AF597968

METHODS IN MOLECULAR BIOLOGY™

For other titles published in this series, go to
www.springer.com/series/7651

Mammalian Chromosome Engineering

Methods and Protocols

Edited by

Gyula Hadlaczky

Institute of Genetics, Biological Research Center, Hungarian Academy of Sciences, Szeged, Hungary

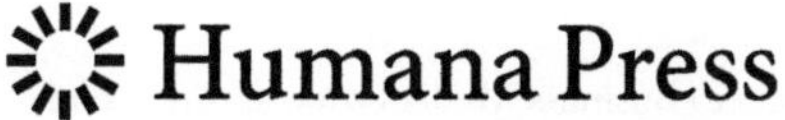

Editor
Gyula Hadlaczky, Ph.D.
Institute of Genetics
Biological Research Center
Hungarian Academy of Sciences
Szeged, Hungary
hgy@brc.hu

ISSN 1064-3745 e-ISSN 1940-6029
ISBN 978-1-61779-098-0 e-ISBN 978-1-61779-099-7
DOI 10.1007/978-1-61779-099-7
Springer New York Dordrecht Heidelberg London

Library of Congress Control Number: 2011923965

Printed on acid-free paper

Humana Press is part of Springer Science+Business Media (www.springer.com)

Preface

The rapid progression of genetics and molecular biology has turned chromosomal engineering from science fiction to reality. Transgenic animals with engineered chromosomes have been produced with success, and chromosomes developed for pharmaceutical protein production are now ready for the medical industry. Engineered chromosomes have also been used in preclinical model experiments for ex vivo stem-cell therapy.

This volume is intended to provide the reader with up-to-date information on this rapidly evolving field, and will attempt to take the reader into the exciting realm of chromosomal engineering from the basic principles to the practical applications of these new technologies. The five overview and ten protocol chapters cover the engineering of chromosomes with extrachromosomal vectors and transposon systems, the manipulation of naturally occurred minichromosomes, the generation and engineering of synthetic artificial chromosomes, and the induced de novo platform artificial chromosome system.

The efforts of the authors and editors will hopefully provide a manual that serves as a bench-side resource for current protocols and help explore prospects for future research and applications.

I am greatly indebted to all contributors, who devoted their precious time to share ideas and expertise that brought about this book, which will be a source of information for anyone interested in new ideas in gene technology.

Szeged, Hungary *Gyula Hadlaczky*

Contents

Contributors

PETER BLAZSO • *Institute of Genetics, Biological Research Center, Hungarian Academy of Sciences, Szeged, Hungary*
ERIKA CSONKA • *Biological Research Center, Institute of Genetics, Hungarian Academy of Sciences, Szeged, Hungary*
IVANA GRABUNDZIJA • *Max Delbrück Center for Molecular Medicine, Berlin, Germany*
AMY L. GREENE • *Department of Biomedical Sciences, Mercer University School of Medicine, Savannah, GA, USA*
OLIVIA HIBBITT • *Department of Anatomy and Genetics, University of Oxford, Oxford, UK*
MASASHI IKENO • *School of Medicine, Keio University, Tokyo, Japan*
ZOLTÁN IVICS • *Max Delbrück Center for Molecular Medicine, Berlin, Germany; University of Debrecen, Debrecen, Hungary*
ZSUZSANNA IZSVÁK • *Max Delbrück Center for Molecular Medicine, Berlin, Germany; University of Debrecen, Debrecen, Hungary*
ROBERT L. KATONA • *Institute of Genetics, Biological Research Center, Hungarian Academy of Sciences, Szeged, Hungary*
MALCOLM L. KENNARD • *Kennard Biologic Consultants, North Vancouver, BC, Canada*
LAJOS MÁTÉS • *Biological Research Center, Hungarian Academy of Sciences, Szeged, Hungary*
MARIANNA PAULIS • *Istituto di Ricerca Genetica e Biomedica, Consiglio Nazionale delle Ricerche, Monserrato, Italy, Istituto Clinico Humanitas, IRCCS, Rozzano, Italy*
CARL F. PEREZ • *Metaara Medical Technologies, Inc., Vancouver, BC, Canada*
EDWARD L. PERKINS • *Department of Biomedical Sciences, Mercer University School of Medicine, Savannah, GA, USA; Curtis and Elizabeth Anderson Cancer Institute, Memorial Health University Medical Center, Savannah, GA, USA*
TÜNDE PRAZNOVSZKY • *Institute of Genetics, Biological Research Center, Hungarian Academy of Sciences, Szeged, Hungary*
ELENA RAIMONDI • *Department of Genetics and Microbiology "A. Buzzati Traverso", University of Pavia, Pavia, Italy*
ILDIKO SINKO • *Institute of Genetics, Biological Research Center, Hungarian Academy of Sciences, Szeged, Hungary*
NOBUTAKA SUZUKI • *School of Medicine, Keio University, Tokyo, Japan*
SANDRA L. VANDERBYL • *Vanderbyl Consulting, Kamloops, BC, Canada*
RICHARD WADE-MARTINS • *Department of Physiology, Anatomy and Genetics, University of Oxford, Oxford, UK*

Chapter 1

Developing Extrachromosomal Gene Expression Vector Technologies: An Overview

Richard Wade-Martins

Abstract

Extrachromosomal, or episomal, vectors offer a number of advantages for therapeutic and scientific applications compared to integrating vectors. Extrachromosomal vectors persist in the nucleus without the requirement to integrate into the host genome, hence avoiding the recent concerns surrounding the genotoxic effects of vector integration. By avoiding integration, episomal vectors avoid vector rearrangement, which can occur at integration, and also avoid any effect of surrounding DNA activity on transgene expression ("position effect"). Extrachromosomal vectors offer a very high transgene capacity, allowing either the incorporation of large promoter and regulatory elements into an expression cassette, or the use of complete genomic loci of up to 100 kb or larger as transgenes. Whole genomic loci transgenes offer an elegant means to express genes under physiological and developmental-stage regulation, to express multiple transcript variants from a single locus, and to express multiple genes from a single tract of genomic DNA. The combined advantages of episomal vectors of prolonged transgene persistence in the absence of vector integration, avoiding silencing by flanking heterochromatin, and high capacity, facilitating delivery and expression of genomic DNA transgenes, will be reviewed here and potential therapeutic and scientific uses outlined.

Key words: Extrachromosomal vectors, Episome, Epstein–Barr virus plasmid, S/MAR plasmid, Mammalian artificial chromosome, High transgene capacity, Genomic DNA, Herpes simplex virus type 1 amplicon

1. Introduction

Gene expression applications typically require efficient retention and long-term expression of a transgene which may be achieved in different ways, determined by the choice of the vector used. Vectors can be divided into two groups based on their ability either to integrate into the host genome, or to persist in the nucleus as an extrachromosomal, or episomal, vector. Integrating

Gyula Hadlaczky (ed.), *Mammalian Chromosome Engineering: Methods and Protocols*, Methods in Molecular Biology, vol. 738, DOI 10.1007/978-1-61779-099-7_1, © Springer Science+Business Media, LLC 2011

vectors have been widely used in gene expression systems as permanent vector integration ensures long-term transgene retention in the absence of selection. However, random integration of a transgene can lead to transgene silencing, if an insertion event occurs in condensed inactive heterochromatin, or to host gene disruption and insertional mutagenesis, when integration occurs a region of active gene expression. Retroviruses and lentiviruses tend to integrate into active regions of the genome, targeting coding or regulatory regions of transcriptionally active genes. Recent adverse events in gene therapy trials to treat X-linked severe combined immuno-deficiency (SCID-X1) using a Molony leukaemia virus (MLV)-based vector carrying the gene coding for the common γ-chain resulted in leukaemias caused by vector integration leading to the activation of the growth promoting proto-oncogene *LMO2*, which caused clonal T-cell proliferation (1, 2). Before these adverse events insertional mutagenesis was of theoretical concern; now, it will prove to be a major obstacle in further clinical application of vectors which undergo insertion into the host genome.

In contrast, extrachromosomal, or episomal, vectors persist in the nucleus in an extrachromosomal state, offering advantages for transgene delivery and expression over integrating vectors. First, the transgene of interest will not be rearranged, concatemerised or disrupted upon integration, nor will it be subject to inappropriate regulation by the chromatin structure of flanking DNA ("position effect"). Second, episomal vectors will not lead to cell transformation caused either by inactivation by insertion and disruption of a tumour suppressor gene, or by overactivation of a proto-oncogene through a vector integrating a strong promoter or enhancer element. Third, episomal vectors often persist in multiple copies per cell, resulting in high expression of the gene of interest. In stable transfection experiments in cell culture, the use of episomal vectors can result in higher efficiency of generating selected cell lines than the conventional plasmid vectors, because integration into the host genome occurs with very low frequency. Fourth, and perhaps of greatest interest and the subject of much work in my laboratory, the high insert capacity of extrachromosomal vectors allows the delivery and expression of entire genomic DNA loci, which has proven an excellent way to achieve physiological levels of transgene expression (3).

Long-term persistent stable maintenance of an episomal vector requires vector replication once per cell cycle, and vector segregation to daughter cells at cell division. A number of episomal systems will provide such mechanisms of maintenance, based either on viral or mammalian DNA sequences, capable of a vector retention efficiency of >95–99% per cell division. Key experimental techniques to analyse such extrachromosomal vectors include plasmid rescue from the mammalian back to the bacterial (or yeast) host, fluorescent in situ hybridization (FISH), and demonstration of vector replication.

2. Epstein–Barr Virus-Based Plasmids

The best-developed of the episomal gene expression vector systems is based on the human Epstein–Barr virus (EBV), a human gamma herpesvirus. Approximately 90% of the world population carries EBV, most of whom are infected in early childhood. Following infection, EBV is carried indefinitely in long-term memory B-lymphocytes in a latent state where the 172 kb genome persists as a replicating extrachromosomal multicopy plasmid (4). The EBV latent origin of replication (*oriP*; origin of plasmid replication) requires the presence of the *trans* acting factor Epstein–Barr virus nuclear antigen-1 (EBNA-1) for replication which binds to *oriP* (5–7). The 1.7 kb *oriP* region contains two components each of which has a specific and distinct function. First, the dyad symmetry (*DS*) element is a 120 bp region containing four EBNA-1 binding sites and is responsible for the replication of EBV during latency (8). Second, the family of repeats (*FR*) element carries 20 copies of a 30 bp repeat each of which contains an EBNA-1 binding site. Binding of EBNA-1 to the *FR* element is responsible for EBV genome retention as an episome and serves the same role in EBV-based plasmids. Removing *FR* from an EBV-derived plasmid prevents the extrachromosomal retention of the plasmid decreasing formation of antibiotic-resistant colonies by almost 1,000-fold (9). It is also known that the *FR* element can act as a transcriptional enhancer and that EBNA-1 can enhance expression by binding to the *FR* element inserted in either direction, located upstream or downstream of a reporter gene (10). The level of transcriptional enhancement depends on the number of EBNA-1 binding sites as reducing the number of 30 bp repeats in *FR* affects the efficiency of transactivation of a reporter gene (11). The *FR*-mediated transcriptional enhancement also works in vivo where the *FR* element in the presence of EBNA-1 provided a 10- to 100-fold increase in expression of Factor IX in mouse liver (11). EBNA-1 binds to metaphase chromosomes and interphase chromatin, especially at newly replicated regions. Dimerisation of EBNA-1 molecules bound to *FR* and host cell chromatin allows the partition of *oriP* plasmids into the daughter cells during mitosis. The high mitotic stability of EBV-based vectors has been termed by Calos as "stability without a centromere" (12).

DNA replication driven by the *DS* element occurs in human and primate cells, but is not supported in rodent cells (5–7).

However, plasmid replication can instead be driven by the insertion of large fragments of mammalian genomic DNA >10 kb in size which provide as-yet uncharacterised origins of replication (13, 14). When large genomic DNA regions are included in plasmids in conjunction with the *FR* element, episomal retention can therefore be achieved in rodent cells (15, 16).

EBV immortalises primary B cells in tissue culture and is found associated with several cancers world-wide, such as Burkitt's lymphoma (BL) and nasopharyngeal carcinoma (NPC). EBNA-1 is the only EBV protein consistently expressed in endemic BL which has led to speculation that EBNA-1 might be an oncogene (17). However, several observations argue against EBNA-1 being oncogenic; for example, the vast majority of the human population (around 95%) carry EBV and do not develop BL or NPC, suggesting an absolute need for other cofactors for tumour formation. In addition, EBNA-1 does not immortalise primary cells in culture. In 1996, Wilson et al. showed that two lines of transgenic mice expressing EBNA-1 developed lymphomas (18), although the susceptibility to tumours did not correlate with the levels EBNA-1 protein expression. The sites of integration were not mapped and the exact role of EBNA-1 in lymphoma development was not determined. A further study on EBNA-1 transgenic mice in 2005 did not find EBNA-1 expression to be oncogenic (19). More recently, a role for EBNA-1 in regulating reactive oxygen species and hence perhaps genomic instability has been suggested (20). Overall, evidence for EBNA-1 being an oncogene is inconclusive, but the EBNA-1-positive state of most people in the world makes it unlikely.

In cells, EBNA-1 binding to the *oriP* element on a plasmid will lead to vector *replication* through the *DS* element, and to vector *retention* by the *FR* element binding to chromosomes, hence tethering the plasmid to chromosomes. The combined efficient *replication* and *retention* will allow stable episomal *maintenance*. The replicating episomal state can be experimentally confirmed by transferring the plasmid back to the bacterial host by plasmid rescue assay to confirm its extrachromosomal presence (14), and by digestion of plasmid DNA extracted from mammalian cells with *Mbo* I to confirm loss of bacterial methylation at the GATC motif through two rounds of replication in the mammalian host (14).

3. S/MAR Vectors

The disadvantage of viral-based episomal vector systems is that they require the long-term expression of a viral protein as a cofactor. Extrachromosomal vector systems which entirely use the mammalian replication machinery would potentially therefore confer significant advantages.

The last 10 years has seen the rapid development of a new episomal vector retention system carrying a *scaffold/matrix attachment region* (*S/MAR*) isolated from the human interferon β-gene (21). *S/MARs* are 70% AT-rich DNA sequence elements which lack a consensus sequence and are instead defined by their affinity for the nuclear matrix/chromosome scaffold. In native mammalian chromosomes *S/MAR* sequences are likely responsible for the organisation of chromatin into independent loops, anchoring DNA to the nuclear matrix (22). Other functions proposed for *S/MAR* elements include the ability to function as insulators (23), and to enhance gene expression by the ability to confer an open, active chromatin state (24, 25), properties which make these sequences particularly attractive for gene therapy vectors.

The first episomal *S/MAR*-based vector (pEPI-1) was generated from an SV40-based vector in which the sequence coding for the large T antigen was replaced with an *S/MAR* sequence isolated from the interferon β-gene (21). Importantly, the *S/MAR* was placed downstream of an active transcription unit without an intervening polyadenylation signal. The vector was able to replicate and be retained as an episome at low copy number for >100 cell generations in the absence of selection, with a mitotic stability of 98%. The same plasmid without the *S/MAR* element could only integrate into the cellular genome, confirming that the presence of *S/MAR* allowed extrachromosomal replication and stable segregation during mitosis (21). It is likely that pEPI binds to the nuclear matrix to facilitate replication (26) through interaction with nuclear matrix protein scaffold attachment factor-A (SAF-A) (27, 28). A defined 180 bp sequence from the *S/MAR* sequence attaches pEPI-1 to SAF-A, and a tetramer of this module can functionally replace the original *S/MAR* sequence in a plasmid (29). pEPI-1 replicates early in S phase in a once-per-cell cycle fashion initiated at several DNA sites in the plasmid (30). Active transcription into the *S/MAR* is essential for autonomous replication (31), probably by generating an accessible chromatin structure enabling pEPI-1 interaction with the nuclear matrix.

S/MAR vectors have now been developed further in vivo as potentially useful gene therapy vectors to provide long-term transgene in the liver (32). Argyros et al. showed long-term transgene expression (6 months) from their *S/MAR*-based vector in the liver, but were unable to confirm vector replication in vivo. Instead the authors suggested that long-term expression was due to the *S/MAR* element maintaining an open chromatin structure through inhibiting promoter methylation, hence facilitating vector expression (32). To develop *S/MAR* vectors as a tool for transgenesis, genetically modified pig foetuses carrying pEPI were generated by sperm-mediated gene transfer. Tissues analysed for pEPI episomal retention and gene expression showed the presence of episomal vectors in 12/18 foetuses at <10 copies per cell. Reporter gene expression was shown in 9/12 episome-positive foetuses (33), opening up an exciting new application for *S/MAR* technology, although further studies are needed.

4. Mammalian Artificial Chromosome Vectors

EBV and S/MAR-based vectors both share the attractive properties of being small and easy to manipulate plasmid-based systems, as the DNA elements required for the episomal maintenance of each are only a few kilobases long. Mammalian artificial chromosomes, on the other hand, rely on the presence of very large and complex centromere sequences >1 Mb in size requiring highly specialised methodologies for vector manipulation and construction. Their potential advantage, however, is that they would confer the very high stability and transgene payload of a native chromosome.

In essence there are two basic approaches to construct a mammalian artificial chromosome. The bottom-up approach exploits our knowledge of the minimum functional elements required for a functioning chromosome and builds a chromosome de novo from the basic elements of a centromere, telomeres and origins of replication. The top-down approach truncates natural chromosomes by radiation or telomere-associated fragmentation methodologies.

The bottom-up approach to building a MAC was largely based on earlier work on the *Saccharomyces cerevisiae* yeast artificial chromosome (YAC) model (34, 35). In *S. cerevisiae*, three elements were required to build a functional YAC: a centromere composed of the 110 bp *CEN* fragment to ensure chromosomal segregation at cell division; a replication origin, composed of the 50 bp *ARS1* element to drive DNA replication in a regulated manner; and a telomere at each end, composed of tandem repeats of a specific sequence to protect chromosomal ends from degradation.

Hoping what was true for *S. cerevisiae* must be true for mammals, several groups attempted to build artificial chromosomes from the three equivalent elements in humans. Human telomeres comprise an array of a tandemly repeated $(TTAGGG)_n$ sequence in length orientated 5′–3′ towards the end of the chromosome. Repeats of the DNA sequence $(TTAGGG)_n$ a few hundred base pairs long can generate de novo *telomeres* when transfected into human chromosomes, seeding new telomeres at the end of the truncation fragments (36). Second, there have been several reports identifying specific replication origins; for example, the origin of directional replication from the hamster *DHFR* locus (37). However, several elegant functional experiments designed to identify mammalian origins of replication were not successful at isolating specific sequences responsible for mammalian genome replication, and it seems likely that any reasonably large (>10–12 kb) of mammalian genomic DNA will drive efficient replication (38, 39). Third, a large highly repetitive sequence termed alpha-satellite, or alphoid, DNA is the key component of centromere structure and function as the integration of alpha-satellite DNA sequences into existing mammalian chromosomes disrupts the normal segregation of chromosomes (40).

The first human artificial chromosomes (HACs) constructed from bottom-up took the approach of co-transfecting synthetic arrays of alphoid DNA, with genomic DNA fragments and telomeric sequences to generate linear artificial chromosomes of 6–10 Mb in size (41). The HACs showed high mitotic stability (close to 100%) and contained functional centromeres, confirmed by immunostaining of centromere proteins C and E (CENP-C, CENP-E).

High-capacity yeast, bacteria and P1-derived artificial chromosomes (YACs, BACs and PACs, respectively) cloning vectors have allowed the manipulation of large genomic constructs to facilitate the bottom-up approach HAC construction (41–44). This allows the more elegant approach of assembling potential artificial chromosome constructs as complete vectors before transfection into mammalian cells. Ikeno et al. cloned alpha-satellite DNA into a YAC vector together with human telomeric sequences and selection markers for mammalian cells (42). The resulting artificial chromosomes were 10–50 times greater than the introduced YAC vector. No acquisition of host DNA had occurred and the HACs were comprised of multimers of the original YAC construct. Clonal cell lines carrying a HAC showed a segregation efficiency of 99.2–99.5% per cell division.

MAC construction can result in either linear or circular chromosomes much larger than the starting construct, likely due to the recombination and amplification of input construct. Circular PACs carrying ~70 kb of alphoid DNA with or without telomeric sequences in the vectors when transfected into HT1080 cells resulted in the establishment of circular minichromosomes carrying alphoid DNA, suggesting that telomeres are not required for the circular MACs. However, telomeres were essential for linear PAC vectors to establish artificial chromosomes (45). The efficiency of establishing MACs carrying only alpha-satellite DNA is ~10%, but this can be increased to ~70% when additional genomic DNA sequences carrying potential replication origins or *S/MAR*s are included (46).

Generating minichromosomes in somatic cell hybrids by the "top-down" approach followed the identification of sequences responsible for telomere formation and allowed the development of telomere-associated chromosome fragmentation. Minichromosomes derived from the X chromosome using this method comprised telomeres flanking a 1.8–2.5 Mb alpha-satellite array and showed full mitotic function (47, 48).

Once a MAC or HAC vector has been constructed, many different techniques such as vector co-transfection followed by homologous recombination, or using site-specific recombinases such as Cre, can be used to insert a transgene into a HAC to exploit their potential as vectors for future gene therapy applications (49, 50).

Several lines of transgenic mice carrying either circular or linear MACs have been generated suggesting that artificial chromosomes may represent a potential transgenic technology. For example, in one study, FISH analyses showed the minichromosome to be present in 90% of tail fibroblasts from the F1 generation and in 10–80% of tissues of the F2 and F3 progeny (51).

Constructing MACs remains an enormous challenge demanding great patience and determination, yet major problems remain regarding their application to gene therapy. Delivery of MACs in vivo remains the principal obstacle to their use, although recent delivery of HACs using a viral vector has now been shown, as discussed below (52).

5. High Capacity Genomic DNA Expression Vectors

Many gene expression technologies have benefited from the use of large genomic DNA transgenes carrying all introns and exons and flanked by non-coding, potentially regulatory DNA. The delivery of a complete genomic DNA locus provides the context for gene expression comparable to that of the native endogenous locus, allowing the physiological regulation of expression and production of multiple splice variants, for example. Clear demonstrations as to the advantages of employing a genomic DNA locus can be found in the production of transgenic mice where delivery of a whole locus provides for correctly regulated tissue-specific expression and production of alternative splice isoforms. BAC-based transgenesis is now regarded as the state-of-the-art in rodent transgenic models, producing, for example a highly respected atlas resource of gene expression (53), and several highly relevant disease models (54).

The high capacity of extrachromosomal vector systems allows the inclusion of very large fragments of genomic DNA for gene delivery and expression. Several studies have developed extrachromosomal high capacity expression plasmid vectors incorporating the EBV episomal retention elements into a BAC or YAC vector. YAC *oriP* vectors of 90 and 660 kb introduced into

293 cells expressing EBNA-1 were shown to exhibit episomal retention and maintenance in the absence of selection at ~97–99% per cell division (55). The 90 kb YAC vector could furthermore be rescued back into yeast (56). The EBV/YAC systems is limited by the requirement for a recircularisation step of the linear YAC vector before introduction into eukaryotic cells, the laborious nature of working with YACs and their frequent instability and chimaerism.

Recent work by my group and by others into developing episomal genomic DNA expression vectors based on BAC vectors has shown the advantages of such technologies for gene expression. In the first demonstration that a genomic DNA locus on an episome can correct a genetic deficiency, a BAC of 115 kb in size containing the whole hypoxanthine phosphoribosyltransferase (*HPRT*) genomic locus and EBV retention elements was constructed and delivered to a *HPRT* deficient cell line. Isolated clonal lines carrying the BAC as an extrachromosomal element showed strong HPRT enzyme activity using hypoxanthine incorporation assays and survived continuous selection in HAT (hypoxanthine/aminopterine/thymidine) medium, confirming the transgene expression and phenotype rescue (57).

Following this initial proof-of-principle study, two further pieces of work sought to demonstrate that the use of such episomal genomic DNA expression vectors can exploit the complexity emerging from our better understanding of mammalian genomes by (a) providing appropriate levels of expression under physiological regulation, and (b) allowing production of alternatively spliced transcripts and promoter usage (16, 58). First, the low-density lipoprotein receptor (*LDLR*) genomic locus was chosen as an excellent example of gene regulation under physiological control. *LDLR* expression is tightly regulated through a negative feedback mechanism in which intracellular cholesterol levels are detected by sterol response elements in the *LDLR* promoter. A BAC containing the entire *LDLR* genomic locus and incorporating the EBV episomal retention elements was delivered to an *ldlr*$^{-/-}$ Chinese hamster ovary cell line (16). Clonal cell lines carrying the *LDLR* BAC were selected and the episomal status confirmed by plasmid rescue of the *LDLR* BAC vector. Functional LDLR expression and correction of the deficiency phenotype was demonstrated in the *ldlr*$^{-/-}$ CHO cells and in primary fibroblasts from familial hypercholesterolaemia patients. High sterol levels inhibited *LDLR* transgene expression, recapitulating the natural control of expression occurring at the chromosomal site.

Second, validation of the BAC-based episomal vector systems to deliver and express genomic loci undergoing complex alternative splicing came from a study expressing the complex genomic region containing the loci *CDKN2A* and *CDKN2B* (58). The region shows high complexity of expression as it contains only six exons, but encodes for five proteins involved in cell cycle regulation. A BAC vector carrying a 132 kb insert containing the *CDKN2* locus showed expression and correct splicing for three of the five genes in the region ($p15^{Ink4b}$, $p16^{Ink4a}$, $p14^{ARF}$) from the single genomic DNA insert, and a growth-arrest response in p53+ glioma cells deleted for the *CDKN2* region (58).

The ability to precisely engineer BACs with base-pair precision opens up exciting possibilities to use BAC-based episomal vectors as tools for functional genomics to better understand the effect of non-coding variation on gene expression. Eeds et al. (59) used an episomal BAC vector to study the effect of non-synonymous mutations, potential splicing defects and promoter variations in the carbamyl phosphate synthetase I gene, characterising the functional importance of DNA changes not possible in conventional cDNA-based systems.

As discussed above there may be advantages in the use of *S/MAR* episomes over EBV systems. To that end our laboratory has developed a new design of high-capacity episomal vector by incorporating the *S/MAR* retention system into a BAC containing a 135 kb genomic DNA insert carrying the *LDLR* gene. The 156 kb vector replicated and persisted extrachromosomally at a low copy number in the $ldlr^{-/-}$ CHO cell line. The vector could be rescued unrearranged after 100 cell generations in the absence of selection, showed a mitotic stability of ~97.6–99.8% per cell division, and was able to rescue the phenotypic deficiency in LDLR expression (60).

6. Vector Delivery

High capacity episomal vectors may be delivered to cells by either non-viral or viral delivery systems, each of which has its own advantages and disadvantages as discussed briefly below.

6.1. Non-viral Delivery

Non-viral delivery has the dual advantages of no upper limit on packaging size and reduced immunogenicity. Any of the usual transfection reagents may be used to deliver episomal vectors to cells in vitro. At larger vector sizes >100 kb there may be advantages in using more complex delivery reagents, such as the LID vector, composed of Lipofectin (L), Integrin (I) and DNA (D) which we have used to deliver intact large BAC-based constructs of up to 242 kb in size in vitro (61). Interestingly, in vitro and in vivo it has been shown that delivery efficiency is independent of the size of a plasmid, so long as a constant number of molecules are used across a range of vector sizes between 12 and 242 kb in size (61, 62).

In vivo our group and others have found hydrodynamic delivery to be a highly efficient method to deliver plasmids to the liver, achieving liver transduction levels of 20–40% with small plasmids. The method uses an injection of a large volume of liquid at a high speed to generate high pressure in the liver, leading to the formation of pores in the cell membranes and allowing entry of plasmid into hepatocytes. This method to deliver episomal vectors to the liver is described in Chapter 2 of the current volume by our laboratory. Using hydrodynamic delivery we have demonstrated plasmid retention (by plasmid rescue) and transgene expression in hepatocytes from EBV-based genomic DNA *LDLR* expression plasmids for 9 months post-injection (62, 63). Another study analysed different non-viral methods for delivery of BACs of different sizes in vitro and in vivo up to 150 kb (64). In vitro, PEI22 resulted in a higher transfection efficiency than Lipofectamine 2000 (~10%), but the latter proved to give a higher proportion of intact DNA. Delivery of an 80 kb BAC in vivo using hydrodynamic injection showed a higher delivery efficiency per copy compared to low volume intravenous injection and intramuscular electroporation of naked DNA (64).

6.2. Viral Delivery

The large size of genomic DNA loci precludes the use of most viral vectors for their delivery. The exception is to be found in vectors based on herpesviruses, which are attractive delivery systems for several reasons. The herpesviruses have large genomes and hence high insert capacity; herpesviruses infect dividing and non-dividing cells, and herpesviruses are naturally retained as episomes during latency. "Amplicon" vectors are plasmid-based systems, which lack any viral coding sequences but contain *cis*-acting DNA sequences which allow the plasmid to be packaged into infectious virions in the presence of viral protein expression, usually provided by a helper virus or plasmid. Amplicon vectors based on a number of herpesviruses have been produced, including herpes simplex virus type 1 (HSV-1), EBV and cytomegalovirus (CMV) (65–67).

EBV amplicon plasmids carry the lytic origin of replication (*oriLyt*), the terminal repeats for vector packaging into virions, and the EBV episomal retention elements described above. Helper-dependent amplicons can be packaged into virus particles, infect B cell, kept in an extrachromosomal state and express a transgene for prolonged periods (66). In 2002, this system was developed further to exploit the potential high-capacity of EBV to deliver genomic loci. EBV BAC amplicons containing 60 and 123 kb genomic DNA inserts were shown to be packaged by a helper virus into virions and delivered intact into Loukes B cells where they were retained as stable extrachromosomal elements, as shown by plasmid rescue (68). The development of a CMV-based amplicon vector is also underway (67). Plasmids containing the CMV lytic origin of replication and the packaging signals were incorporated into CMV virions when transfected in fibroblasts previously infected with a helper CMV virus (67).

Although the large packaging capacities of the EBV and CMV amplicon vectors (170 and 230 kb, respectively) are attractive, the systems are limited as the resulting virions can infect only a limited range of cell types and the viral stocks produced are heavily contaminated by helper viruses, although this has been partially solved using early-generation helper virus-free EBV packaging systems (69, 70).

Amplicon systems based on HSV-1 represent the best-developed viral system to deliver extrachromosomal genomic DNA vectors. HSV-1 is a 152 kb double-stranded DNA virus widespread in the human population capable of infecting a wide range of cell types and is maintained episomally during latency. HSV-1 amplicons are bacterial plasmids carrying an HSV-1 origin of replication (*oriS*) and the DNA cleavage/packaging terminal repeats (the "a" sequence or *pac*) (65). As "gut-less" vectors carry only a few kilobases of *cis*-acting DNA, HSV-1 amplicons have a potential maximum transgene capacity of ~150 kb, although they do require packaging by a helper virus or helper plasmid. The recent development of an improved helper-virus free packaging system in which a BAC plasmid carrying the HSV-1 viral genome lacking the packaging signals provides the viral protein expression for amplicon replication and packaging into virion particles with no detectable helper virus (71).

The HSV-1 virus can enter a state of latency in infected neurons, but only as a non-replicating episome. Therefore, to generate a replicating HSV-1 based episome, additional extrachromosomal retention elements need to be included. For example, HSV-1/EBV hybrid vector have been developed to include EBV retention elements in an amplicon of HSV-1 vector (72). In 2001 the full transgene capacity potential of the HSV-1 amplicon system was achieved with the delivery of BAC vectors containing large inserts of genomic DNA (73). Cre–*loxP* recombination was used to incorporate ("retrofit") an amplicon vector carrying EBV retention elements and HSV-1 packaging sequences to BAC and PAC clones from human genomic DNA libraries. Several vectors of different sizes were shown to be packaged into HSV-1 particles (as determined by Southern blot) and delivered intact and efficiently into cells (as determined by plasmid rescue). One amplicon insert contained the *HPRT* genomic locus within a 115 kb insert and was used to transduce *HPRT* deficient human fibroblasts. The vector was demonstrated to be retained episomally in stable clonal lines and to provide functional complementation of the *HPRT* deficiency (73). This system, termed the infectious BAC (iBAC) exploits the high-capacity of the HSV-1 amplicon system in a genomic DNA expression system capable of in vitro functional screening to study genes involved in genetic deficiencies and has been used extensively by my group and others (16, 58, 60, 73–76). For example, Xing et al. used an iBAC vector carrying the bone morphogenic protein-2 (*BMP-2*) locus.

Introduction of the *BMP-2* BAC in MC3T3-E1 cells resulted in transgene expression which induced the transduced cell line to differentiate into osteoblast (76). A modification of this approach has been recently described with iBAC-*S/MAR*, an HSV-1 amplicon-based high-capacity BAC vector in which episomal retention properties are mediated by an *S/MAR* sequence (60).

Finally, the HSV-1 amplicon system has been used to achieve viral delivery of mammalian artificial chromosomes (52), which may serve to overcome the major limitation of the use of MACs in vivo. BAC-based vectors carrying human alphoid DNA, HSV-1 packaging elements and an EGFP reporter gene cassette were constructed. Transduction of cell lines using these amplicon vectors resulted in a high percentage of transduction, measured by EGFP expression, and high efficiency of stable clones formation, estimated to be 10,000-fold greater than previously obtained by transfection of BACs using lipofection. Analysis of stable clones revealed the formation of functional HACs in two cell lines, which showed a mitotic stability >99% and binding of centromeric proteins. These data suggest that the HSV-1 amplicon vector system can deliver intact HACs at high efficiency, thus making somatic gene therapy with HACs a feasible gene therapy application.

Adenoviral vectors have a mid-range transgene capacity (~37 kb) and have been used to deliver EBV-based replicating systems (77, 78). However, since the genome of adenovirus remains linear in cells after infection and the EBV episomal system requires a circular plasmid for replication, an extra circularisation step needs to be built into the vectors to function post-infection. This may be achieved using *loxP* recombination, requiring a second vector to deliver and express Cre recombinase to circularise the first adenoviral vector carrying EBNA-1 and *oriP* elements (77). This methodology has been extended to helper-dependent adenovirus (HDA) vectors which lack viral genes and hence provide a greater transgene capacity (78).

7. Conclusions

Recent concerns over the safety of integrating viral vectors have provided an excellent opportunity for the further development and application of extrachromosomal, or episomal, vector systems. A further advantage of extrachromosomal systems lies in their high capacity and ability to deliver and express whole genomic DNA loci transgenes. Here, I have tried to review this exciting class of vector system and provide an overview of its potential applications for therapeutic and laboratory applications. Later chapters, including one from my own laboratory, provide details of key methodologies to work successfully with many of the systems described above.

Acknowledgements

I would like to thank colleagues, M. James, R. White, Y. Saeki, E.A. Chiocca, Z. Larin-Monaco, M. Lufino and O. Hibbitt with whom I have worked on the development of episomal systems, for all their help, advice and support over the years. I also acknowledge funding support from the Wellcome Trust, the Medical Research Council, the Biotechnology and Biological Sciences Research Council and the British Heart Foundation.

References

1. Hacein-Bey-Abina, S, von Kalle, C, Schmidt, M, Le Deist, F, Wulffraat, N, McIntyre, E, Radford, I, Villeval, JL, Fraser, CC, Cavazzana-Calvo, M *et al.* (2003) A serious adverse event after successful gene therapy for X-linked severe combined immunodeficiency. *N Engl J Med* 348:255–256.
2. Hacein-Bey-Abina, S, Von Kalle, C, Schmidt, M, McCormack, MP, Wulffraat, N, Leboulch, P, Lim, A, Osborne, CS, Pawliuk, R, Morillon, E *et al.* (2003) LMO2-associated clonal T cell proliferation in two patients after gene therapy for SCID-X1. *Science* 302:415–419.
3. Lufino MM, Edser PA, Wade-Martins R. (2008) Advances in High-capacity Extrachromosomal Vector Technology: Episomal Maintenance, Vector Delivery, and Transgene Expression. *Mol Ther.* 16(9):1525–38.
4. Lindahl, T, Adams, A, Bjursell, G, Bornkamm, GW, Kaschka-Dierich, C and Jehn, U. (1976) Covalently closed circular duplex DNA of Epstein-Barr virus in a human lymphoid cell line. *J Mol Biol* 102:511–530.
5. Yates, J, Warren, N, Reisman, D and Sugden, B. (1984) A cis-acting element from the Epstein-Barr viral genome that permits stable replication of recombinant plasmids in latently infected cells. *Proc Natl Acad Sci USA* 81: 3806–3810.
6. Yates, JL, Warren, N and Sugden, B. (1985) Stable replication of plasmids derived from Epstein-Barr virus in various mammalian cells. *Nature* 313:812–815.
7. Rawlins, DR, Milman, G, Hayward, SD and Hayward, GS. (1985) Sequence-specific DNA binding of the Epstein-Barr virus nuclear antigen (EBNA-1) to clustered sites in the plasmid maintenance region. *Cell* 42:859–868.
8. Yates, JL, Camiolo, SM and Bashaw, JM. (2000) The minimal replicator of Epstein-Barr virus oriP. *J Virol* 74:4512–4522.
9. Wysokenski, DA and Yates, JL. (1989) Multiple EBNA1-binding sites are required to form an EBNA1-dependent enhancer and to activate a minimal replicative origin within oriP of Epstein-Barr virus. *J Virol* 63: 2657–2666.
10. Reisman, D and Sugden, B. (1986) trans activation of an Epstein-Barr viral transcriptional enhancer by the Epstein-Barr viral nuclear antigen 1. *Mol Cell Biol* 6:3838–3846.
11. Sclimenti, CR, Neviaser, AS, Baba, EJ, Meuse, L, Kay, MA and Calos, MP. (2003) Epstein-Barr virus vectors provide prolonged robust factor IX expression in mice. *Biotechnol Prog* 19:144–151.
12. Calos MP. (1998) Stability without a centromere. *Proc Natl Acad Sci USA.* 95(8): 4084–5.
13. Krysan, PJ, Haase, SB and Calos, MP. (1989) Isolation of human sequences that replicate autonomously in human cells. *Mol Cell Biol* 9:1026–1033.
14. Wade-Martins, R, Frampton, J and James, MR. (1999) Long-term stability of large insert genomic DNA episomal shuttle vectors in human cells. *Nucleic Acids Res* 27:1674–1682.
15. Krysan, PJ and Calos, MP. (1993) Epstein-Barr virus-based vectors that replicate in rodent cells. *Gene* 136:137–143.
16. Wade-Martins, R, Saeki, Y and Chiocca, EA. (2003) Infectious delivery of a 135-kb LDLR genomic locus leads to regulated complementation of low-density lipoprotein receptor deficiency in human cells. *Mol Ther* 7:604–612.
17. Schulz TF, Cordes S. (2009) Is the Epstein-Barr virus EBNA-1 protein an oncogene? *Proc Natl Acad Sci USA.* 106(7):2091–2.
18. Wilson JB, Bell JL, Levine AJ (1996) Expression of Epstein–Barr virus nuclear antigen-1 induces B cell neoplasia in transgenic mice. *EMBO J* 15:3117–3126.

19. Kang MS, *et al.* (2005) Epstein–Barr virus nuclear antigen 1 does not induce lymphoma in transgenic FVB mice. *Proc Natl Acad Sci USA* 102:820–825.
20. Gruhne B, *et al.* (2009) The Epstein–Barr virus nuclear antigen-1 promotes genetic instability via induction of reactive oxygen species. *Proc Natl Acad Sci USA* 106: 2313–2318.
21. Piechaczek, C, Fetzer, C, Baiker, A, Bode, J and Lipps, HJ. (1999) A vector based on the SV40 origin of replication and chromosomal S/MARs replicates episomally in CHO cells. *Nucleic Acids Res* 27:426–428.
22. Vogelstein, B, Pardoll, DM and Coffey, DS. (1980) Supercoiled loops and eucaryotic DNA replicaton. *Cell* 22:79-85.
23. Bode, J, Benham, C, Knopp, A and Mielke, C. (2000) Transcriptional augmentation: modulation of gene expression by scaffold/matrix-attached regions (S/MAR elements). *Crit Rev Eukaryot Gene Expr* 10:73–90.
24. Zahn-Zabal, M, Kobr, M, Girod, PA, Imhof, M, Chatellard, P, de Jesus, M, Wurm, F and Mermod, N. (2001) Development of stable cell lines for production or regulated expression using matrix attachment regions. *J Biotechnol* 87:29–42.
25. McKnight, RA, Shamay, A, Sankaran, L, Wall, RJ and Hennighausen, L. (1992) Matrix-attachment regions can impart position-independent regulation of a tissue-specific gene in transgenic mice. *Proc Natl Acad Sci USA* 89:6943–6947.
26. Cook, PR. (1999) The organization of replication and transcription. *Science* 284: 1790–1795.
27. Jenke, BH, Fetzer, CP, Stehle, IM, Jonsson, F, Fackelmayer, FO, Conradt, H, Bode, J and Lipps, HJ. (2002) An episomally replicating vector binds to the nuclear matrix protein SAF-A in vivo. *EMBO Rep* 3:349–354.
28. Baiker, A, Maercker, C, Piechaczek, C, Schmidt, SB, Bode, J, Benham, C and Lipps, HJ. (2000) Mitotic stability of an episomal vector containing a human scaffold/matrix-attached region is provided by association with nuclear matrix. *Nat Cell Biol* 2:182–184.
29. Jenke, AC, Stehle, IM, Herrmann, F, Eisenberger, T, Baiker, A, Bode, J, Fackelmayer, FO and Lipps, HJ. (2004) Nuclear scaffold/matrix attached region modules linked to a transcription unit are sufficient for replication and maintenance of a mammalian episome. *Proc Natl Acad Sci USA* 101:11322–11327.
30. Schaarschmidt, D, Baltin, J, Stehle, IM, Lipps, HJ and Knippers, R. (2004) An episomal mammalian replicon: sequence-independent binding of the origin recognition complex. *Embo J* 23:191–201.
31. Stehle, IM, Scinteie, MF, Baiker, A, Jenke, AC and Lipps, HJ. (2003) Exploiting a minimal system to study the epigenetic control of DNA replication: the interplay between transcription and replication. *Chromosome Res* 11:413–421.
32. Argyros O, Wong SP, Niceta M, Waddington SN, Howe SJ, Coutelle C, Miller AD, Harbottle RP (2008) Persistent episomal transgene expression in liver following delivery of a scaffold/matrix attachment region containing non-viral vector. *Gene Ther.* 15(24):1593–605.
33. Manzini, S, Vargiolu, A, Stehle, IM, Bacci, ML, Cerrito, MG, Giovannoni, R, Zannoni, A, Bianco, MR, Forni, M, Donini, P *et al.* (2006) Genetically modified pigs produced with a nonviral episomal vector. *Proc Natl Acad Sci USA* 103:17672–17677.
34. Stinchcomb, DT, Struhl, K and Davis, RW. (1979) Isolation and characterisation of a yeast chromosomal replicator. *Nature*, 282: 39–43.
35. Murray, AW and Szostak, JW. (1983) Construction of artificial chromosomes in yeast. *Nature* 305:189–193.
36. Farr, CJ, Stevanovic, M, Thomson, EJ, Goodfellow, PN and Cooke, HJ. (1992) Telomere-associated chromosome fragmentation: applications in genome manipulation and analysis. *Nat Genet* 2:275–282.
37. Burhans WC, Selegue JE, Heintz NH. (1986) Isolation of the origin of replication associated with the amplified Chinese hamster dihydrofolate reductase domain. *Proc Natl Acad Sci USA.* 83(20):7790–4.
38. Heinzel, SS, Krysan, PJ, Tran, CT and Calos, MP. (1991) Autonomous DNA replication in human cells is affected by the size and the source of the DNA. *Mol Cell Biol* 11: 2263–2272.
39. Krysan, PJ and Calos, MP. (1991) Replication initiates at multiple locations on an autonomously replicating plasmid in human cells. *Mol Cell Biol* 11:1464–1472.
40. Haaf, T, Warburton, PE and Willard, HF. (1992) Integration of human alpha-satellite DNA into simian chromosomes: centromere protein binding and disruption of normal chromosome segregation. *Cell* 70:681–696.
41. Harrington, JJ, Van Bokkelen, G, Mays, RW, Gustashaw, K and Willard, HF. (1997) Formation of de novo centromeres and construction of first-generation human artificial microchromosomes. *Nat Genet* 15:345–355.

42. Ikeno, M, Grimes, B, Okazaki, T, Nakano, M, Saitoh, K, Hoshino, H, McGill, NI, Cooke, H and Masumoto, H. (1998) Construction of YAC-based mammalian artificial chromosomes. *Nat Biotechnol* 16:431–439.

43. Grimes, BR, Schindelhauer, D, McGill, NI, Ross, A, Ebersole, TA and Cooke, HJ. (2001) Stable gene expression from a mammalian artificial chromosome. *EMBO Rep* 2:910–914.

44. Mejia, JE, Willmott, A, Levy, E, Earnshaw, WC and Larin, Z. (2001) Functional complementation of a genetic deficiency with human artificial chromosomes. *Am J Hum Genet* 69:315–326.

45. Ebersole, TA, Ross, A, Clark, E, McGill, N, Schindelhauer, D, Cooke, H and Grimes, B. (2000) Mammalian artificial chromosome formation from circular alphoid input DNA does not require telomere repeats. *Hum Mol Genet* 9:1623–1631.

46. Basu, J, Compitello, G, Stromberg, G, Willard, HF and Van Bokkelen, G. (2005) Efficient assembly of de novo human artificial chromosomes from large genomic loci. *BMC Biotechnol* 5:21.

47. Farr, CJ, Bayne, RA, Kipling, D, Mills, W, Critcher, R and Cooke, HJ. (1995) Generation of a human X-derived minichromosome using telomere-associated chromosome fragmentation. *Embo J* 14:5444–5454.

48. Mills, W, Critcher, R, Lee, C and Farr, CJ. (1999) Generation of an approximately 2.4 Mb human X centromere-based minichromosome by targeted telomere-associated chromosome fragmentation in DT40. *Hum Mol Genet* 8:751–761.

49. Kotzamanis, G, Cheung, W, Abdulrazzak, H, Perez-Luz, S, Howe, S, Cooke, H and Huxley, C. (2005) Construction of human artificial chromosome vectors by recombineering. *Gene*, 351, 29–38.

50. Mejia, JE and Larin, Z. (2000) The assembly of large BACs by in vivo recombination. *Genomics* 70:165–170.

51. Suzuki, N, Nishii, K, Okazaki, T and Ikeno, M. (2006) Human artificial chromosomes constructed using the bottom-up strategy are stably maintained in mitosis and efficiently transmissible to progeny mice. *J Biol Chem* 281:26615–26623.

52. Moralli, D, Simpson, KM, Wade-Martins, R and Monaco, ZL. (2006) A novel human artificial chromosome gene expression system using herpes simplex virus type 1 vectors. *EMBO Rep* 7:911–918.

53. Gong, S, Zheng, C, Doughty, ML, Losos, K, Didkovsky, N, Schambra, UB, Nowak, NJ, Joyner, A, Leblanc, G, Hatten, ME *et al.* (2003) A gene expression atlas of the central nervous system based on bacterial artificial chromosomes. *Nature* 425:917–925.

54. Heintz N. (2001) BAC to the future: the use of BAC transgenic mice for neuroscience research. *Nat Rev Neurosci.* 2(12):861–70.

55. Simpson, K, McGuigan, A and Huxley, C. (1996) Stable episomal maintenance of yeast artificial chromosomes in human cells. *Mol Cell Biol* 16:5117–5126.

56. Simpson, K and Huxley, C. (1996) A shuttle system for transfer of YACs between yeast and mammalian cells. *Nucleic Acids Res* 24: 4693–4699.

57. Wade-Martins, R, White, RE, Kimura, H, Cook, PR and James, MR. (2000) Stable correction of a genetic deficiency in human cells by an episome carrying a 115 kb genomic transgene. *Nat Biotechnol* 18:1311–1314.

58. Inoue, R, Moghaddam, KA, Ranasinghe, M, Saeki, Y, Chiocca, EA and Wade-Martins, R. (2004) Infectious delivery of the 132 kb *CDKN2A/CDKN2B* genomic DNA region results in correctly spliced gene expression and growth suppression in glioma cells. *Gene Ther* 11:1195–1204.

59. Eeds AM, Mortlock D, Wade-Martins R, Summar ML. (2007) Assessing the functional characteristics of synonymous and nonsynonymous mutation candidates by use of large DNA constructs. *Am J Hum Genet.* 80(4):740–50.

60. Lufino, MM, Manservigi, R and Wade-Martins, R. (2007) An S/MAR-based infectious episomal genomic DNA expression vector provides long-term regulated functional complementation of LDLR deficiency. *Nucleic Acids Res* 35, e98.

61. White, RE, Wade-Martins, R, Hart, SL, Frampton, J, Huey, B, Desai-Mehta, A, Cerosaletti, KM, Concannon, P and James, MR. (2003) Functional delivery of large genomic DNA to human cells with a peptide-lipid vector. *J Gene Med* 5:883–892.

62. Hibbitt, OC, Harbottle, RP, Waddington, SN, Bursill, CA, Coutelle, C, Channon, KM and Wade-Martins, R. (2007) Delivery and long-term expression of a 135 kb LDLR genomic DNA locus in vivo by hydrodynamic tail vein injection. *J Gene Med* 9:488–497.

63. Hibbitt, O.C., McNeil, E., Lufino, M.M.P., Seymour, L. Channon, K. and Wade-Martins, R. (2009) Long-term physiologically regulated expression of the low-density lipoprotein receptor in vivo using genomic DNA minigene constructs. 2009 Oct 27. [Epub ahead of print].

64. Magin-Lachmann, C, Kotzamanis, G, D'Aiuto, L, Cooke, H, Huxley, C and Wagner, E. (2004) In vitro and in vivo delivery of intact BAC DNA- comparison of different methods. *J Gene Med* 6, 195–209.
65. Spaete, RR and Frenkel, N. (1982) The herpes simplex virus amplicon: a new eucaryotic defective-virus cloning-amplifying vector. *Cell* 30:295–304.
66. Banerjee S, Livanos E, Vos JM. (1995) Therapeutic gene delivery in human B-lymphoblastoid cells by engineered non-transforming infectious Epstein-Barr virus. Nat Med. 1(12):1303–8.
67. Borst, EM and Messerle, M. (2003) Construction of a cytomegalovirus-based amplicon: a vector with a unique transfer capacity. *Hum Gene Ther* 14:959–970.
68. White, RE, Wade-Martins, R and James, MR. (2002) Infectious delivery of 120-kilobase genomic DNA by an epstein-barr virus amplicon vector. *Mol Ther* 5:427–435.
69. Delecluse, HJ, Pich, D, Hilsendegen, T, Baum, C and Hammerschmidt, W. (1999) A first-generation packaging cell line for Epstein-Barr virus-derived vectors. *Proc Natl Acad Sci USA* 96:5188–5193.
70. Hettich E, Janz A, Zeidler R, Pich D, Hellebrand E, Weissflog B, Moosmann A, Hammerschmidt W. (2006) Genetic design of an optimized packaging cell line for gene vectors transducing human B cells. *Gene Ther.* 13(10):844–56.
71. Saeki, Y, Fraefel, C, Ichikawa, T, Breakefield, XO and Chiocca, EA. (2001) Improved helper virus-free packaging system for HSV amplicon vectors using an ICP27-deleted, oversized HSV-1 DNA in a bacterial artificial chromosome. *Mol Ther* 3:591–601.
72. Wang, S and Vos, JM. (1996) A hybrid herpesvirus infectious vector based on Epstein-Barr virus and herpes simplex virus type 1 for gene transfer into human cells in vitro and in vivo. *J Virol* 70:8422–8430.
73. Wade-Martins, R, Smith, ER, Tyminski, E, Chiocca, EA and Saeki, Y. (2001) An infectious transfer and expression system for genomic DNA loci in human and mouse cells. *Nat Biotechnol* 19, 1067–1070.
74. Gomez-Sebastian, S., Gimenez-Cassina, A., Diaz-Nido, J. Lim, F. and Wade-Martins, R. (2007) Infectious delivery and expression of a 135 kb human FRDA genomic DNA locus complements Friedreich s ataxia deficiency in human cells. *Mol. Ther.* 15:248–254.
75. Peruzzi, P.P. Lawler, S., Senior, S.L., Dmitrieva, N., Edser, P.A.H., Gianni, D., Chiocca, E.A. and Wade-Martins, R. (2009) Physiological gene regulation and functional complementation of a neurological disease gene deficiency in neurons *Mol. Ther.* 17(9):1517–26.
76. Xing, W, Baylink, D, Kesavan, C and Mohan, S. (2004) HSV-1 amplicon-mediated transfer of 128-kb BMP-2 genomic locus stimulates osteoblast differentiation in vitro. *Biochem Biophys Res Commun* 319, 781–786.
77. Tan, BT, Wu, L and Berk, AJ. (1999) An adenovirus-Epstein-Barr virus hybrid vector that stably transforms cultured cells with high efficiency. *J Virol* 73:7582–7589.
78. Dorigo, O, Gil, JS, Gallaher, SD, Tan, BT, Castro, MG, Lowenstein, PR, Calos, MP and Berk, AJ. (2004) Development of a novel helper-dependent adenovirus-Epstein-Barr virus hybrid system for the stable transformation of mammalian cells. *J Virol* 78: 6556–6566.

Chapter 2

High Capacity Extrachromosomal Gene Expression Vectors

Olivia Hibbitt and Richard Wade-Martins

Abstract

Extrachromosomal gene expression vectors that contain native genomic gene expression elements have numerous advantages over traditional integrating mini-gene vectors. In this protocol chapter we describe our work using episomal vectors where expression of a cDNA is controlled by a 10 kB piece of genomic DNA encompassing the promoter of the low density lipoprotein receptor. We explain methods to subclone large genomic inserts into gene expression vectors. We also illustrate various methods employed to ascertain whether expression from these vectors is robust and physiologically relevant by investigating their sensitivity to changes in cellular milieu. Delivery of gene expression vectors in vivo is also described using hydrodynamic tail vein injection, a high pressure, high volume tail vein injection used for liver-directed gene transfer.

Key words: Episomal, Hydrodynamic tail vein injection, LDLR, Familial hypercholesterolaemia, Live imaging, Cholesterol, Genomic promoter, Luciferase

1. Introduction

High capacity genomic DNA gene expression vectors where transgene expression is controlled by native expression elements have a strong advantage over mini-gene vectors where a heterologous promoter drives cDNA expression. When using an entire genomic locus it is possible to deliver a complete gene including all introns, exons and regulatory elements in the correct genomic context. This is important for many applications that require systems that do not lead to transgene over-expression (1, 2).

Working with gene expression vectors that deliver transgenic DNA into cells without integrating into the genome is becoming increasingly attractive. Integration of a vector does ensure long-term retention of the transgene, however, it can also lead to gene silencing through positional effects, and cellular transformation (3, 4).

Gyula Hadlaczky (ed.), *Mammalian Chromosome Engineering: Methods and Protocols*, Methods in Molecular Biology, vol. 738,
DOI 10.1007/978-1-61779-099-7_2,

In this chapter we describe the use of high capacity extrachromosomal vectors in vitro and in vivo in the context of our work in gene therapy for familial hypercholesterolaemia (FH) (5–8). FH is a condition caused by mutations in the low density lipoprotein receptor (*LDLR*) gene and is characterised by high circulating levels of cholesterol (9, 10). The condition represents a unique challenge in gene therapy as over-expression of LDLR leads to toxic intracellular accumulations of LDL (11, 12). In addition, any transduced population of cells will be required to clear large amounts of cholesterol from the plasma continuously as cholesterol synthesis is constitutively active in the liver. This means that the therapeutic *LDLR* transgene has to complement the loss of function of the endogenous gene by expressing the LDLR in a physiologically regulated manner. In this protocol chapter we include descriptions of functional analysis of LDLR transgene expression including expression of reporter genes from genomic promoter regions and analysis of LDL binding and internalisation by quantitative cell culture assay.

2. Materials

Unless otherwise stated all chemicals were obtained from Sigma (Dorset, UK).

2.1. Vector Design

2.1.1. BAC DNA Maxi-Prep

1. LB agar (e.g. Calbiochem) prepared as per manufacturer's instructions and autoclaved.
2. Antibiotics: Ampicillin (Amp): 50 mg/ml solution made up in MilliQ water and filtered through a 0.22 μm filter. Kanamycin (Kan): 25 mg/ml solution made up as for Ampicillin. Chloramphenicol (Chl): 15 mg/ml solution made up in 70% ethanol. All antibiotic solutions were stored in aliquots at –20°C.
3. LB Broth Miller (e.g. Novagen, VWR, Leighton Buzzard) made up as per manufacturer's instructs and autoclaved.
4. Qiagen Tip 50 Maxiprep kit (Qiagen, Crawley): all buffers are included with kit. Buffer P1 should have RNAse added before storing at 4°C. Buffer P3 should also be stored at 4°C, buffer QF should be heated to 55°C before use.
5. Kimwipe tissues (Kimberly Clark, Fisher Scientific, UK).
6. All centrifugations were performed in a Beckman Avanti J-E centrifuge – Rotors: J10.5 and J17.
7. 250-ml centrifuge bottles (Beckman, High Wycombe, UK).
8. Oakridge tubes (Beckman, High Wycombe, UK).
9. Isopropanol (VWR, Leighton Buzzard, UK).

10. Tris–EDTA: TE, 10 mM Tris–HCl at pH 8, 1 mM EDTA.
11. Materials required for pulsed field gel electrophoresis.

2.1.2. Sub-cloning Genomic Fragments

1. pSC101-BAD-gbA-tet plasmid (Gene Bridges, Heidelberg). This plasmid contains tetracycline resistance (tet), has a temperature sensitive origin of replication and the genes required for homologous recombination (recE, an exonuclease and recT) are under the control of an arabinose inducible promoter.
2. BioRad Gene Pulser Controller (BioRad, Hemel Hempsted, UK); unless otherwise stated all electroporation into bacterial host cells was performed at settings: 25 μF, 1.8 kV and 200 Ω.
3. Electrocompetent cells containing genomic expression plasmid (e.g. BAC).
4. SOC medium (Invitrogen, Paisley, UK).
5. LB agar plates containing tetracycline (9 μg/ml, pSC101) and chloramphenicol (15 μg/ml, BAC plasmid).
6. L-arabinose.
7. Glycerol.
8. BioXact Long, long range polymerase (Bioline, London, UK).
9. Qiagen PCR purification kit (Qiagen, Crawley, UK).

2.1.3. Retrofitting Expression Plasmids with Episomal Maintenance Plasmids

1. Cre enzyme/buffer (NEB, Hitchin, UK).
2. Dialysis membrane (Millipore, Watford, UK).
3. DH10B electrocompetent cells (Invitrogen, Paisley, UK).

2.2. Cell Culture

1. CHO a7 Ldlr −/− cell line.
2. Hams F12 medium (Invitrogen, Paisley, UK).
3. L-glutamine (L/G, Invitrogen, Paisley, UK).
4. Penicillin/streptomycin (P/S, Invitrogen, Paisley, UK).
5. Foetal bovine serum (FBS, Invitrogen, Paisley, UK).
6. Lipid depleted foetal bovine serum (LPDS, Biomedical Technologies, Stoughton, MA).
7. Tissue culture plasticware, e.g. 75 cm^2, 25 cm^2 flasks, 96/24/12/6 well plates.

2.2.1. Establishment of Episomal Clonal Cell Lines

1. Lipofectamine (Invitrogen, Paisley, UK).
2. Opti-MEM, serum free medium (Invitrogen, Paisley, UK).
3. Trypsin/EDTA (Invitrogen, Paisley, UK).
4. G418 (neomycin analogue, Invitrogen, Paisley, UK).
5. Selection medium: Hams F12, 1% L/G, 1% P/S, 10% FBS, 600 μM G418.

2.2.2. Confirmation of Plasmid Maintenance (Plasmid Rescue)

1. 10-cm cell culture plates.
2. STET buffer: 8% sucrose, 5% triton X-100, 50 mM EDTA, 50 mM Tris–HCl at pH 8.
3. Alkaline SDS: 1% SDS and 0.2 N NaOH.
4. 7.5 M ammonium acetate.
5. 1.5-ml Heavy Phase Lock Gel tubes (Eppendorf, Hamburg, Germany).
6. Phenol:chloroform:isoamyl alcohol 25:24:1 saturated with 10 mM Tris–HCl at pH 8 and 1 mM EDTA.
7. Chloroform.
8. TE + RNase: 10 mM Tris–HCl at pH 8, 1 mM EDTA and 5 μg/ml RNase A.

2.3. Functional Assays In Vitro

2.3.1. Luciferase Assay

1. Dynex luciferase plate reader with dual injectors (or similar).
2. Hams F12 medium P/S, L/G plus lipid depleted serum.
3. Transfection reagents (as in Subheading 2.2.1).
4. Cholesterol, make a 12 μg/ml working solution in 70% ethanol.
5. 25-Hydroxycholesterol, make a 0.6 μg/ml working solution in 70% ethanol.
6. Mevastatin (Merck, Nottingham, UK), working solution 1 mM made up in ethanol.
7. Sterol incubation medium: Hams F12 + LPDS, 1:1,000 cholesterol and 1:2,000 25-hydroxycholesterol.
8. Statin incubation medium: Hams F12 + LPDS and 1:1,000 Statin.
9. Luciferase lysis buffer: 25 mM Tris–PO_4 at pH 7.8, 0.2 mM 1,2-diaminocyclohexane tetraacetic acid, 1:10 glycerol, 1:100 Triton X-100 and 2 μM dithiothretol.
10. Luciferase Solution A – Luciferin solution: 0.3 mg/ml (Caliper Life Sciences, Hopkinton, MA).
11. Luciferase Solution B – Luciferase assay buffer: 15 mM $MgSO_4$, 15 mM KPO_4 at pH 7.8, 0.04 mM ethylene glycol tetraacetic acid at pH 7.8, 2 μM dithiothretol, 50 μl β-mercaptoethanol and 200 mg/ml adenosine triphosphate.
12. O-nitrophenyl-β-galactopyranoside (ONPG) assay buffer: 6 mM Na_2HPO_4, 4 mM $Na_2H_2PO_4$, 10 mM KCl, 1 mM $MgSO_4$, 20 mg/ml ONPG, 2 μM dithiothretol and 50 μl β-mercaptoethanol; 100 μl.
13. ONPG assay stop solution: 50 mM Na_2CO_3.

2.3.2. Dil-LDL Assay

1. Spectrofluorimeter, plate reader or similar. Excitation wavelength 520 nm and emission wavelength 580 nm.

2. HamsF12 medium P/S, L/G plus lipid depleted serum.
3. Transfection reagents (as in Subheading 2.2.1).
4. Cholesterol, make a 12 μg/ml working solution in 70% ethanol.
5. 25-Hydroxycholesterol, make a 0.6 μg/ml working solution in 70% ethanol.
6. Mevastatin (Merck), working solution 1 mM made up in ethanol.
7. Sterol incubation medium: HamsF12 + LPDS, 1:1,000 cholesterol and 1:2,000 25-hydroxycholesterol.
8. Statin incubation medium: HamsF12 + LPDS and 1:1,000 Statin.
9. DiI-LDL (AbD Serotech, Abingdon, UK).
10. Unlabelled human LDL (AbD Serotech, Abingdon, UK).
11. DiI-LDL medium: HamsF12 + LPDS and 10 μg/ml DiI-LDL.
12. DiI-LDL plus cold medium: HamsF12 + LPDS, 10 μg/ml DiI-LDL and 500 μg/ml human LDL.
13. DiI-LDL lysis buffer: 25 mM Tris–PO_4 at pH 7.8, 0.2 mM 1,2-diaminocyclohexane tetraacetic acid, 1:10 glycerol, 1:100 Triton X-100 and 2 μmol/l dithiothretol.
14. DiI-LDL standard curve solutions ranging from 0.016 to 2 μg/ml made up in lysis buffer.

2.4. Delivery and Analysis In Vivo

2.4.1. Liver-Specific Plasmid Delivery

1. Adult mice, 25–30 g.
2. Prewarmed sterile PBS.
3. Plasmid DNA 20–50 μg/animal in 2.5 ml PBS.
4. 27 g needles.
5. A 38–40°C heating box suitable for mice.
6. Isofluorane.
7. Oxygen.
8. Anaesthetic machine with an isofluorane vaporiser.
9. Warming pad set to 37°C.

2.4.2. Transfection Efficiency Analysis (Immunohistochemistry)

1. Cannula needle attached to a 50-ml syringe.
2. Phosphate buffered saline tablets (Sigma 79382).
3. 4% Paraformaldehyde w/v in PBS.
4. Ethanol: 70, 95 and 100%.
5. Xylene/histoclear.
6. Paraffin wax.
7. Automatic Tissue Processor.

8. Paraffin embedder.
9. Microtome.
10. Polysine slides (VWR, Leighton Buzzard, UK).
11. PAP pen (Abcam, Cambridge, UK) – creates a hydrophobic barrier around section keeping staining reagents on the section. This reduces the amount of reagent needed and also reduces cross-contamination between sections on the same slide.
12. Endogenous biotin blocking kit (Invitrogen, Paisley, UK) consisting of blocking solution A (streptavidin reagent) and blocking solution B (biotin reagent).
13. Immunohistochemistry blocking solution: 1% fish gelatin, 0.1% Triton X-100, 10% goat serum in Tris buffered saline – 50 mM Tris–HCl (pH 7.5) and 150 mM NaCl.
14. Biotinylated anti-human LDLR monoclonal primary antibody (Fitzgerald Industries International, North Acton, MA).
15. Anti-β-galactosidase secondary antibody (Invitrogen, Paisley, UK).
16. 4′,6-Diamidino-2-phenylindole (DAPI, Invitrogen, Paisley, UK) nuclear material counterstain that emits a blue fluorescence.
17. Mounting medium such as glycerol or Clearmount (Invitrogen, Paisley, UK).

2.4.3. Transfection Efficiency Analysis (Live Imaging)

1. IVIS 100 live imaging camera and software (Caliper Life Sciences, Hopkinton, MA).
2. Luciferin (Caliper Life Sciences, Hopkinton, MA).

2.4.4. Transfection Efficiency Analysis (Plasmid Rescue)

1. Genomic lysis buffer: 0.6% SDS, 100 mM NaCl, 50 mM Tris–HCl (pH 8), 20 mM EDTA.
2. Proteinase K 10 mg/ml working solution.
3. Phase lock gel (light, Eppendorf).
4. Ethanol 70 and 100%.
5. TE.
6. DH10B bacteria.
7. LB agar plates containing an appropriate antibiotic.

3. Methods

3.1. Vector Design

3.1.1. Construction of Retrofitting Plasmids

Extrachromosomal vector maintenance requires the inclusion of elements that will promote the maintenance of a plasmid vector as a replicating, episomal gene expression unit. Mammalian cells being either; the Epstein–Barr virus (EBV) derived episomal system, the S/MAR system or human artificial chromosomes (2).

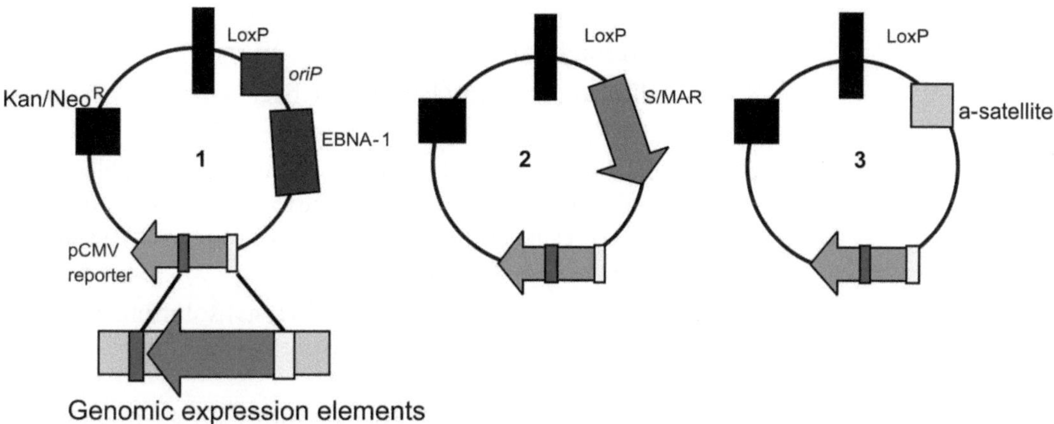

Fig. 1. Gene expression plasmid vectors for the promotion of extrachromosomal vector maintenance may contain one of three systems. (1) The EBV system requires the inclusion of *OriP* and EBNA-1. (2) The S/MAR system requires the inclusion of S/MAR sequences. (3) Human artificial chromosomes are produced from vectors containing alpha-satellite DNA. Also included in these vectors are pCMV-reporter gene expression cassettes which can be converted to genomic promoter–reporter gene expression cassettes through homologous recombination. Kanamycin/neomycin resistance (Kan/neo^R) is essential for selection in bacterial and mammalian cells. A *LoxP* site is included so that plasmids can be retrofitted to BAC plasmids or other *LoxP* containing plasmids.

Each maintenance system will require specific modifications to any plasmid. The EBV system requires the addition of the *trans*-acting Epstein–Barr virus Nuclear Antigen-1 (EBNA-1) protein and the *cis*-acting *oriP* origin of replication. The *S/MAR* system requires the *S/MAR* sequence from the pEPI-based vectors. Human artificial chromosomes require the inclusion of α-satellite DNA (Fig. 1). It is also important for expression analysis and clonal cell establishment to include a reporter gene under a constitutive promoter and a mammalian selection cassette (Fig. 1). For in vivo use a reporter gene such as luciferase is particularly useful if you are able to utilise live imaging technology. If this is not possible β-galactosidase is an excellent and versatile reporter with very little background expression. All vectors need to include a *loxP* site, which facilitates Cre-mediated recombination.

3.1.2. BAC DNA Maxi-Prep

The following protocol uses Qiagen Tip 500 maxi-prep kits with a modified protocol. It is highly efficient for the purification of large plasmids, but can also be used to obtain high yields from smaller plasmids.

1. On a LB agar plate containing the appropriate antibiotics, streak a small amount of bacterial stock and incubate overnight at 37°C.
2. Grow a small starter culture of a single colony in 1.5 ml LB containing the appropriate antibiotics for a minimum of 6 h shaking at 37°C. At this point, the media should appear slightly cloudy.

3. Tip the 1.5 ml culture into 250 ml of LB + antibiotics and grow overnight at 37°C shaking at 225 rpm.
4. The following morning harvest the bacterial cells by centrifugation; 6,000 × *g* for 10 min.
5. Tip off the media and resuspend the bacterial pellet with 15 ml of cold (4°C) P1 (resuspension) solution containing RNAse. To resuspend, completely leave the tubes to shake in the incubator for 10 min at 225 rpm.
6. Lyse the bacteria with 15 ml of P2 (lysis) solution. Incubate the cells for precisely 5 min, mixing every 1 min by gentle swirling.
7. Neutralise the lysis by adding 15 ml of P3 (neutralisation) solution and swirl to mix.
8. Incubate in P3 for 20 min on ice by gently inverting the tube at 2-min intervals. This step is important because the BAC DNA can be precipitated with the bacterial genomic DNA and is a major cause of low yields.
9. Pellet the flocculate by centrifugation; 15,000 × *g* for 35 min at 4°C.
10. Prepare the Tip-500 columns for DNA binding
    - Equilibrate with 15 ml of QBT (equilibration) buffer
    - Insert a double layer of "kimwipe" tissue into the column by pushing in with a finger. This tissue acts a filter to prevent bacterial flocculate from clogging the column.
11. Pour supernatant containing the plasmid DNA through the tissue and let it run through the column. Gently squeeze out the tissue being careful to avoid any precipitate falling onto the column.
12. Wash the column twice with 30 ml of QC (wash) buffer.
13. Elute DNA into Oakridge tubes using 15 ml of prewarmed (55°C) QF (elution) buffer.
14. Precipitate the DNA by adding 10.5 ml isopropanol and pellet by centrifugation at 27,000 × *g* for 30 min at 4°C. Be aware that isopropanol pellets tend to be glassy in appearance and so may not be visible.
15. Carefully tip off supernatant into a clean 50-ml plastic tube. At this point check to make sure the supernatant does not contain anything that looks like a pellet.
16. Wash the pellet in 3.5 ml of 70% ethanol without mixing. Centrifuge at 27,000 × *g* for 30 min at 4°C.
17. Very carefully decant the supernatant and leave to air dry.
18. Resuspend the pellet by gentle flicking in 250 μl of TE buffer overnight at 4°C.

19. The following day, flick the tube again and spin briefly to collect the solution. Transfer to a fresh microcentrifuge tube and store at 4°C.
20. Check the quality of DNA preparation by restriction enzyme digestion of 400–500 ng of DNA. Digest should be separated using pulsed field gel electrophoresis.

3.1.3. Sub-cloning Genomic Fragments Using Homologous Recombination

Here we describe the sub-cloning of large genomic fragments into plasmids using RecE/RecT or ET recombination. In our work we used recombination to create a plasmid that contained a 10 kB piece of genomic DNA encompassing the LDLR genomic promoter driving either luciferase or LDLR cDNA (8).

1. Generate an expression plasmid containing (Fig. 2); transgene expression cassette driven by heterologous promoter, human origin of replication, a polyadenylation site and antibiotic resistance. Kanamycin/neomycin is useful as it allows for selection in both bacterial and mammalian cells and a loxP site for retrofitting.
2. Identify genomic region for subcloning.
3. Design recombination primers (Fig. 2); at least 55 bp homologous to genomic sequence and at least 25 bp homologous to the vector sequence

Primer A – 55 bp arm homologous to the genomic DNA 10 kB down stream of LDLR start codon. 25 bp arm homologous

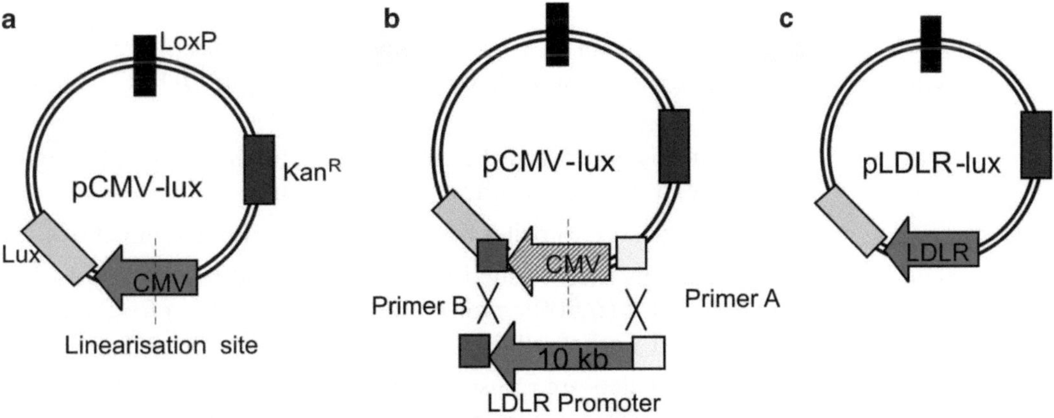

Fig. 2. Subcloning of genomic inserts into plasmid vectors using ET recombination is shown here schematically. The plasmid is designed that contains an expression cassette (**a**). The plasmid is linearised in the middle of the pCMV sequence. Primer A and Primer B are designed with homology arms that equate to 25 bp homology to the plasmid and 55 bp to genomic DNA (**b**). A PCR is performed using the linearised pCMV-lux plasmid. This results in a PCR product that incorporates the entire plasmid minus the pCMV and containing 5′ and 3′ homology arms homologous to genomic DNA. This product is electroporated into bacteria containing the *LDLR* BAC. Following recombination the resulting plasmid has the *LDLR* promoter in place of the pCMV promoter (**c**).

to the region immediately upstream of CMV promoter on the expression plasmid.

Primer B – 55 bp arm homologous to 55 bp up stream of LDLR start codon. The 25 bp arm is homologous to either LDLR or luciferase cDNA and includes the start codon.

Production of ET recombination electrocompetent cells

4. Thaw on ice a vial of electrocompetent DH10B cells containing specific genomic expression plasmid (e.g. LDLR BAC).
5. Electroporate 10 ng of pSC101-BAD-gbA-tet plasmid.
6. Add 450 µl of SOC medium; mix. Transfer cell suspension to a 5-ml tube and shake at 30°C for 1 h.
7. Plate out bacteria on Chl/Tet LB agar plates and grow overnight at 30°C.
8. To confirm presence of pSC101-BAD-gbA-tet plasmid and BAC plasmid in bacteria, pick a single colony using a sterile inoculation needle. Dip the needle first into 1.5 ml of LB media containing chl then the same needle into 1.5 ml of LB media containing tet. Grow shaking overnight at 30°C (tet), or 37°C (chl). The following day extract plasmid DNA using an appropriate mini-prep method and check for intact plasmids using restriction enzyme digestion.
9. To create recombination ready cells; pick single clones from LB (tet/chl) plates and grow overnight in 1.5 ml LB (tet/chl) at 30°C.
10. The following morning tip the small starter culture into 100 ml LB (tet/chl) media and grow (30°C) to an OD of 0.1–0.15.
11. Add 1.5 ml of 10% L-arabinose to 100 ml culture and continue to grow with shaking at 37°C to an OD of 0.35–0.40 at A_{600}.
12. At this point stop the cells growing further by incubating on ice in the cold room (4°C) for 40 min.
13. Centrifuge to pellet bacteria (6,000 × *g* for 15 min at 4°C).
14. Pellet is then washed three times in 100 ml of ice cold 10% glycerol (6,000 × *g* for the first wash, 8,000 × *g* for the subsequent washes, all for 15 min at 4°C).
15. The pellets are resuspended in about 0.5–1 ml of the remaining supernatant from the final wash, and aliquoted into 50-µl aliquots. Aliquots are snap frozen and stored at –80°C.

Sub cloning (see Note 1)

It is important to optimise annealing temperature for the recombination primers. A gradient of temperatures between 55 and 72°C should be sufficient to obtain the most efficient annealing temperature.

16. Set-up 6× long range PCR reactions as follows:

10×[a] OPTi buffer	2.5 µl
$MgCl_2$[a] (50 mM)	0.875 µl
dNTP (8 mM)	1.6 µl
DNA[b]	50 ng
Forward primer (1 µM)	2.5 µl
Reverse primer (1 µM)	2.5 µl
BioXact long	0.25 µl
MilliQ Water	Up to 25 µl

[a]Reagents included with BioXact long polymerase
[b]Linearised at an appropriate site located between the 25 bp homology arms (see Note 2)

17. Perform the polymerase chain reaction on the six reactions using the following protocol.
18. Once PCR programme has run, make two pools of PCR mix containing reactions 1–3 and 4–6. Purify DNA from reaction components using a PCR purification kit (Qiagen) eluting DNA in 50 µl of milliQ water. Elute a second time in 40 µl of milliQ water and pool with 50 µl of eluate.

95°C 95°C 55–65°C	15 min 30 s 30 s	30–50 Cycles
72°C	1 min/kB to amplify	
72°C	1 min/kB to amplify	

19. Digest 85 µl of DNA with Dpn1.
20. PCR purify using PCR purification kit eluting in 20 µl.
21. Electroporate 8 µl into ET recombination electrocompetent cells (produced in step 4).
22. Following electroporation add 550 µl of LB media with no antibiotics, transfer cell suspension to a 5-ml tube and shake at 37°C for 75 min.
23. On a 10-cm agar plate containing appropriate antibiotics, spread 2 µl of bacteria. On a second 15-cm agar plate, spread the remaining bacteria and grow at 37°C overnight. The antibiotics used here should correspond to the plasmid, not the BAC, i.e. if the plasmid is kanamycin resistant then grow the cells on kanamycin plates that do not contain chloramphenicol.
24. Pick single clones for mini-prep analysis of recombination.

3.1.4. Retrofitting Expression Plasmids with Episomal Maintenance Plasmids

In this section we describe Cre/loxP mediated retrofitting. This is a highly efficient way of combining expression cassettes with episomal maintenance elements.

1. Prepare the following recipe (see Note 3).

Sub-cloned plasmid	1 μg
Retrofitting plasmid	50 ng
10× Cre buffer	3 μl
Cre enzyme	1 unit
MilliQ water	Up to 30 μl

2. Incubate samples at 37°C for 30 min followed by 10 min at 75°C to inactivate the Cre enzyme.
3. Dialyse sample against water for 3 h to remove salts.
4. Electroporate 15 μl into DH10B cells.
5. Plate onto LB agar containing the appropriate antibiotic combination and incubate overnight at 37°C.
6. Analyse retrofitting using restriction enzyme digestion.

3.2. Cell Culture

3.2.1. Establishment of Episomal Clonal Cell Lines

Described here is a protocol for the establishment of clonal cell lines in CHO a7 *Ldlr–/–* cells (8).

1. Seed 1×10^5–1×10^6 cells per well of a 6-well dish. Leave to grow for 24 h.
2. Make up the transfection mix in a 15-ml tube. For a 6-well dish use up to 4 μg plasmid DNA and 10 μl of lipofectamine in a total of 1.5 ml Opti-MEM. Leave mix to complex for about 10 min.
3. While waiting for the DNA/lipofectamine to complex, wash cells three times with Opti-MEM.
4. Apply 1.5 ml of transfection mix to cells and swirl gently.
5. Incubate cells in transfection mix for 4–6 h.
6. Remove transfection mix and wash cells three times in Opti-MEM.
7. After the final wash apply 3 ml of normal growth media to each well and leave cells for 48 h.

For each transfected well:

8. Wash 1× with PBS.
9. Apply 0.75 ml of trypsin and leave for 2 min.
10. Apply 0.75 ml of selection media (Subheading 2.2.1) and mix up and down to dislodge cells from plate.
11. Dispense all 1.5 ml of media plus cells into 13.5 ml of selection media (tube A).

12. Perform a serial dilution of cells; take 1 ml from tube A and dispense into tube B containing 14 ml of selection media and mix. Take 1 ml from tube B and dispense into tube C containing 14 ml of selection media.
13. Seed the cells from the serial dilutions into 6-well plates. Three wells per dilution should be sufficient.
14. In addition, seed a control well containing untransfected cells in selection media.
15. The cells should now be left until single clones have grown. It is important that cells are left as long as it takes for the untransfected cells to be completely killed by the selection antibiotics and for large, well defined colonies to form. It is normal for this to take up to 15 days.

    Clones should be reasonably large, about 2 mm, and completely isolated from surrounding cells to avoid contamination of clonal populations. Once clones have reached a reasonable size that is, they are visible as small dots on the base of the plate, they can be picked. Clones can be picked using plastic clone rings (Sigma); however, in our experience, it is easier to pick the clones by hand as we will describe.
16. Looking at the plate from underneath, circle clones to be picked with a marker.
17. Check circled clones under the microscope; they should be discreet cell clones.
18. Remove selection media and wash cells with PBS before applying 1.5 ml of trypsin.
19. Take the plate to the microscope and using a 4× objective identify clone to be picked and using a P20 pipette, aspirate clonal cells from the plate (see Note 4).
20. Dispense cells into a single well of a 96-well plate containing 100 μl of selection media. Leave cells to grow to confluency.
21. Once cells are confluent, transfer them to progressively larger wells until they are growing in 25 cm^2 flasks.

3.2.2. Confirmation of Plasmid Maintenance (Plasmid Rescue)

1. Plate $2–5 \times 10^6$ clonal cells into 10-cm tissue culture dishes.
2. When confluent, extract episomal plasmid DNA using alkaline lysis. Scrape cells into 1.5 ml of PBS and centrifuge for 3 min at $5{,}000 \times g$.
3. Resuspend the cell pellet in 60 μl of STET buffer.
4. Lyse cells with 130 μl of alkaline SDS.
5. Neutralise with 110 μl of ammonium acetate and incubate on ice for 5 min.
6. Centrifuge at $13{,}000 \times g$ for 30 min at 4°C.

7. Transfer supernatant to Phase Lock gel Heavy tube that has been centrifuged at maximum speed for 1 min.
8. Pipette 500 μl of Phenol:Chloroform onto supernatant and mix well.
9. Centrifuge at 13,000×g for 2 min at room temperature. Remove the upper aqueous phase and repeat.
10. Extract twice with 400 μl of Chloroform using phase lock Eppendorfs.
11. Precipitate DNA using 2.5 times the volume of absolute ethanol. Centrifuge at 13,000×g for 30 min at 4°C to pellet DNA.
12. Wash pellet with 70% ethanol.
13. Resuspend DNA in 20–50 μl of TE/RNase.
14. Confirm circular plasmid status by restriction enzyme digestion.

3.3. Functional Assays In Vitro

3.3.1. Luciferase Assay

Functional analysis of expression can be undertaken using luciferase reporter gene expression. Here we describe an assay that is used to assess the expression from a 10 kB piece of genomic DNA encompassing the LDLR promoter. We use sterols and statins to investigate the expression dynamics from the promoter region (Fig. 3) (8).

1. Seed 1×10^4 cells per well in a 24-well plate allowing for 4 wells per condition and leave for 24 h.

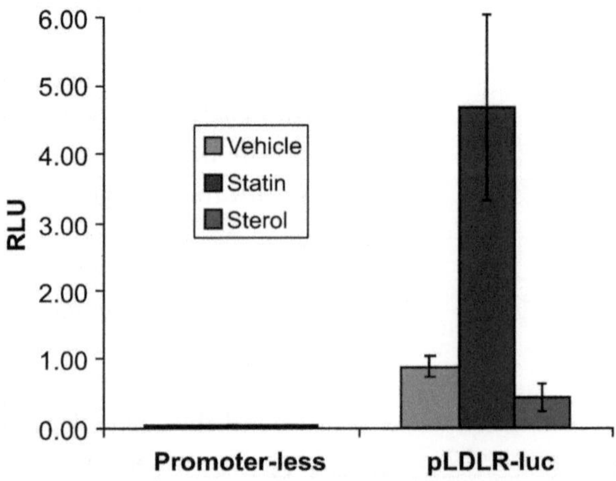

Fig. 3. Luciferase expression in CHO A7 Ldlr –/– cells expressing luciferase under the control of the LDLR genomic promoter is sensitive to regulation by sterols and statins. The CHO a7 *Ldlr –/–* cells respond in a physiologically relevant manner to cellular stimuli with a 50% reduction in luciferase expression seen with the addition of sterols and a fivefold increase in luciferase expression seen with the addition of statin.

2. The following day transfect plasmid DNA as described in Subheading 2.2.1 with the exception that the amount of DNA should be no more than 1 μg and using 1 μl lipofectamine.
3. Following the incubation period wash off transfection mix and incubate the cells in 250 μl of either sterol incubation media, statin incubation media, or HamsF12+LPDS with 10 μl of vehicle (ethanol) control for 24–72 h.
4. Wash cells twice with PBS and apply 100 μl of luciferase lysis buffer, incubate for 20 min at room temperature. Luciferase lysis buffer can be made up in advance and stored at room temperature which the exception of the DTT, which should be added just before use from frozen aliquots.
5. Make up Solution A and Solution B. Solution A should be made up fresh each time. Solution B can be made up and stored at room temperature with the exception of the addition of ATP and DTT, which should be added fresh from frozen aliquots.
6. Set-up luciferase plate reader such that 50 μl of Solution A and 100 μl of Solution B will be dispensed into each well immediately prior to the luciferase value being read.
7. Transfer the entire contents of each well into a well of a black 96-well assay plate. Remove 3 μl of lysate to a separate colourless assay plate.
8. Run luciferase assay.
9. In the separate assay plate apply 100 μl of ONPG assay buffer and incubate at 37°C for 1–10 min checking at regular intervals.
10. Once the colour has changed to a very light yellow, apply 50 μl of ONPG stop solution and read the OD at A_{460}.
11. The luciferase value divided by the ONPG value gives a luciferase expression level corrected for transfection efficiency. Alternatively, luciferase can be normalised to total well protein using established protein assay methods.

3.3.2. DiI-LDL Assay

Functional analysis of LDL receptor activity in vitro was undertaken using a fluorescently labelled LDL analogue called DiI-LDL (7, 8).

1. Prepare cells as in items 1–3 of Subheading 2.3.1.
2. For each condition incubate 3 wells with DiI-LDL media and one well with DiI-LDL plus cold media for 5 h.
3. Wash cells twice with PBS (1% Bovine Serum Albumin).
4. Wash cells three times with PBS (Fig. 4).

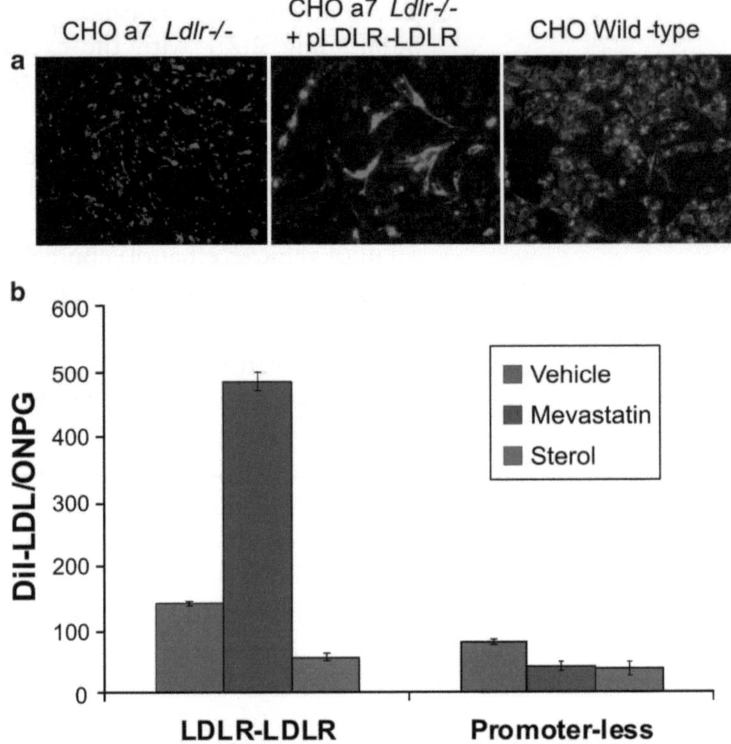

Fig. 4. Binding and internalisation of LDL can be analysed using a fluorescent analogue of LDL, DiI-LDL. (**a**) Incubation of CHO cells with fluorescently labelled LDL (DiI-LDL) demonstrates efficiency of expression of *LDLR* cDNA from the genomic promoter. In CHO a7 *Ldlr–/–* cells no binding of LDL is seen. CHO a7 *Ldlr–/–* cells expressing LDLR from the genomic promoter exhibit binding and internalisation of DiI-LDL that is comparable to the untransfected wild-type CHO cells. (**b**) The binding and internalisation seen in CHO cells expressing LDLR from the genomic promoter is quantifiable and is sensitive to cellular stimuli. Incubation of infected cells with sterols leads to a reduction in functional LDL receptors and therefore a 50% reduction in binding and internalisation of DiI-LDL. Whereas incubation with statins leads to a twofold increase in binding and internalisation of DiI-LDL.

5. Lyse cells with lysis buffer and transfer to spectrofluorimeter plate/cuvette.
6. Remove 3 μl of sample and perform ONPG assay as per items 9–11 of Subheading 2.3.1.
7. Read DiI-LDL fluorescence levels using spectrofluorimeter and normalise to ONPG (Fig. 4).

3.4. Delivery and Analysis In Vivo

3.4.1. Liver-Specific Plasmid Delivery

Here we describe the protocol for liver-specific plasmid delivery using hydrodynamic tail vein injection. This is a particularly efficient means of transfecting the liver in vivo using plasmid DNA (13–15). The premise of the hydrodynamic injection is a combination of high pressure and large volume. A bolus of fluid equalling the total blood volume is injected into the tail vein of a mouse in less than 10 s. This bolus of fluid travels up the vena cava and is stopped by the heart causing retrograde flow into the hepatic portal vein. The large volume of fluid flowing into the capillaries of the liver causes them to swell, which results in the breaking apart of cell adhesions thus making holes in the membranes allowing the DNA to flow in. The transfection efficiency can be as high as around 60% of hepatocytes and is very well tolerated by the mice.

1. Warm the animals in a heating unit for 5–10 min at 38–40°C taking care to not let the animals get too hot (see Note 5).
2. Induce animal with 5% isofluorane in 2 l/min oxygen.
3. Maintain anaesthesia with 2% isofluorane in 2 l/min oxygen.
4. Locate the tail vein and inject 2.5 ml of injection solution (see Note 6).
5. Allow the animal to recover in home cage.

3.4.2. Transfection Efficiency Analysis (Immunohistochemistry)

Here we describe immunohistochemical protocol to detect β-galactosidase and human LDLR in injected mice (6, 8).

1. Administer an overdose of anaesthetic to the animal.
2. Once death is confirmed, through cessation of respiration and absence of reflexes, pin out the animal exposing its abdomen. Make a mid-line incision in the skin on the abdomen and fold back.
3. Cut through the abdominal muscle layer exposing the organs. Cut up to the thorax.
4. Cut through the diaphragm and the ribcage on either side exposing the thoracic cavity.
5. Using a cannula needle pierce the apex of the left ventricle and make a nick in the right atrium (see Note 7).
6. Using a large syringe attached to the cannula needle push through at least 10 ml of PBS until the fluid leaving the heart at the atrium is clear.
7. Once the fluid is running clear switch to PFA, pushing through at least 10 ml (see Note 8).
8. Remove required organs, chop into 5 mm^2 pieces for processing and incubate in PFA for at least 48 h, but no more than 1 week.
9. The tissue is now ready for processing. The tissue is dehydrated by incubation in increasing concentrations of ethanol, cleared of ethanol using Xylene and infiltrated with paraffin wax (see Note 9).
10. Embed the tissue in paraffin blocks and cut 5 μm thin sections.
11. Float onto coated slides and store at room temperature.
12. Rehydrate sections to prepare for staining; Histoclear (2 × 1 min), 100% ethanol (2 × 1 min), 95% ethanol (1 × 1 min), 70% ethanol (1 × 1 min), water (2 × 1 min), PBS (2 × 5 min) and leave sections in PBS.

13. Using a PAP pen draw circles around each section and apply blocking solution A. Place slide in humidified container and incubate at 37°C for 10 min.
14. Flick off solution A, but do not wash before applying a drop of solution B to the section. Place in a humidified container and incubate at 37°C for 10 min.
15. Wash 3 × 5 min in PBS.
16. Apply 50 μl of IHC solution containing 1:50 dilution of anti-LDLR and a 1:500 dilution of anti-β-galactosidase to each section and incubate overnight at 4°C. Control sections – anti-LDLR only, anti-β-galactosidase only, primary antibody only and secondary antibody only.
17. Wash 3 × 5 min in PBS.
18. Apply secondary antibody for not more than 1 h. Do not expose to light.
19. Wash 3 × 5 min in PBS (in a light-tight box).
20. Incubate in DAPI solution for 10 min (in a light-tight box).
21. Wash 3 × 5 min in PBS (in a light-tight box).
22. Perform a wet mount and analyse using a fluorescence microscope (Fig. 5).

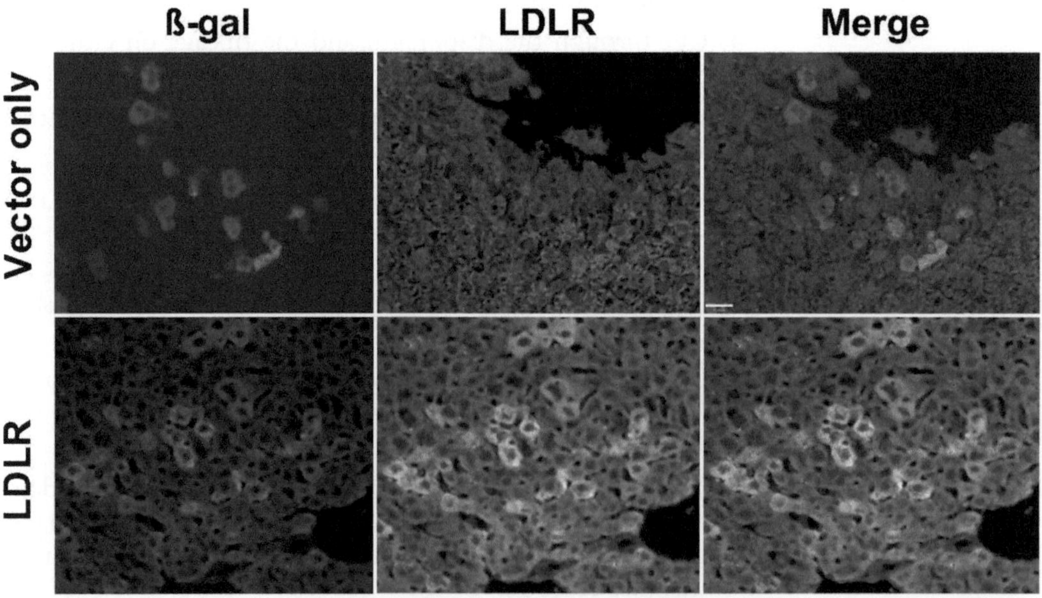

Fig. 5. Hydrodynamic tail vein injection of a plasmid expressing β-galactosidase from a CMV promoter and LDLR from the *LDLR* genomic promoter results in efficient transfection of hepatocytes in vivo. Here we show that expression of human LDLR protein is detectable and co-localises with β-galactosidase. Sections are co-stained with antibodies specific to β-galactosidase and human LDLR and counterstained with the DAPI nuclear stain. The livers show co-localisation of staining for human LDLR and β-galactosidase. This co-localisation is absent in livers from animals injected with a plasmid that only expresses β-galactosidase. This liver is only positive for β-galactosidase expression. For colour image, see Ref. (8).

3.4.3. Transfection Efficiency Analysis (Live Animal Imaging)

This protocol describes live animal imaging using an IVIS 100 luciferase imaging camera (Caliper Life Sciences, Hopkinton, MA) (8).

1. Anaesthetise animals with 5% isofluorane in 2 l/min oxygen.
2. Once induced dose animals with 100 μl of luciferin via intraperitoneal injection.
3. Place the animals inside the chamber and maintain anaesthesia at 2% isofluorane in 2 l/min oxygen.
4. After a 4-min incubation period from the time of luciferin injection, image the animals for luciferase expression (Fig. 6).

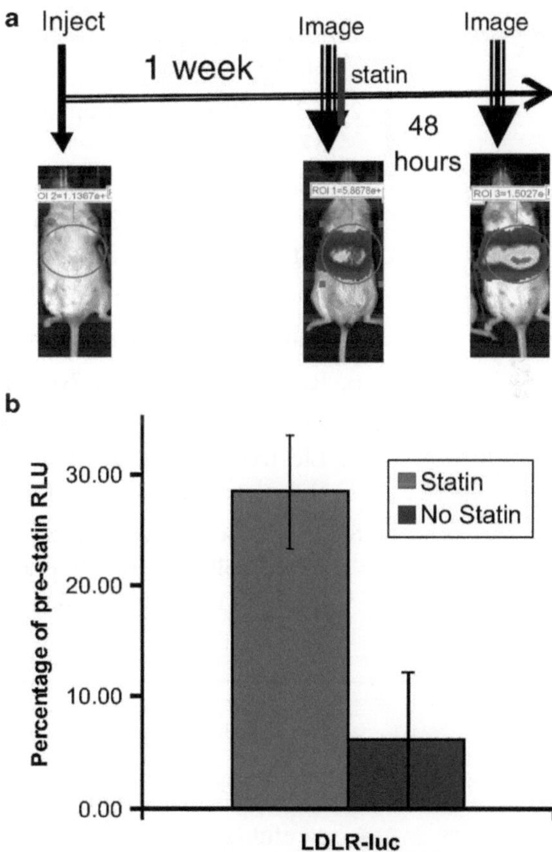

Fig. 6. Luciferase expression in vivo following hydrodynamic tail vein injection is robust and sensitive to drug administration. (**a**) Experimental time-line of a statin administration protocol showing representative luciferase activity images, hydrodynamic injection (*black line*), luciferase expression imaging (*triple line*) and statin administration (*large arrow*). (**b**) Administration of a single dose of 600 mg/kg of pravastatin resulted in five-fold more luciferase expression. Luciferase levels are expressed as a percentage of the luciferase levels calculated from the pre-statin administration imaging.

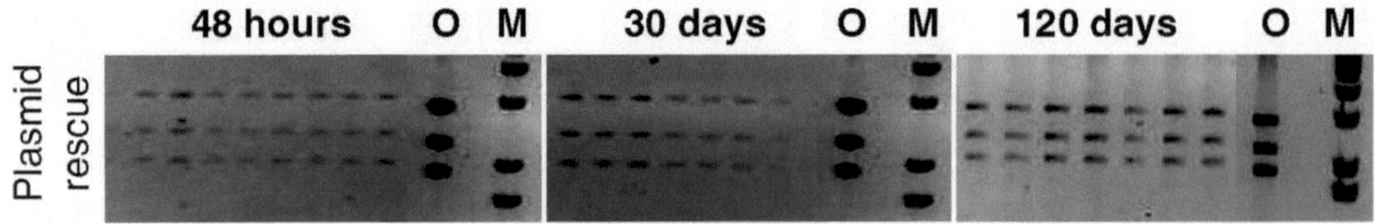

Fig. 7. Plasmid rescue from genomic DNA isolated from animals injected with an EBV containing plasmid. Plasmid retention is demonstrated by the presence of the plasmid as a circular episomal element that can be rescued 48 h, 30 days and 120 days following injection. 0, original injected plasmid; M, marker.

5. Analyse luciferase expression levels using LivingImage software that allows you to quantify amount of luciferase in photons/s (Fig. 6).
6. Allow animals to recover in home cage.

3.4.4. Plasmid Rescue from Tissue

1. Homogenise small pieces of frozen liver in genomic lysis buffer.
2. Add 100 μl of proteinase K and incubate for at least 24 h at 37°C.
3. Perform a phenol:chloroform extraction as per items 6 and 7 in Subheading 2.2.1 using light phase lock gel tubes (Eppendorf).
4. Precipitate DNA in absolute ethanol.
5. Wash in 70% ethanol and leave DNA pellets to air dry.
6. Resuspend DNA in 200 μl of Tris–EDTA at room temperature for about 48 h.
7. Electroporate 1 μl of genomic DNA into DH10B bacteria and plate onto appropriate LB agar plates.
8. To confirm circular plasmid DNA make small cultures from single colonies, purify plasmid DNA using alkaline lysis, and perform restriction enzyme digestion (Fig. 7).

4. Notes

1. This protocol can be particularly problematic and requires careful optimisation. There are a variety of things that can be altered to optimise the protocol. Annealing temperature is a key variable, also extension time; most protocols recommend 1 min per kilobase. Addition of agents such as Q solution (Qiagen) or glycerol can result in better long-range products.
2. We have found that linearising the plasmid between the homology arms results in a more efficient recombination. This step can however, be left out.

3. The amount of each plasmid used in the retrofitting protocol will be dictated by the relative sizes of the plasmids. If you are retrofitting a large BAC plasmid that is 100 kB with a small retrofitting plasmid that is 10 kB you will want to have a tenfold excess of BAC plasmid.
4. The easiest way to pick a clone is to identify the cell population down the microscope using the 4× objective. Then, depress the plunger of the pipette and, still looking down the microscope, position the tip of the pipette tip in the media over the top of the clone. Once you can see the shadow of the tip down the microscope slowly manoeuvre it to be immediately adjacent to the cells. Slowly release the plunger aspirating the clonal cells.
5. It is very important that the animals are adequately heated before injection. Heating the animals properly will not only allow easy visualisation of the vein, in our experience the injection is more effective and the recovery of the animal is quicker. The animals should not however be allowed to overheat. For this reason it is important to not allow the heating chamber to get above 40°C and to not leave the animals in the chamber for longer than 10 min. If an animal does start to show signs of overheating such as sweating around the neck and face, remove the animal from the chamber to its home cage, allow free access to water and wait at least 24 h before reattempting injection.
6. Hydrodynamic tail vein injection is a high speed high pressure tail vein injection. Once the needle is in the vein the entire 2.5 ml injection solution should be injected in no more than 10 s. Slow injection speed is a major cause for poor transfection efficiency.
7. It is important that the cannula is correctly placed in the ventricular cavity and not in the wall or pushed too far through into the atrium. If the needle is poorly placed, the perfusion will not work. If the needle is placed correctly when the PBS is pushed through there should be an obvious flow from the nick in the right atrium and you will notice organs such as the liver turning increasingly pale.
8. Once the flow of PFA begins it is normal for the muscles to involuntarily contract.
9. The process of tissue processing is lengthy and takes about 8 h. For this reason it is easier to use an automatic tissue processor such as the Leica TP1020 tissue processor (Leica, Milton Keynes, UK). However, if this is not available, use the following protocol; 70% ethanol (2 × 20 min), 95% ethanol (2 × 20 min), 100% ethanol (2 × 20 min), Xylene (2 × 20 min), paraffin (65°C, 2 × 30 min). Take care not to leave the tissue blocks in hot paraffin for longer than the specified time as this can dry out the block and make it very difficult to cut.

References

1. Hibbitt OC, Wade-Martins R. (2006) Delivery of large genomic DNA inserts >100 kb using HSV-1 amplicons. Curr Gene Ther;6(3):325–36.
2. Lufino MM, Edser PA, Wade-Martins R. (2008) Advances in high-capacity extrachromosomal vector technology: episomal maintenance, vector delivery, and transgene expression. Mol Ther;16(9):1525–38.
3. Hacein-Bey-Abina S, von Kalle C, Schmidt M, et al. (2003) A serious adverse event after successful gene therapy for X-linked severe combined immunodeficiency. N Engl J Med;348(3):255–6.
4. Bushman F, Lewinski M, Ciuffi A, et al. (2005) Genome-wide analysis of retroviral DNA integration. Nat Rev Microbiol;3(11): 848–58.
5. Wade-Martins R, Saeki Y, Chiocca EA. (2003) Infectious delivery of a 135-kb LDLR genomic locus leads to regulated complementation of low-density lipoprotein receptor deficiency in human cells. Mol Ther;7(5 Pt 1): 604–12.
6. Hibbitt OC, Harbottle RP, Waddington SN, et al. (2007) Delivery and long-term expression of a 135 kb LDLR genomic DNA locus in vivo by hydrodynamic tail vein injection. J Gene Med;9(6):488–97.
7. Lufino MM, Manservigi R, Wade-Martins R. (2007) An S/MAR-based infectious episomal genomic DNA expression vector provides long-term regulated functional complementation of LDLR deficiency. Nucleic Acids Res;35(15):e98.
8. Hibbitt OC, McNeil E, Lufino MM, Seymour L, Channon K, Wade-Martins R. (2010) Long-term Physiologically Regulated Expression of the Low-density Lipoprotein Receptor In Vivo Using Genomic DNA Mini-gene Constructs. Mol Ther 18(2): 317–26. Epub 2009 Oct 27.
9. Goldstein JL, Kita T, Brown MS. (1983) Defective lipoprotein receptors and atherosclerosis. Lessons from an animal counterpart of familial hypercholesterolemia. N Engl J Med;309(5):288–96.
10. Tolleshaug H, Hobgood KK, Brown MS, Goldstein JL. (1983) The LDL receptor locus in familial hypercholesterolemia: multiple mutations disrupt transport and processing of a membrane receptor. Cell;32(3):941–51.
11. Cichon G, Willnow T, Herwig S, et al. (2004) Non-physiological overexpression of the low density lipoprotein receptor (LDLr) gene in the liver induces pathological intracellular lipid and cholesterol storage. J Gene Med;6(2):166–75.
12. Heeren J, Steinwaerder DS, Schnieders F, Cichon G, Strauss M, Beisiegel U. (1999) Nonphysiological overexpression of low-density lipoprotein receptors causes pathological intracellular lipid accumulation and the formation of cholesterol and cholesteryl ester crystals in vitro. J Mol Med;77(10):735–43.
13. Zhang G, Gao X, Song YK, et al. (2004) Hydroporation as the mechanism of hydrodynamic delivery. Gene Ther;11(8):675–82.
14. Zhang G, Song YK, Liu D. (2000) Long-term expression of human alpha1-antitrypsin gene in mouse liver achieved by intravenous administration of plasmid DNA using a hydrodynamics-based procedure. Gene Ther;7(15): 1344–9.
15. Zhang X, Dong X, Sawyer GJ, Collins L, Fabre JW. (2004) Regional hydrodynamic gene delivery to the rat liver with physiological volumes of DNA solution. J Gene Med; 6(6):693–703.

Chapter 3

Naturally Occurring Minichromosome Platforms in Chromosome Engineering: An Overview

Elena Raimondi

Abstract

Artificially modified chromosome vectors are non-integrating gene delivery platforms that can shuttle very large DNA fragments in various recipient cells: theoretically, no size limit exists for the chromosome segments that an engineered minichromosome can accommodate. Therefore, genetically manipulated chromosomes might be potentially ideal vector systems, especially when the complexity of higher eukaryotic genes is concerned.

This review focuses on those chromosome vectors generated using spontaneously occurring small markers as starting material. The definition and manipulation of the centromere domain is one of the main obstacles in chromosome engineering: naturally occurring minichromosomes, due to their inherent small size, were helpful in defining some aspects of centromere function. In addition, several distinctive features of small marker chromosomes, like their appearance as supernumerary elements in otherwise normal karyotypes, have been successfully exploited to use them as gene delivery vectors. The key technologies employed for minichromosome engineering are: size reduction, gene targeting, and vector delivery in various recipient cells. In spite of the significant advances that have been recently achieved in all these fields, several unsolved problems limit the potential of artificially modified chromosomes. Still, these vector systems have been exploited in a number of applications where the investigation of the controlled expression of large DNA segments is needed. A typical example is the analysis of genes whose expression strictly depends on the chromosomal environment in which they are positioned, where engineered chromosomes can be envisaged as epigenetically regulated expression systems. A novel and exciting advance concerns the use of engineered minichromosomes to study the organization and dynamics of local chromatin structures.

Key words: Small supernumerary marker chromosomes, Centromere, Chromosome engineering, Size reduction, Gene targeting, Chromosome transfer

1. Introduction

To date, two main approaches have been followed to establish higher eukaryotic chromosome vectors, these are currently known as "bottom up" and "top down" respectively. The "bottom up"

Gyula Hadlaczky (ed.), *Mammalian Chromosome Engineering: Methods and Protocols*, Methods in Molecular Biology, vol. 738, DOI 10.1007/978-1-61779-099-7_3, © Springer Science+Business Media, LLC 2011

approach, as first demonstrated for the construction of yeast artificial chromosomes (YACs) in *Saccharomyces cerevisiae* (1), consists in the assembly of simple DNA sequences in order to obtain minimal functional chromosomes. Conversely, the "top down" approach is a process of step-by-step size reduction of naturally occurring chromosomes, aiming at the conservation of only those sequence elements that are necessary and sufficient for proper chromosome function.

Artificially built up and artificially modified chromosomes to be used for genetic engineering in higher eukaryotes should possess all the features that characterize natural chromosomes. This means that they should have functional telomeric ends or otherwise they should be circular; moreover they should correctly segregate during mitosis and meiosis (they should contain a functional centromere capable to recruit kinetochore proteins) and finally they should undergo the canonical condensation process, in other words they should be large enough to contain the cis-acting elements that regulate chromatin condensation. In addition, engineered chromosomes should have a defined genetic content and should behave as highly stable non-integrating vectors able to accommodate large DNA fragments. Further points, critically affecting artificial chromosomes as valuable chromosomal vectors for gene transfer in higher eukaryotic cells, are the presence of an effective homologous recombination system, acting as a cloning site, and the development of high efficiency transduction procedures. Here the discussion will be limited to artificially engineered minichromosomes derived from supernumerary small markers occasionally occurring in wild populations. However, for a better understanding of the following arguments, some general points must be addressed.

2. Non-integrating Large Capacity Vectors

Studies on gene delivery systems based on chromosome vectors were strongly stimulated by the need to find appropriate tools for the genetic modification of the human genome, not only because they could provide novel gene therapy approaches, but also because they should improve the knowledge of the organization and function of complex genomes, particularly when the genes or chromosomal regions of interest are too large to be handled with the currently available cloning and delivery systems. In this respect, chromosomal vectors could offer a number of advantages compared to viral vectors. First, engineered chromosomes are nonintegrating thus, on the one hand, they are not affected by site-specific position effects and on the other hand they are not associated to deleterious effects caused by insertional mutagenesis.

Second, the size of the DNA segments that they can accommodate is theoretically unlimited and therefore they represent promising tools to generate animal models for contiguous gene syndromes and chromosomal diseases. Third, most eukaryotic genes are very large and their regulated expression often depends on genomic control elements that can reside at a substantial distance from the coding sequences themselves. Finally, proper expression of most eukaryotic genes strictly depends on the chromosomal environment in which they are positioned, therefore engineered chromosomes might be used as epigenetically regulated expression systems as well as model systems to investigate chromatin architecture.

2.1. The Centromere Is the Key Component of Chromosome Vectors

Any effort aimed at building a chromosome vector requires as a prerequisite the exact knowledge of the sequence elements that are crucial for chromosome function. Approaches to engineer higher eukaryotic chromosomes have been hampered for a long time by the absence of detailed knowledge on the exact nature of the centromere. The only eukaryotic centromere that could be defined at the sequence level using a shot-gun cloning approach was the point centromere of budding yeasts (2); indeed point centromeres are quite small and compact (about 125 bp DNA) and support a kinetochore with only one microtubule attachment; they contain three well defined DNA sequence domains that serve as binding sites for essential kinetochore proteins. Conversely, in the case of regional centromeres, which are very large and support many kinetochore-microtubule attachments, the picture becomes much more complex; as an example, *Schizosaccharomyces pombe*, despite being a close relative of the budding yeast, presents regional centromeres consisting of 40–100 kb of DNA organized into distinct classes of centromere-specific repeats (3–7). The scenario becomes even more complicated when higher eukaryotes are considered, as most eukaryotes have widespread regional centromeres that contain large amounts of DNA, typically organized in very large arrays of repetitive sequences. An intriguing point is that, despite the essential function of the centromere, the DNA that comprises it is rapidly evolving and varies from species to species; in addition, examples of eukaryotic centromeres completely devoid of clusters of repetitive DNA sequenced have been described (named neo-centromeres, see the next paragraph); ultimately, satellite DNA contributes to, but does not define centromere function (reviewed in (8–11)).

A further point that deserves discussion is the existence of transcribed sequences at the centromeres of the majority of eukaryotes. As a matter of fact, the silenced heterochromatin should not be transcribed by definition (reviewed in (12)), still there have been frequent reports of low-level transcription in heterochromatic regions ((13, 14), for a review see Ref. (15)) and some examples of transcriptional activation of centromeric specific sequences (reviewed in (16)), the most studied one being that involved in cell stress response (17, 18). Furthermore, recent investigations implicate RNA interference mechanisms in targeting and maintaining heterochromatin, and these mechanisms are inherently dependent on transcription (reviewed in (19)). In some cases, transcription of the region to be silenced seems to be required for silencing itself. As it will be explained later on, when building an artificially modified minichromosome via chromosome fragmentation, the majority of single copy and pericentromeric repetitive sequences can be removed, but the centromere itself has not to be affected to obtain correctly segregating end products. Therefore, centromeric transcriptional competence might have detrimental effects when the engineered chromosome is transferred into the final recipient.

Neo-centromeres are particularly interesting in this respect and have been actually used as platforms for the construction of engineered human chromosomes (20, 21). Neo-centromeres are fully functional ectopic centromeres first described in human pathology (22). Two intriguing features of neo-centromeres are the complete absence of satellite DNA sequences and the localization at euchromatic chromosome regions. Different hypotheses have been proposed to explain the mechanism of neo-centromere formation; however they all share the assumption that epigenetic signals might be occasionally recruited by non centromeric DNA sequences. As an example, chromosome rearrangements could trigger neo-centromere formation, possibly through a change in the epigenetic state of the chromatin following DNA repair (10). As neo-centromeres are of small size and contain single copy genomic DNA along with interspersed repeated sequences, they are amenable to full sequence characterization; thus, it should be possible to identify genes that may have negative effects and also to investigate neo-centromere formation outcome on gene expression.

Another significant example of satellite-free natural centromeres has been recently reported in the genus *Equus* (23, 24). Equids show an extraordinarily high evolution rate that has been accompanied by rapid karyotypic evolution, with a high frequency of centromere repositioning (centromere shift generating evolutionary neo-centromeres) (25). The phylogenetic reconstruction of centromere repositioning events showed that the acquisition of satellite DNA occurs after the formation of the centromere during evolution and that centromeres can function over millions of years without detectable satellite DNA. Specifically, it was shown, by a detailed sequence analysis, that the centromere of horse chromosome 11 is completely devoid of repetitive DNA sequences (23). In addition, FISH data strongly suggested that eight donkey and one Burchell's zebra evolutionary neo-centromeres lack satellite DNA as well (24). Thus, equids might provide new resources for the construction of artificially modified minichromosomes.

2.2. Naturally Occurring Minichromosomes

A number of animal, plant, and fungi species contain B chromosomes (also termed supernumerary or accessory chromosomes), in addition to the normal chromosome set (A chromosomes). In most of the individuals of wild populations B chromosomes are missing, therefore they should be nonessential for the survival of the species. Generally speaking, B chromosomes are largely noncoding, but some supernumeraries may contain euchromatic segments and they seem to be able to persist in a species; it is likely that these extra chromosomes may provide some positive adaptive advantage (26).

In humans, the presence of supernumerary chromosomes is an unusual finding, which may or may not be associated with developmental abnormalities and malformations. Most human supernumerary chromosomes derive from the A chromosome set and are mitotically stable. In some cases, transmission in families has also been observed. Human supernumerary marker chromosomes have been used as platforms for the construction of engineered minichromosomes (20, 21, 27–29), because they display distinctive features that would make them suitable as gene therapy vectors.

Small marker chromosomes are present in 0.075% of prenatal cases and occur in about one out of 3,000 live newborns. It has been shown that approximately 70% of de novo cases have no phenotypic effects (reviewed in (30–32)) and all the clinical cytogenetic data available, concerning the behavior during cell division and the clinical effects of small supernumerary marker chromosomes, suggest that human cells may be tolerant of the presence of a supernumerary centromere; as a consequence it is possible to foresee the in vivo effect of the engineered end product.

Another feature of naturally occurring minichromosomes that makes them convenient in view of the construction of chromosomal vectors, is the intrinsic enrichment in the set of sequences required to fulfill chromosome function; this is a prerequisite, which simplifies the following manipulation steps.

On average, the use of spontaneously occurring minichromosome platforms should be extremely favored, primarily when the knowledge of the effects of the chromosome vector on the host genome is crucial.

2.3. Size Reduction of Natural Minichromosomes

Size reduction is a key step towards the generation of engineered minichromosomes, the final goal being to obtain a "minimal" normally behaving chromosome that does not impair cell functions. Stepwise removal of chromosome segments unnecessary for

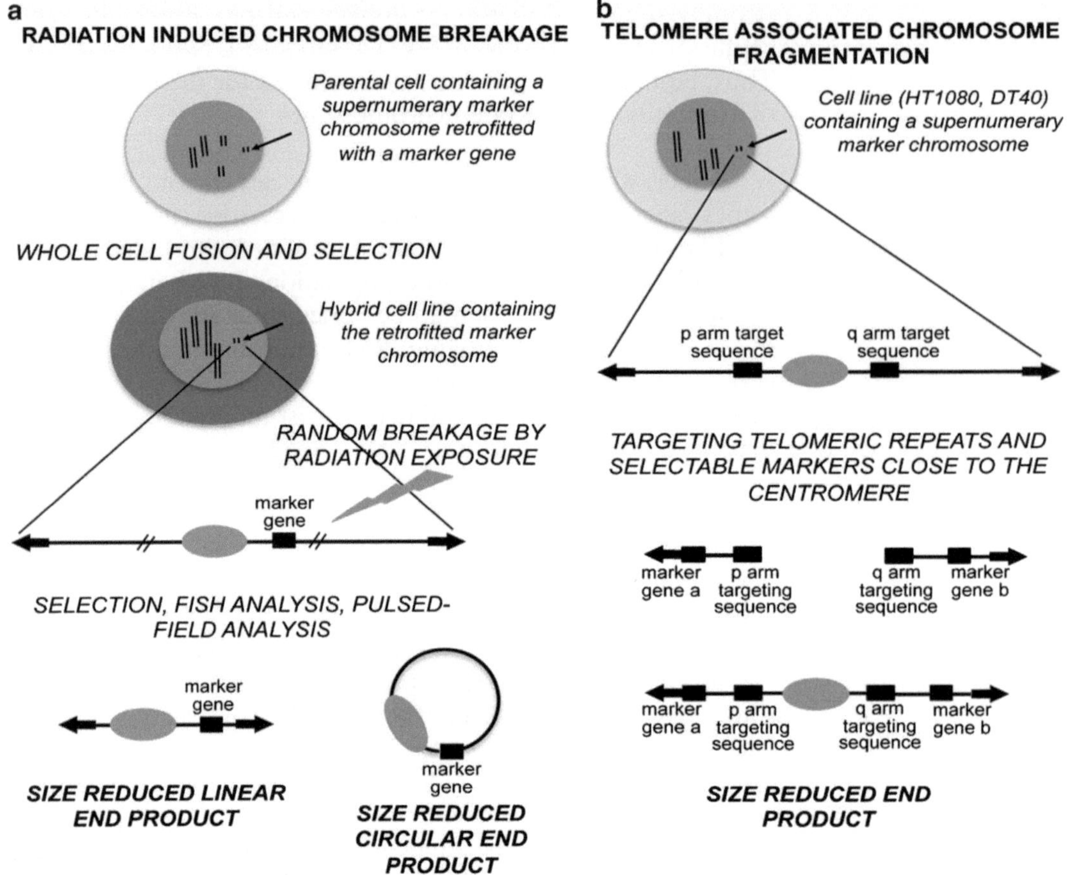

Fig. 1. Schematic representation of the key steps for minichromosome size reduction. Two alternative routes can be followed: in (**a**), radiation induced chromosome breakage, through the generation of an intermediate hybrid host cell, is represented; the (**b**) panel shows telomere associated chromosome fragmentation through targeting of telomeric repeats.

proper chromosome function can be ultimately attained following two main procedures (Fig. 1): radiation induced chromosome breakage and telomere associated fragmentation (TACF).

The well-known clastogenic effect of ionizing radiations has been exploited to break natural chromosomes in a number of applications. This technique relies on the uncontrolled radiation-induced breakage of chromosomal arms and is the method of choice for the production of radiation hybrids (33). As sketched in Fig. 1a, chromosome manipulation is generally performed into an intermediate host cell. To this purpose, the marker chromosome is transferred into an hybrid cell line by means of whole cell fusion; for the subsequent manipulation steps, it is mandatory to have a selectable marker gene as close as possible to the centromere of the minichromosome. Thus, in most cases, the marker chromosome is retrofitted with a drug resistance gene.

After radiation exposure, different deletion derivatives are generated and maintained as free elements in the selected hybrid colonies; these can be both circular and linear deletion products, with telomere healing performed by the host cell machinery (Fig. 1a). However, since chromosome breakage is completely random, the selection of suitable deletion derivatives can be performed only a posteriori, which represents a serious drawback in the application of radiation induced size reduction.

Telomere associated chromosome fragmentation is by far the most employed strategy (Fig. 1b). The technique relies on the integration of cloned human telomeric DNA into an endogenous chromosome arm. Functional activation of the new telomere, which leads to the loss of the chromosome segment downstream of the insertion site, can occasionally occur, giving rise to the de novo formation of a stable, size reduced chromosome. In general, telomere sequences integrate at random, but targeting procedures and effective selection strategies were designed to isolate the deletion products (34, 35). Homologous recombination occurs sporadically in mammalian somatic cells, therefore various strategies to insert the telomeric repeat as close as possible to the centromere have been developed. In the first attempts, chromosome specific satellite DNA was used as a carrier in co-transformation experiments (27, 35). The targeting efficiency observed in these pilot experiments was surprisingly high, but random integration could not be prevented and there was no way to control the number of integrated copies at each target site. An alternative approach made use of recombination proficient chicken DT40 cells as intermediate hosts for chromosome manipulation (36). More recently, other targeting designs have been described that require an exact knowledge of the sequence of the chosen integration region and are performed knocking down selectable marker genes (21, 37). A promising improvement in gene targeting has been recently achieved by means of "recombineering," a technique that utilizes the homologous recombination functions encoded by gamma phage to construct knockout vectors (reviewed in (38)).

The first paper which described the construction of an engineered "minimal" minichromosome by means of size reduction of a supernumerary small precursor, was performed in the fission yeast *S. pombe* (39). The authors used as starting material an aneuploid strain, which carried a partial disomy for chromosome III. Deletions in the left or the right arm of the minichromosome were induced by gamma ray exposure. Deletion derivatives, for loci placed on each chromosome arm close to the centromere, were isolated after phenotypic selection. The size of the minichromosomes obtained was then tested by pulsed-field gel electrophoresis, while functional tests allowed the verification of the mitotic and meiotic behavior of the deletion derivatives.

Yeasts have a number of characteristics (ease and economy in laboratory handling, haplontic life cycle, and high number of selectable phenotypic mutants) that make them ideal organisms for studies of cell cycle and chromosome segregation. Unfortunately, it was realized very soon that they are inappropriate model systems for higher eukaryotic chromosome engineering. Indeed, metazoan chromosomes are much larger and more complex than the yeast ones, with major differences concerning the condensation cycle and the size and identity of centromeric domains. Besides, in higher eukaryotes, the number of phenotypic selectable markers is generally exiguous and functional tests are laborious.

A significant advancement in minichromosome engineering was represented by the localization of the DNA elements responsible for centromere activity in *Drosophila melanogaster* (40). *Drosophila* is a model organism in genetics and cytogenetics since over a century, therefore many genetic and cytological assays for chromosome functions in vivo exist. In addition, a *D. melanogaster* strain was isolated that contained a fully functional X-chromosome-derived minichromosome, which was used for molecular dissection of chromosome functions. Deleted minichromosomes were generated by irradiation mutagenesis, and their molecular structures were determined by pulsed-field electrophoresis and Southern blot analysis. Minichromosomes, retaining only the indispensable elements controlling chromosome function, were produced (40).

An experimental strategy very similar to that outlined for the construction of artificially modified chromosomes in *S. pombe* and in *D. melanogaster* was designed to generate human size reduced minichromosomes (27). The authors used as starting material a chromosome-9-derived accessory minichromosome (estimated size 20 Mb) found in a human karyotype. Thus, in this case, as in the previously reported examples, the original chromosome was per se enriched in sequence elements necessary for proper function; moreover the marker chromosome did not seem correlated with pathological symptoms and was apparently devoid of any detrimental effect on chromosome mechanics (41, 42). This was an important prerequisite for further manipulation as it strongly suggested that the marker chromosome did not carry extra copies of DNA sequences that could impair normal cell functions. The authors developed a co-transformation procedure, based on the use of chromosome specific subcentromeric satellite DNA as a transformation carrier, to target the *neo* gene to the marker chromosome. This minichromosome, retrofitted with the selectable gene, was subsequently size reduced by X-ray exposure (smallest end product estimated size 4.7 Mb) and modified with a *lox-P* site (see Subheading 2.4), enabling the recombinogenic introduction of a reporter cassette (29). The molecular structure and the size of the deleted derivatives were determined by pulsed-field electrophoresis, their mitotic stability was checked by anaphase analysis and finally, the ability to bind centromeric proteins was verified by immunofluorescence.

Other variations of the "spontaneously arising minichromosome" platform have been described (20, 21, 28). Of particular interest are those cases in which chromosomes containing a neo-centromere were used to develop chromosome vectors (20, 21). Minichromosomes lacking centromeric satellite DNA, but harboring a functional neo-centromere, might be especially attractive for the construction of engineered chromosomes with relatively small centromeric domain that can be characterized at the sequence level.

More recently, engineered minichromosomes have been generated in plants. B chromosomes occur at high frequency in plants, and it is known that plants well tolerate gene dosage unbalance, therefore these organisms are particularly suitable for chromosome engineering. Plant artificially modified minichromosomes have been built by radiation-induced chromosomal breakage, by breakage-fusion-bridge (BFB) cycles and by telomere mediated chromosomal truncation (reviewed in (43)). Yu and colleagues (44, 45) produced engineered minichromosomes, derived from both A and B chromosomes, in maize. The authors developed an efficient method of telomere associated chromosomal truncation. *Agrobacterium*-mediated gene transformation was employed to insert constructs with multiple copies of the telomere sequence and at the same time to integrate site-specific recombination cassettes in maize embryos. Transgenes expression was obtained as well as transmission from one generation to the next. The construction of plant chromosome vectors could have a number of applications in plant genetic engineering, especially in those instances where the simultaneous expression of multiple genes is needed.

2.4. Site-Specific Recombination in Engineered Minichromosomes

An artificially modified minichromosome cannot be used as a vector for gene transfer unless it carries a cloning site which enables to insert a controlled number of copies of a given sequence into a known site, thus providing the stable expression of the transgene/s. In addition, since the ability to deliver large inserts is inherent by the nature of chromosome vectors, a system to target large DNA fragments needs to be developed.

Effective homologous recombination systems used in minichromosome engineering are mainly based on site-specific recombination mediated by *Cre–loxP*. The *Cre–loxP* approach allows the control of both the site-specific integration and copy number of the transgene, as well as the replacement or deletion of any specific segment within a target sequence (Fig. 2). The *Cre* recombinase from bacteriophage P1 catalyzes the exchange

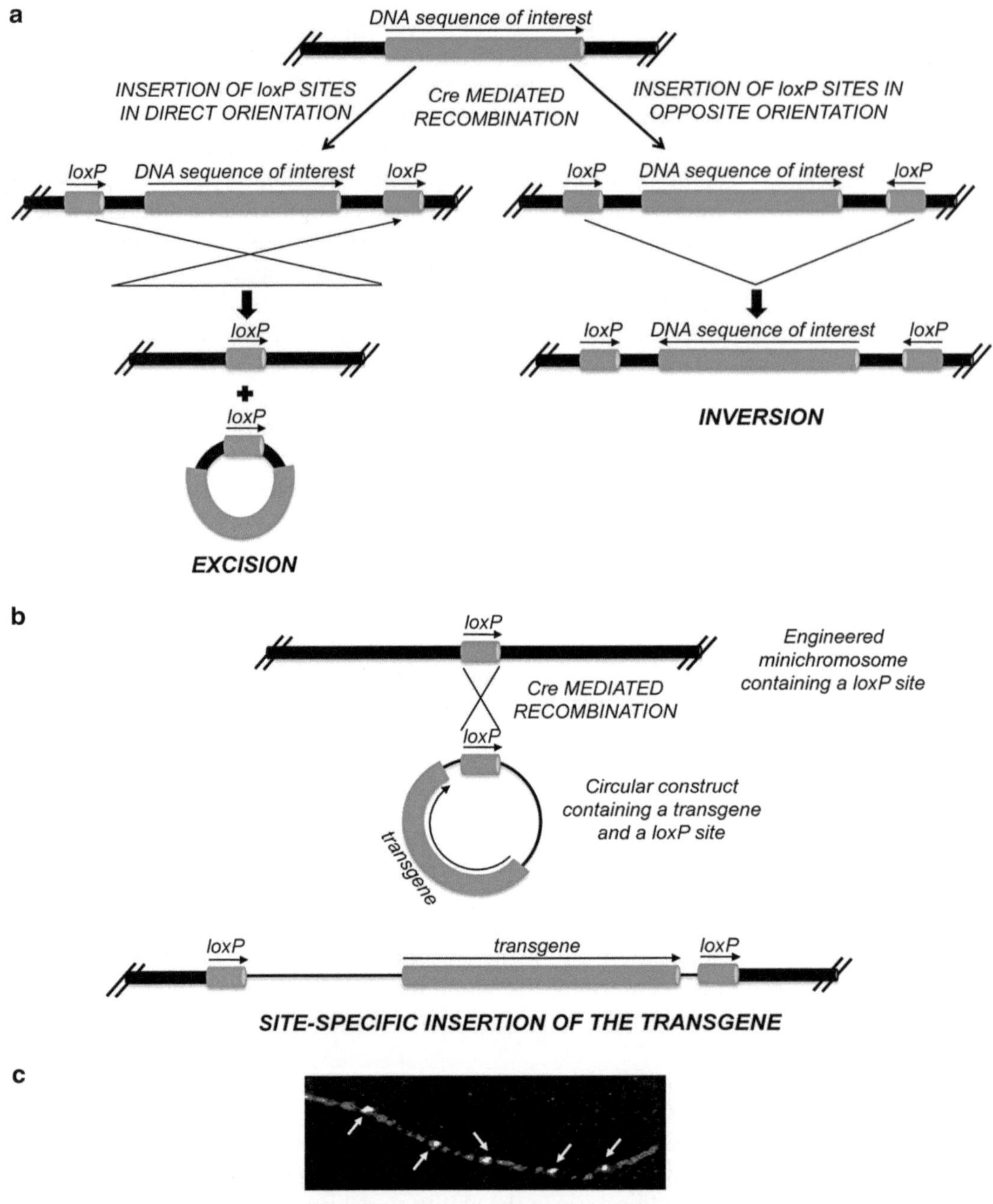

Fig. 2. *Cre–loxP* mediated homologous recombination. In (**a**), the mechanism of *loxP* mediated excision and inversion of chromosome segments is outlined. The (**b**) panel sketches the mechanism of site-specific insertion of a given exogenous DNA sequence. An example of site-specific gene targeting is shown in (**c**). Two color FISH on chromatin fibers enables to demonstrate that a single copy of a reporter gene (*bright fluorescent spots*), is inserted into each of the *loxP* sites (*darker fluorescent signals*) carried by a small size-reduced supernumerary marker chromosome (29). In all, five copies of the *loxP* sequence and of the reporter gene are present (*white arrows*).

between specific boxes within two *loxP* sites by concerted cleavage and rejoining reactions. The recombination between two *loxP* sites inserted in direct orientation on the same DNA molecule results in the deletion of the framed DNA (Fig. 2a, left side), while recombination between two *loxP* sites in opposite orientation leads to its inversion (Fig. 2a, right side) (46, 47). The recombination between two *loxP* sites placed on different DNA molecules mediates the translocation or site-specific insertion of an exogenous DNA sequence (Fig. 2b) (46, 47). In Fig. 2b, the mechanism by which a transgene can be targeted to an engineered minichromosome vector is shown: the *Cre–loxP* mediated recombination allows the insertion of exactly one copy of a given sequence at each *loxP* site (Fig. 2c), thus offering a very effective copy number control procedure.

The limiting step in minichromosome modification with recombinase recognition sites is the targeting of the recombination cassette to the minichromosome itself. However, this problem can be bypassed exploiting methods similar to those used for the targeting of telomeric repeats to minichromosomes (see previous section).

Site-specific homologous recombination schemes have also been designed to engineer minichromosomes in recombination proficient chicken DT40 cells. These procedures can be coupled with TACF and *Cre–loxP* mediated targeting, obtaining, at the same time, size reduction and genetic modification of the chromosome vector (36, 48, 49).

Once the minichromosomes have been modified by *loxP* integration, they can be easily retrofitted with protein expression cassettes, reaching site-specific insertion rates close to 50% and a stringent control of the number of copies of the inserted transgenes. Such modified minichromosomes could be theoretically used as platforms for the site-specific integration of any exogenous sequence. Indeed, site-specific integration and long-term expression has been obtained in a number of cases, the most recent improvements concerning the tissue specific control of transgenes expression and the delivery of very large DNA segments such as whole chromosome regions. In the latter case, *Cre–loxP* mediated targeting was employed to develop a chromosome-cloning system in which a defined human chromosomal region can be cloned into a stable minichromosome vector. This technique also allowed the introduction of defined rearrangements into mouse ES (Embryonic Stem) cells (50, 51). Transchromosomal mice that carry engineered chromosomal deletions were successfully used to model the human chromosomal deletions that are responsible for the DiGeorge, the Prader–Willi, and the Smith-Magenis syndromes (reviewed in (52)). Animal models for deletion syndromes should lead to a better understanding of the molecular basis of gene dosage alterations opening new therapeutic frontiers. A highly promising recent breakthrough was the delivery of the entire *dystrophin* gene, which spans 2.4 Mb, into a minichromosome vector derived from a truncated human chromosome 21 through *loxP* recombination (53). The engineered construct was shown to be stable in human immortalized mesenchymal stem cells. In addition, tissue specific expression of the appropriate isoforms of dystrophin was obtained in chimeric mice. These studies are particularly significant because they provide direct evidence that minichromosome engineering technology can be successfully employed to deliver and express large genomic constructs. An exciting new development, exploiting engineered chromosome vectors to understand the comprehensive mechanisms by which chromatin organization modulates gene regulation, was recently published by Ikeno and colleagues (54). These technologies might have significant applications in basic research as well as in applied research in a number of fields of study ranging from regenerative medicine to plant transgenesis.

2.5. Transfer of Engineered Minichromosomes Between Cell Types

One of the main reasons for creating engineered minichromosomes is that they can vehicle and maintain exogenous genes into the cells of an organism. Chromosome manipulation is currently performed in intermediate recipient cells such as HT1080 (human fibrosarcoma), CHO (Chinese Hamster Ovary), or DT40 (recombination proficient chicken cells) that can be easily handled; however, these are only transient hosts and the end products need to be transferred to other recipients, such as primary cells or stem cells. Given the typical chromosome vectors size and ability to accommodate very large DNA fragments, conventional transfection approaches are inappropriate. Consequently, ad hoc strategies have been developed to deliver intact chromosome vectors into the final recipient cells, although it should be stressed that transfer efficiency remains the main and largely unsolved problem in minichromosome engineering.

Microcell mediated chromosome transfer (MMCT) is the method of choice to shuttle engineered chromosomes between cells. MMCT was originally developed in the 1970s to transfer exogenous chromosome material into host cells forming hybrid cell lines (55, 56); this procedure was first used to generate transgenic animals carrying freely segregating human chromosomes or fragments of human chromosomes in 1997 by Tomizuka and coworkers (57). Tomizuka's studies were fundamental to highlight the potential of chromosome engineering combined with MMCT that was used successfully ever since to deliver artificially modified chromosomes into different hosts. The rationale of the MMCT technique is to centrifuge cells arrested in metaphase in the presence of drugs that destroy the cytoskeleton, thus favoring micronucleation. Microcells are then filtered through membranes to collect cells containing single chromosomes. The final step is cell fusion with the chosen recipient and selection for a marker carried by the engineered chromosome.

In spite of the low transfer frequencies achieved with MMCT (10^{-5} to 10^{-7}), the approach has been used in transferring engineered minichromosomes into various recipients including stem cells. Alternative delivery procedures, such as modifications of the whole cell fusion platform, have been attempted (58). Whole cell fusion is simplified with respect to MMCT, however it is of limited applicability since the results strictly depend on the parental cells that are used for cell fusion.

3. Concluding Remarks

Naturally occurring accessory minichromosomes show some unique features that have been successfully exploited to generate engineered chromosome vectors. First of all, they are of small size per se, which means that they represent platforms enriched in the functional elements that define chromosome functions. Secondly, small marker chromosomes generally appear as extra elements in otherwise normal karyotypes and often do not have any phenotypic effect; this is a crucial prerequisite since consequences of gene dosage unbalance on cell functions can be envisaged. Therefore further manipulations of these minicromosomes can be performed being somehow guaranteed of the low perturbing effects of the engineered minichromosome on the host cell genome. A number of size reduction, site-specific recombination, and chromosome transfer strategies have been developed, so that nowadays several natural minichromosome platforms exist and have been used to transfer exogenous genes in primary cells as well as in stem cells for the creation of transchromosomal mice.

In this scenario, particularly promising, are neo-centromere based minichromosome platforms because they have small centromeric domains devoid of satellite DNA sequences, therefore they are prone to complete sequence characterization. A new example of satellite DNA free centromeres has been recently described in *Equus* species; in the next future these centromeres might be used as novel chromosome engineering tools.

Acknowledgements

I wish to acknowledge Luca Ferretti (Dept. of Genet. and Microbiol., University of Pavia), Chiara Mondello (IGM, CNR Pavia) and Marianna Paulis (ITB, CNR Milano) for critical reading of the manuscript. Research in my lab is funded by Ministero dell'Università e della Ricerca (PRIN 2006).

References

1. Murray, A. W., Szostak, J. W. (1983) Construction of artificial chromosomes in yeast *Nature* **305**, 189–93.
2. Clarke, L., Carbon, J. (1980) Isolation of a yeast centromere and construction of functional small circular chromosomes *Nature* **287**, 504–9.
3. Clarke, L., Amstutz, H., Fishel, B., Carbon, J. (1986) Analysis of centromeric DNA in the fission yeast *Schizosaccharomyces pombe Proc Natl Acad Sci USA* **83**, 8253–7.
4. Carbon, J., Clarke, L. (1990) Centromere structure and function in budding and fission yeasts *New Biol* **2**, 10–9.
5. Clarke, L. (1998) Centromeres: proteins, protein complexes, and repeated domains at centromeres of simple eukaryotes. *Curr Opin Genet Dev* **8**, 212-8.
6. Pluta, A. F., Mackay, A. M., Ainsztein, A. M., Goldberg, I. G., Earnshaw, W. C. (1995) The centromere: hub of chromosomal activities *Science* **270**, 1591–4.
7. Allshire, R. C. (1997) Centromeres, checkpoints and chromatid cohesion *Curr Opin Genet Dev* 7, 264–73.
8. Henikoff, S., Ahmad, K., Malik, H. S. (2001) The centromere paradox: stable inheritance with rapidly evolving DNA *Science* **293**, 1098–102.
9. Allshire, R. C., Karpen, G. H. (2008) Epigenetic regulation of centromeric chromatin: old dogs, new tricks? *Nat Rev Genet* **9**, 923–37.
10. Marshall, O. J., Chueh, A. C., Wong, L. H., Choo, K. H. (2008) Neo-centromeres: new insights into centromere structure, disease development, and karyotype evolution *Am J Hum Genet* **82**, 261–82.
11. Vagnarelli, P., Ribeiro, S. A., Earnshaw, W. C. (2008) Centromeres: old tales and new tools *FEBS Lett* **582**, 1950–9.
12. John, B. (1988) The biology of heterochromatin. In: Verma RS (ed) Heterochromatin: molecular and structural aspects. *Cambridge University Press, Cambridge*, pp 1–128.
13. Maison, C., Bailly, D., Peters, A. H., Quivy, J. P., Roche, D., Taddei, A., Lachner, M., Jenuwein, T., Almouzni, G. (2002) Higher-order structure in pericentric heterochromatin involves a distinct pattern of histone modification and an RNA component *Nat Genet* **30**, 329–34.
14. Muchardt, C., Guilleme, M., Seeler, J. S., Trouche, D., Dejean, A., Yaniv, M. (2002) Coordinated methyl and RNA binding is required for heterochromatin localization of mammalian HP1alpha *EMBO Rep* **3**, 975–81.
15. Eymery, A., Callanan, M., Vourc'h, C. (2009) The secret message of heterochromatin: new insights into the mechanisms and function of centromeric and pericentric repeat sequence transcription *Int J Dev Biol* **53**, 259–68.
16. Dimitri, P., Caizzi , R., Giordano, E., Accardo, M. C., Lattanzi, G., Biamonti, G. (2009) Constitutive heterochromatin: a surprising variety of expressed sequences *Chromosoma* **118**, 419–35.
17. Jolly, P. D., Smith, P. R., Heath, D. A., Hudson, N. L., Lun, S., Still, L. A., Watts, C. H., McNatty, K. P. (1997) Morphological evidence of apoptosis and the prevalence of apoptotic versus mitotic cells in the membrana granulosa of ovarian follicles during spontaneous and induced atresia in ewes *Biol Reprod* **56**, 837–46.
18. Denegri, M., Moralli, D., Rocchi, M., Biggiogera, M., Raimondi, E., Cobianchi, F., De Carli, L., Riva, S., Biamonti, G. (2002) Human chromosomes 9, 12, and 15 contain the nucleation sites of stress-induced nuclear bodies *Mol Biol Cell* **13**, 2069–79.
19. Kloc, A., Martienssen, R. (2008) RNAi, heterochromatin and the cell cycle *Trends Genet* **24**, 511–7.
20. Saffery, R., Wong, L. H., Irvine, D. V., Bateman, M. A., Griffiths, B., Cutts, S. M., Cancilla, M. R., Cendron, A. C., Stafford, A. J., Choo, K. H. (2001) Construction of neocentromere-based human minichromosomes by telomere-associated chromosomal truncation *Proc Natl Acad Sci USA* **98**, 5705–10.
21. Wong, L. H., Saffery, R., Choo, K. H. (2002) Construction of neo-centromere-based human minichromosomes for gene delivery and centromere studies *Gene Ther* **9**, 724–6.
22. Voullaire, L. E., Slater, H. R., Petrovic, V., Choo, K. H. (1993) A functional marker centromere with no detectable alpha-satellite, satellite III, or CENP-B protein: activation of a latent centromere? *Am J Hum Genet* **52**, 1153–63.
23. Wade, C. M., Giulotto, E., Sigurdsson, S., Zoli, M., Gnerre, S., Imsland, F., Lear, T. L., Adelson, D. L., Bailey, E., Bellone, R. R., Blöcker, H., Distl, O., Edgar, R. C., Garber, M., Leeb, T., Mauceli, E., MacLeod, J. N., Penedo, M. C., Raison, J. M., Sharpe, T., Vogel, J., Andersson, L., Antczak, D. F., Biagi, T., Binns, M. M., Chowdhary, B. P., Coleman, S. J., Della Valle, G., Fryc, S., Guérin, G.,

Hasegawa, T., Hill, E. W., Jurka, J., Kiialainen, A., Lindgren, G., Liu, J., Magnani, E., Mickelson, J. R., Murray. J., Nergadze, S. G., Onofrio, R., Pedroni, S., Piras, M. F., Raudsepp, T., Rocchi, M., Røed, K. H., Ryder, O. A., Searle, S., Skow, L., Swinburne, J. E., Syvänen, A. C., Tozaki, T , Valberg, S. J., Vaudin, M., White, J. R., Zody, M. C.; Broad Institute Genome Sequencing Platform; Broad Institute Whole Genome Assembly Team, Lander, E. S., Lindblad-Toh, K. (2009) Genome sequence, comparative analysis, and population genetics of the domestic horse *Science* **326**, 865–7.

24. Piras, M. F., Nergadze, S. G., Magnani, E., Bertoni, L., Attolini, C., Khoriauli, L., Raimondi, E., Giulotto, E. (2010) Uncoupling of satellite DNA and centromeric function in the genus *Equus Cytogenet Genome Res PLOS Genet*, **6**, 1000–845.
25. Carbone, L., Nergadze, S. G., Magnani, E., Misceo, D., Cardone, F. M, Roberto, R., Bertoni, L., Attolini, C., Piras, F. M., de Jong, P., Raudsepp, T., Chowdhary, B. P., Guérin, G., Archidiacono, N., Rocchi, M., Giulotto, E. (2006) Evolutionary movement of centromeres in horse, donkey, and zebra *Genomics* **87**, 777–82.
26. Camacho, J. P. M. (ed.) (2004) B Chromosomes in the Eukaryote Genome *Special issue of Cytogenet Genome Res.* **106**, nos. 2–4.
27. Raimondi, E., Balzaretti, M., Moralli, D., Vagnarelli, P., Tredici, F., Bensi, M., De Carli, L. (1996) Gene targeting to the centromeric DNA of a human minichromosome *Hum Gene Ther* 7, 1103–9.
28. Guiducci, C., Ascenzioni, F., Auriche, C., Piccolella, E., Guerrini, A. M., Donini, P. (1999) Use of a human minichromosome as a cloning and expression vector for mammalian cells *Hum Mol Genet* **8**, 1417–24.
29. Moralli, D., Vagnarelli, P., Bensi, M., De Carli, L., Raimondi, E. (2001) Insertion of a *loxP* site in a size-reduced human accessory chromosome *Cytogenet Cell Genet* **94**, 113–20.
30. Crolla, J. A. (1998) FISH and molecular studies of autosomal supernumerary marker chromosomes excluding those derived from chromosome 15: II. Review of the literature *Am J Med Genet* **75**, 367–81.
31. Liehr ,T., Claussen, U., Starke, H. (2004) Small supernumerary marker chromosomes (sSMC) in humans *Cytogenet Genome Res* **107**, 55–67.
32. Liehr, T., Ewers, E., Kosyakova, N., Klaschka, V., Rietz, F., Wagner, R., Weise, A. (2009) Handling small supernumerary marker chromosomes in prenatal diagnostics *Expert Rev Mol Diagn* **9**, 317–24.
33. Walter, M. A., Goodfellow, P. N. (1993) Radiation hybrids: irradiation and fusion gene transfer *Trends Genet* **9**, 352–6.
34. Farr, C. J., Bayne, R. A., Kipling, D., Mills, W., Critcher, R., Cooke, H. J. (1995) Generation of a human X-derived minichromosome using telomere-associated chromosome fragmentation *EMBO J* **14**, 5444–54.
35. Heller, R., Brown, K. E., Burgtorf, C., Brown, W. R. (1996) Minichromosomes derived from the human Y chromosome by telomere directed chromosome breakage *Proc Natl Acad Sci USA* **93**, 7125–30.
36. Mills, W., Critcher, R., Lee, C., Farr, C. J. (1999) Generation of an approximately 2.4 Mb human X centromere-based minichromosome by targeted telomere-associated chromosome fragmentation in DT40 *Hum Mol Genet* **8**, 751–761.
37. Katoh, M., Ayabe, F., Norikane, S., Okada, T., Masumoto, H., Horike, S., Shirayoshi, Y., Oshimura, M. (2004) Construction of a novel human artificial chromosome vector for gene delivery *Biochem Biophys Res Commun* **321**, 280–90.
38. Lee, S. C., Wang, W., Liu, P. (2009) Construction of gene-targeting vectors by recombineering *Methods Mol Biol* **530**, 15–27.
39. Niwa, O., Matsumoto, T., Chikashige, Y., Yanagida, M. (1989) Characterization of Schizosaccharomyces pombe minichromosome deletion derivatives and a functional allocation of their centromere *EMBO J* **8**, 3045–52.
40. Murphy, T. D., Karpen, G. H. (1995) Localization of centromere function in a *Drosophila* minichromosome *Cell* **82**, 599–609.
41. Ferretti, L., Raimondi, E., Gamberi, C., Young, B. D., De Carli, L., Sgaramella V. (1991) Molecular cloning of DNA from a sorted human minichromosome *Gene* **99**, 229–34.
42. Raimondi, E., Ferretti, L., Young, B. D., Sgaramella, V., De Carli, L. (1991) The origin of a morphologically unidentifiable human supernumerary minichromosome traced through sorting, molecular cloning, and in situ hybridisation *J Med Genet* **28**, 92–6.
43. Yu, W., Han, F., Birchler, J. A. (2007) Engineered minichromosomes in plants *Curr Opin Biotechnol* **18**, 425–31.
44. Yu, W., Lamb, J. C., Han, F., Birchler, J. A. (2006) Telomere-mediated chromosomal truncation in maize *Proc Natl Acad Sci USA* **103**, 17331–6.

45. Yu, W., Han, F., Gao, Z., Vega, J. M., Birchler, J. A. (2007) Construction and behaviour of engineered minichromosomes in maize *Proc Natl Acad Sci USA* **104**, 8924–9.
46. Fukushige, S., Sauer, B. (1992) Genomic targeting with a positive-selection lox integration vector allows highly reproducible gene expression in mammalian cells *Proc Natl Acad Sci USA* **89**, 7905–9.
47. Baubonis, W., Sauer, B. (1993) Genomic targeting with purified *Cre* recombinase *Nucleic Acids Res* **21**, 2025–29.
48. Kuroiwa, Y., Shinohara, T., Notsu, T., Tomizuka, K., Yoshida, H., Takeda, S., Oshimura, M., Ishida, I. (1998) Efficient modification of a human chromosome by telomere-directed truncation in high homologous recombination-proficient chicken DT40 cells *Nucleic Acids Res* **26**, 3447–8.
49. Kuroiwa, Y., Tomizuka, K., Shinohara, T., Kazuki, Y., Yoshida, H., Ohguma, A., Yamamoto, T., Tanaka, S., Oshimura, M., Ishida, I. (2000) Manipulation of human minichromosomes to carry greater than megabase-sized chromosome inserts *Nat Biotechnol* **18**, 1086–90.
50. Smith, A. J. H., de Sousa, M. A., Kwabi-Addo, B., Heppell-Parton, A., Impey, H., Rabbits. P. (1995) A site-directed chromosomal translocation induced in embryonic stem cells by *Cre-loxP* recombination *Nature Genet* **9**, 376–85.
51. Van Deursen J, Fornerod M, Van Rees B & Grosveld G (1995) *Cre*-mediated site-specific translocation between nonhomologous mouse chromosomes. Proc Natl Acad Sci USA 92: 7376–7380.
52. Van der Weide, L., Bradley, A. (2006) Mouse Chromosome Engineering for Modeling Human Diseases *Annu Rev Genomics Hum Genet* 7, 247–76.
53. Hoshiya, H., Kazuki, Y., Abe, S., Takiguchi, M., Kajitani, N., Watanabe, Y., Yoshino, T., Shirayoshi, Y., Higaki, K., Messina, G., Cossu, G., Oshimura, M. (2009) A highly stable and nonintegrated human artificial chromosome (HAC) containing the 2.4 Mb entire human dystrophin gene *Mol Ther* **17**, 309–17.
54. Ikeno, M., Suzuki, N., Hasegawa, Y., Okazaki, T. (2009) Manipulating transgenes using a chromosome vector *Nucleic Acids Res* **37**, e44.
55. Schor, S. L., Johnson, R. T., Mullinger, A. M. (1975) Perturbation of mammalian cell division. II. Studies on the isolation and characterization of human mini segregant cells *J Cell Sci* **19**, 281–303.
56. Fournier, R. E., Ruddle, F. H. (1977) Microcell-mediated transfer of murine chromosomes into mouse, Chinese hamster, and human somatic cells *Proc Natl Acad Sci USA* **74**, 319–23.
57. Tomizuka, K., Yoshida, H., Uejima, H., Kugoh, H., Sato, K., Ohguma, A., Hayasaka, M., Hanaoka, K., Oshimura, M., Ishida, I. (1997) Functional expression and germline transmission of a human chromosome fragment in chimaeric mice *Nat Genet* **16**, 133–43.
58. Paulis, M., Bensi, M., Orioli, D., Mondello, C., Mazzini, G., D'Incalci, M., Falcioni, C., Radaelli, E., Erba, E., Raimondi, E., De Carli, L. (2007) Transfer of a human chromosomal vector from a hamster cell line to a mouse embryonic stem cell line *Stem Cells* **25**, 2543–50.

Chapter 4

Chromosome Transfer Via Cell Fusion

Marianna Paulis

Abstract

Intact chromosomes as well as chromosome fragments can be vehicled into various recipient cells without perturbing their ability to segregate as free elements; chromosome transfer can be performed both in cultured cells and in living animals. The method of choice to shuttle single chromosomes between cells is microcell fusion named microcell mediated chromosome transfer (MMCT). The use of MMCT is mandatory in a number of applications where alternative chromosome transfection procedures are ineffective; however, the main drawback is the extremely low efficiency of the technique. Recently, we developed a new procedure to shuttle an engineered human minichromosome from a Chinese hamster ovary hybrid cell line to a mouse embryonic stem cell line. This technology ultimately consists in micronucleated whole cell fusion (MWCF) without microcell isolation. Therefore, MWCF is much more simple than MMCT; moreover, chromosome transfer efficiency is higher. The main limit of the MWCF approach is that it can be employed only with parental cells of different species, while the MMCT protocol can be adapted to any donor and recipient cell line.

This chapter will describe both the protocols that we currently use for MMCT and MWCF. The efficiency of the two protocols strictly depends on the parental cell lines to be used for cell fusion.

Key words: Polyethylene glycol, PEG, Chromosome transfer, Cell fusion, Microcell fusion, MMCT, Cell hybrid, Micronucleated whole cell fusion

1. Introduction

Somatic cell fusion is the key technology in somatic cell genetics; since over 40 years intraspecific and interspecific somatic cell hybrids have been used in a number of fields (1–6). Two significant examples of somatic cell genetics applications are the production of monoclonal antibodies and high resolution gene localization via radiation reduced cell hybrids (7, 8). The most recent developments concern the adaptation of cell fusion procedures to single chromosome transfer aimed at genetically modifying primary or stem cell lines to be used in gene therapy trials as well as in regenerative medicine (9).

Gyula Hadlaczky (ed.), *Mammalian Chromosome Engineering: Methods and Protocols*, Methods in Molecular Biology, vol. 738, DOI 10.1007/978-1-61779-099-7_4, © Springer Science+Business Media, LLC 2011

When cells derived from two different species are fused, the resultant hybrid cell lines tend to selectively loose most chromosomes of one of the two parental cells. This process may require a long culture time to isolate single chromosomes of a given donor species on the genomic background of the recipient cell line. Generally speaking, human/rodent hybrid cell lines selectively eliminate human chromosomes. Chromosome loss occurs at random both as regards the rate and as regards the chromosome identity; therefore, while in some instances the majority of human chromosomes are rapidly lost, in other cases a large number of human chromosomes are retained in all the hybrid colonies. This feature critically hinders the production of monochromosomal cell hybrids. Thus, ad hoc technologies have been developed, aimed at transferring single chromosomes into various recipients.

At present, microcell fusion is the approach most commonly used to transfer single or few chromosomes into different host cells, despite its low efficiency (approximately 10^{-6}) (10–12).

The rationale of the technique is quite simple. After a prolonged exposure of the donor cell to colchicine, which inhibits tubulin polymerization, metaphase cells accumulate and the nuclear membrane is reorganized around single chromosomes or small groups of chromosomes, thus giving rise to "micronuclei." Cells can be "micro-enucleated" by centrifugation in the presence of the microfilament disrupting agent, cytochalasin B; in this way microcells, consisting of micronuclei surrounded by the plasma membrane, can be obtained (13). Finally, microcells can be fused with an appropriate recipient cell line.

Recently, we set up a novel protocol for interspecific cell fusion aimed at delivering an engineered human minichromosome, contained in a hamster cell line, to mouse ES cells (14). This procedure allowed us to transfer the minichromosome alone to the recipient cell line in a few culture passages.

We named this new strategy micronucleated whole cell fusion (MWCF). The first step was to micronucleate the donor cells by exposing them to colchicine. Multi-micronucleated cells were then treated with cytochalasin B and fused with the recipient cells, thus skipping the microcells isolation step. The combined treatment of the donor cell line with colcemid and cytochalasin B, followed by stringent selection, presumably induce a rapid loss of hamster chromosomes.

Using MMCT as well as MWCF as chromosome transfer procedure, only a fraction of the cells yields the hybrids. The remaining cells must be eliminated, otherwise they will overgrow; therefore it is mandatory to set up an efficient selection system.

Both the chromosome transfer procedures briefly outlined above are being described here. While on the one hand the MMCT is of general use since it can be adapted to any donor and recipient cell line, on the other hand the MWCF is much easier and more efficient, but can be employed only for interspecific chromosome transfer.

2. Materials

2.1. Preparation of Donor Cells for Cell Fusion (MMCT and MWCF)

1. RPMI 1640 medium (EuroClone).
2. Fetal bovine serum (FBS; Hyclone).
3. Knockout Dulbecco's modified Eagle's medium (KO-DMEM; Invitrogen).
4. Sterile phosphate buffered saline without Ca^{2+} and Mg^{2+} (PBS), pH 7.2 (Lonza Walkersville, Inc.).
5. Trypsin-Versene (EDTA) mixture, containing 0.05% trypsin and 0.02% EDTA (Lonza Walkersville, Inc.).
6. Percoll (Pharmacia) stored at 4°C.
7. KaryoMAX Colcemid Solution, liquid (10 μg/ml) in HBSS (Invitrogen) stored at 4°C.
8. Cytochalasin B (Sigma) dissolved at 10 mg/ml in DMSO and stored at –20°C.
9. 25-cm Tissue culture flask (Falcon).
10. 15-mL and 50-mL Round-bottom centrifuge tubes (Falcon).
11. 0.8–0.5 μm Polycarbonate filter (Millipore).
12. Swinnex adaptor (Millipore).
13. 30-mL Polycarbonate centrifuge tubes (Nalgene).
14. Heraeus multifuge 3S-R centrifuge and Heraeus fixed-angle rotor.

2.2. Preparation of Recipient Cells for Cell Fusion (MMCT and MWCF)

1. Knockout Dulbecco's modified Eagle's medium (KO-DMEM; Invitrogen).
2. Knockout Serum Replacement (KO-serum replacement, Invitrogen).
3. Glutamax-I (Invitrogen).
4. Nonessential amino acids (Invitrogen).
5. 50 U/mL Penicillin and 50 μg/mL streptomycin (Invitrogen).
6. β-mercaptoethanol (Invitrogen).
7. Leukemia Inhibitory Factor (ESGRO; Chemicon).
8. Gelatin, Type A from porcine skin (Sigma) dissolved at 0.1% in tissue-culture water and stored at 4°C.
9. Sterile phosphate buffered saline without Ca^{2+} and Mg^{2+} (PBS), pH 7.2 (Lonza Walkersville, Inc.).
10. Trypsin-Versene (EDTA) mixture, containing 0.05% trypsin and 0.02% EDTA (Lonza Walkersville, Inc.).
11. 60-mm Tissue culture dishes (NUNC).
12. 15-mL and 50-mL Round-bottom centrifuge tubes (Falcon).

2.3. Polyethylene Glycol Mediated Fusion

1. Knockout Dulbecco's modified Eagle's medium (KO-DMEM; Invitrogen).
2. Knockout Serum Replacement (KO-serum replacement, Invitrogen).
3. Glutamax-I (Invitrogen).
4. Nonessential amino acids (Invitrogen).
5. 50 U/mL Penicillin and 50 μg/mL streptomycin (Invitrogen).
6. β-mercaptoethanol (Invitrogen).
7. Leukemia Inhibitory Factor (ESGRO; Chemicon).
8. Gelatin, Type A from porcine skin (Sigma) dissolved at 0.1% in tissue-culture water and stored at 4°C.
9. Sterile phosphate buffered saline without Ca^{2+} and Mg^{2+} (PBS), pH 7.2 (Lonza Walkersville, Inc.).
10. Trypsin-Versene (EDTA) mixture, containing 0.05% trypsin and 0.02% EDTA (Lonza Walkersville, Inc.).
11. 50% Polyethylene glycol (PEG) 1500 in PBS (Roche) stored at 4°C.
12. 60-mm Tissue culture dishes (NUNC).
13. 100-mm Tissue culture dishes (Falcon).
14. 15-mL and 50-mL Round-bottom centrifuge tubes (Falcon).

2.4. Selection and Hybrid Colonies Characterization

1. Knockout Dulbecco's modified Eagle's medium (KO-DMEM; Invitrogen).
2. Knockout Serum Replacement (KO-serum replacement, Invitrogen).
3. Glutamax-I (Invitrogen).
4. Nonessential amino acids (Invitrogen).
5. 50 U/mL Penicillin and 50 μg/mL streptomycin (Invitrogen).
6. β-mercaptoethanol (Invitrogen).
7. Leukemia Inhibitory Factor (ESGRO; Chemicon).
8. Gelatin, Type A from porcine skin (Sigma) dissolved at 0.1% in tissue-culture water and stored at 4°C.
9. Hypoxanthine 100 μM, aminopterin 0.4 μg/ml, and thymidine 16 μM (HAT; Sigma) dissolved at 50× in sterile cell culture medium and stored at −20°C.
10. Puromycin (Sigma) dissolved at 50 mg/ml in tissue-culture water and stored at −20°C.
11. 6-multiwell plates (Falcon).

3. Methods

To illustrate both the following protocols, MClox cells as donors and ES-E14 cells as recipients are being used.

MClox is a monochromosomic human/hamster somatic hybrid originally obtained by cell fusion between HPRT–CHO cells and a lymphoblastoid line from a patient carrying a minichromosome (MC) (15). The MC contains a *loxP* site and is marked with the neomycin gene (*neo*) and the puromycin acetyltransferase resistance gene (*pac*) (15). The cells grow in RPMI 1640 medium, supplemented with 10% fetal calf serum.

E14 is a mouse embryonic stem cell line (ES) from 129/Ola mice (16). The cells grow on Petri dishes coated with 0.1% gelatin in KO-DMEM containing 15% KO-serum replacement and 1,000 U/ml Leukemia Inhibitory Factor; the medium is conditioned with mouse embryonic fibroblasts (MEFs) grown for 2 days in log phase.

Parental cell lines and hybrid clones have been regularly tested for the presence of *Mycoplasma*, since *Mycoplasma* contamination reduces fusion efficiency.

The fusion experiments described below were aimed at transferring the MC from MClox cells into ES-E14 cells. The transfer can be obtained both by the MMCT and MWCF protocol. The main steps of the two protocols are outlined in Fig. 1.

Briefly, donor and recipient cells are prepared and fused with PEG. The two protocols mainly differ in the preparation of the donor cells: in the MMCT (Fig. 1b) the cells are micronucleated, the micronuclei are ejected from the cells by centrifugation forming microcells and finally microcells are purified by filtration; in the MWCF (Fig. 1a) the cells are micronucleated and directly fused with the recipient cells.

3.1. Donor Cells

3.1.1. Preparation of Microcells for MMCT (See Note 1)

1. Seed 1×10^7 donor cells in ten 25-cm tissue culture flasks.
2. Feed cells with fresh medium 16–20 h before colcemid treatment.
3. Add colcemid (0.02 μg/mL) to each flask.
4. Incubate cells for further 24 h at 37°C.
5. Remove medium.
6. Wash the cells with 10 mL of PBS.
7. Add 1 mL of trypsin and incubate at 37°C for 5 min.
8. Harvest micronucleated cells and pellet these cells at $400 \times g$ for 10 min.
9. Wash the pellet with 10 mL of prewarmed serum-free DMEM.

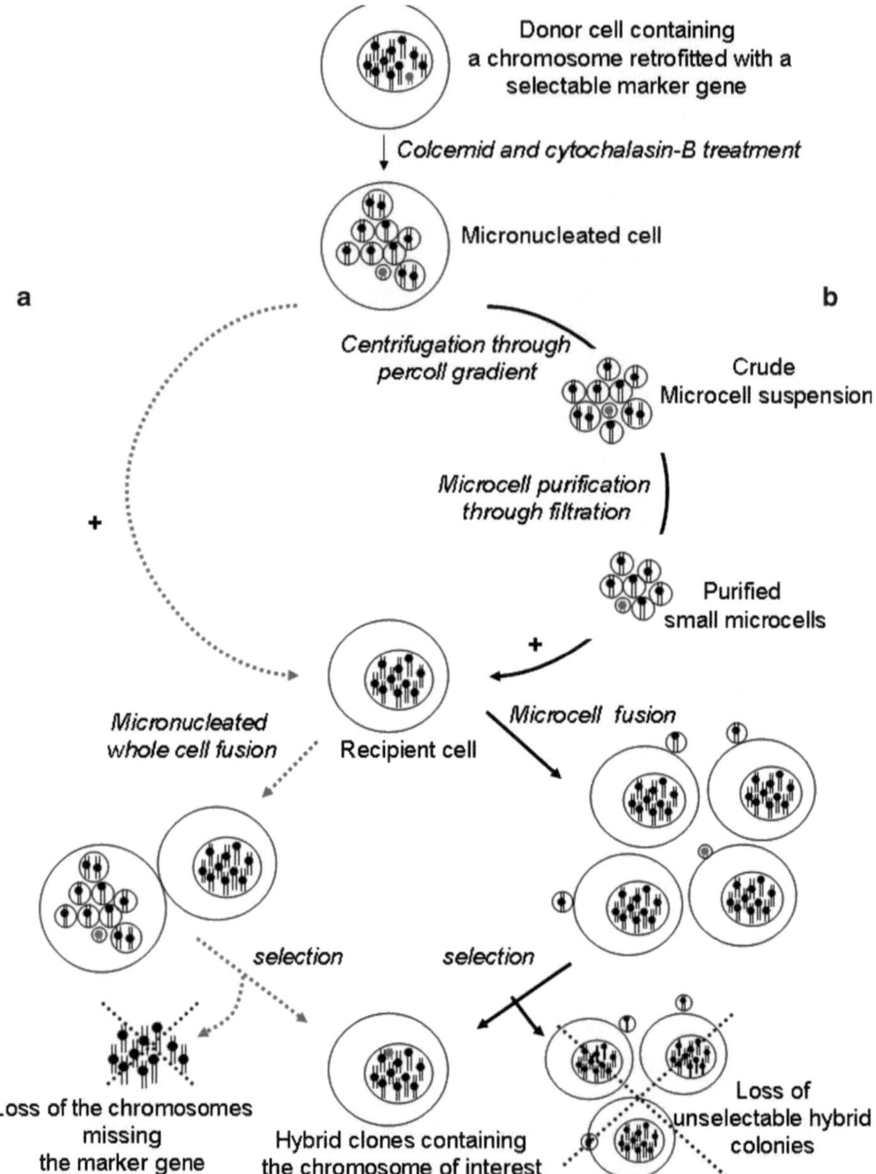

Fig. 1. Schematic representation of the key steps required in transferring single chromosomes from a donor to a recipient cell line. Two alternative approaches can be followed: micronucleated whole cell fusion is shown in (**a**) and microcell mediated chromosome transfer is shown in (**b**). The chromosome to be delivered is indicated in *grey*.

10. Resuspend cells collected in a 40 mL solution containing prewarmed Percoll/serum-free KO-DMEM (1/1) and 10 μg/mL cytochalasin B.
11. Transfer cells into two 30-mL polycarbonate tubes (20 mL/tube).
12. Centrifuge cells at 16,000 × *g* in the prewarmed Heraeus multifuge 3S-R centrifuge at 34°C for 75 min.

13. Two bands may be visible after centrifugation. Collect about 15 mL microcells by gentle aspiration or collect the whole supernatant without touching the gel pellet at the bottom if no bands are visible.
14. Mix 15 mL of microcells with 35 mL of prewarmed serum-free DMEM.
15. Pellet microcells at $500 \times g$ for 10 min.

3.1.2. Purification of Microcells (See Note 2)

1. Resuspend microcells in 20 ml of serum-free KO-DMEM and divide in two aliquots.
2. Filter the two microcell suspensions under gravity or under very light positive pressure through polycarbonate filters mounted in 25-mm Swinnex adaptor by using sterile syringes. Employ two filters in series mounted in separate adaptors and with pore size of 8 and 5 μm, respectively.
3. Wash the collected microcells with 50 mL of prewarmed serum-free KO-DMEM and pellet them at $400 \times g$ for 5 min.
4. Repeat step 3.
5. Resuspend microcells in 1 mL of prewarmed serum-free DMEM.

3.1.3. Micronucleation of Cells for MWCF

1. Seed 4×10^6 MClox donor cells in four 25-cm tissue culture flasks.
2. Feed cells with fresh medium 16–20 h before colcemid treatment.
3. Add colcemid (0.02 μg/mL) to each flask.
4. Incubate cells for further 24 h at 37°C.
5. Wash the cells with 10 mL of PBS.
6. Add 1 mL of trypsin and incubate at 37°C for 5 min.
7. Harvest micronucleated cells and pellet these cells at $400 \times g$ for 10 min.
8. Resuspend cells in 15 ml of prewarmed growth medium containing 20 μg/ml cytocalasin B.
9. Incubate cells for 3–4 h at 37°C with constant stirring.
10. Pellet cells at $500 \times g$ for 10 min.
11. Resuspend microcells in 1 mL of prewarmed serum-free KO-DMEM.

3.2. Preparation of Recipient Cells for Cell Fusion

1. Seed 5×10^6 ES-E14 cells in five 60-mm tissue culture dishes or 1×10^7 ES-E14 cells in ten 60-mm tissue culture dishes for MWCF and MMCT, respectively.
2. Add fresh conditioned KO-DMEM medium to the cell culture 16–20 h before fusion.

3. Remove medium.
4. Wash cells with 10 mL of PBS.
5. Add 1 mL of trypsin and incubate at 37°C for 5 min.
6. Resuspend cells in 15 ml of KO-DMEM.
7. Pellet cells at 400 × *g* for 10 min.
8. Wash cells with 10 mL of prewarmed serum-free KO-DMEM.
9. Pellet cells again.

3.3. PEG-Mediated Fusion (See Note 3)

1. Mix recipient cell pellets (ES-E14) with 5 mL micronucleated donor cells or microcells.
2. Pellet the cell mixture at 500 × *g* for 10 min.
3. Resuspend cells in 1 mL of serum-free KO-DMEM and incubate for 10 min at room temperature.
4. Pellet the mixture again.
5. Resuspend the mixture in 1 mL of 2.5% of PEG 1500 in KO-DMEM.
6. Keep the mixture in this solution for 20 min at 4°C.
7. Pellet the cell mixture at 400 × *g* for 10 min.
8. Resuspend cells in 1 mL of a prewarmed solution of 50% PEG 1500 over 2 min.
9. Gradually add 10 ml of serum-free KO-DMEM to the cell suspension over 10 min.
10. Wash fused sample twice in 50 mL of serum-free KO-DMEM.
11. Add 10 ml (for MWCF fusion protocol) or 30 mL (for the MMCT fusion protocol) of MEF-conditioned KO-DMEM to the fused cells.
12. Transfer cell suspensions into 60-mm tissue culture dishes (5 ml/dish) and incubate for a recovery time of 48 h at 37°C.
13. Plate 10^5 cells per 100-mm tissue culture dish.

3.4. Selection and Hybrid Colonies Characterization (See Note 4)

1. After 24 h apply double selection to the culture: HAT, final concentration of 1×, to eliminate whole donor cells; puromycin at different concentrations (0.3–10 μg/ml) to select the recipient cells containing the MC.
2. Incubate the culture for about 10–15 days at 37°C.
3. Pick up hybrid colonies from individual culture dish.
4. Transfer the colonies into 6-multiwell plates for culture expansion.
5. Use a small fraction of cells from each colony for chromosome analysis by means of FISH analysis on metaphase

chromosomes. Briefly, to detect the MC and the residual CHO chromosomal material, metaphase spreads are hybridized in situ with digoxigenin labeled MC specific satellite DNA and biotin labeled CHO whole genomic DNA. The probes are then revealed by immunofluorescence.

4. Notes

1. Microcell formation. Micronucleation and microcell formation efficiency strictly depend on the cell type. For example, it is much easier to induce micronucleation in mouse LA9 fibroblasts and in CHO cells than in MEF or mouse ES cells. To obtain micronucleation it is essential to have exponentially growing cells. For different cell lines it is necessary to titrate both the concentration of colcemid, that usually ranges from 0.05 to 0.10 μg/ml, and the exposure time that ranges from 10 to 48 h for most mammalian cell lines. Micronucleation can be easily scored by phase contrast microscopic examination.
2. Purification. After centrifugation, the crude microcell preparation contains different kinds of particles: whole cells, cytoplasts, karyoplasts and microcells of various size. To select the smallest microcells, which are likely to contain single chromosomes, the preparation is purified by serial filtration, usually with pore size from 8 to 3 μm. The filter is mounted in a swinnex adaptor, sterilized by autoclaving, and attached to sterile syringes. It is important during filtration not to force through the filter, but to use the gravity force only or eventually a very light positive pressure. Excessive pressure could break the filter. To check filtration efficiency, aliquots can be sampled before and after the procedure and analysed under a bright field microscope after acridine orange staining.
3. PEG-fusion. PEG is a highly toxic detergent. PEG molecular weight, concentration, and time of cell exposure are critical points in fusion experiments. PEG with a molecular weight ranging from 1,000 to 6,000 can be used. The optimal PEG concentration ranges from 35 to 50%. Lower concentrations are less toxic, but also less fusogenic. With respect to incubation time, long exposure is toxic while too short exposure is ineffective: in the majority of cases 1–2 min are enough to generate hybrids.
4. Selectable markers. In order to select the chromosome of interest contained in the donor cell line, a selectable marker gene must be carried by the relevant chromosome itself. A number of selectable markers, suitable for mammalian cell selection, are available. Neomycin (*neo*) and the puromycin

acetyltransferase (*pac*) genes are examples of dominant positive markers. These genes confer to mammalian cells resistance to the antibiotics Geneticin and puromycin, respectively. Since microcells can not grow in culture as free elements, a counter selection, should not be required. Still, whole donor cells, escaped from micronucleation, might be present; therefore, a negative selection procedure is needed as well. Genes involved in the endogenous and salvage metabolic pathways for DNA synthesis, can be used for negative selection of hybrid cells. The method of choice exploits the so called HAT medium, which contains Aminopterin (A in HAT) that acts as a folate metabolism inhibitor by inhibiting dihydrofolate reductase, and hypoxanthine (H in HAT, a purine derivative) and thymidine (T in HAT, a deoxynucleoside), which are intermediates in DNA synthesis. In the presence of Aminopterin, the cells are forced to use the salvage pathway for DNA synthesis, thus the exogenous bases precursors (H and T) that are present in HAT medium are their sole source of purines and pyrimidines. Consequently, hypoxanthine-guanine phosphoribosyl transferase (HGPRT) deficient cells, as well as thymidine kinase (TK) deficient cells, are unable to grow in HAT medium. HGPRT- and TK- cell lines can be obtained via selection for resistance to the nucleoside analogues 6-thioguanine (6TG) and 5-bromo-2-deoxyuridine (BrdU), respectively. An alternative approach for the negative selection of hybrid colonies lies in the use of parental cells transformed with the TK gene of the herpes simplex virus (HSV). This procedure was originally developed in gene therapy as a potential way to selectively eliminate cancer cells via the so called "suicide gene" (17). Contrary to the TK gene carried by mammalian cells, the HSV TK gene is able to activate the prodrug ganciclovir to a phosphorylated form, which acts as an effective cytostatic agent. Thus mammalian cells, transfected with the viral TK gene, become sensitive to ganciclovir.

References

1. Kamarck, M. E., Barker, P. E., Miller, R. L., and Ruddle, F. H. (1984) Somatic cell hybrid mapping panels *Exp. Cell Res.* **152**, 1–14.
2. Cho, M. S., Yee, H., and Chan, S. (2002) Establishment of a human somatic hybrid cell line for recombinant protein production *J. Biomed. Sci.* **9**, 631–38.
3. Kuroiwa, Y., Kasinathan, P., Choi, Y. J., Naeem, R., Tomizuka, K., Sullivan, E. J., Knott, J.G., Duteau, A., Goldsby, R. A., Osborne, B.A., Ishida, I., Robl, J. M. (2002) Cloned transchromosomic calves producing human immunoglobulin *Nat Biotechnol.* **9**, 889–94.
4. Tomizuka, K., Shinohara, T., Yoshida, H., Uejima, H., Ohguma, A., Tanaka, S., Sato, K., Oshimura, M., Ishida, I. (2000) Double trans-chromosomic mice: maintenance of two individual human chromosome fragments containing Ig heavy and kappa loci and expression of fully human antibodies *Proc Natl Acad Sci USA* **2**, 722–27.

5. O'Doherty, A., Ruf, S., Mulligan, C., Hildreth, V., Errington, M. L., Cooke, S., Sesay, A., Modino, S., Vanes, L., Hernandez, D., Linehan, J.M., Sharpe, P.T., Brandner, S., Bliss, T.V., Henderson, D.J., Nizetic, D., Tybulewicz, V.L., Fisher, E.M. (2005) An aneuploid mouse strain carrying human chromosome 21 with Down syndrome phenotypes *Science* **309**, 2033–37.
6. Irvine, D.V., Shaw, M. L., Choo, K. H., (2005) Saffery R.Engineering chromosomes for delivery of therapeutic genes *Trends Biotechnol.* **12**, 575–83.
7. Steinitz, M. (2009) Three decades of human monoclonal antibodies: past, present and future developments *Hum Antibodies* **18**, 1–10.
8. Walter, M. A., Goodfellow P., N. (1993) Radiation hybrids: irradiation and fusion gene transfer *Trends Genet.* **9**, 352–6.
9. Oshimura, M., Katoh, M. (2008) Transfer of human artificial chromosome vectors into stem cells *Reprod Biomed Online* **16**, 57–69.
10. Katoh, M., Ayabe, F., Norikane, S., Okada, T., Masumoto, H., Horike, S., Shirayoshi, Y., Oshimura M. (2004) Construction of a novel human artificial chromosome vector for gene delivery *M.Biochem Biophys Res Commun.* **321**, 280–90.
11. Kakeda, M., Hiratsuka, M., Nagata, K., Kuroiwa, Y., Kakitani, M., Katoh, M., Oshimura, M., Tomizuka, K. (2005) Human artificial chromosome (HAC) vector provides long-term therapeutic transgene expression in normal human primary fibroblasts *Gene Ther.,* **12**, 852–56.
12. Fourier, R. E. K. (1981) A general high-efficiency procedure for production of micro-cell hybrids *Proc Natl Acad Sci USA* **78**, 6349–53.
13. Ege, T., Ringertz, N. R. (1974) Preparation of microcells by enucleation of micronucleate cells *Exp Cell Res.* **87**, 378–82.
14. Paulis, M., Bensi, M., Orioli, D. Mondello, C., Mazzini, G., D'Incalci, M., Falcioni, C., Radaelli, E., Erba, E., Raimondi, E., De Carli, L. (2007) Transfer of a Human Chromosomal Vector from a Hamster Cell Line to a Mouse Embryonic Stem Cell Line *Stem Cell* **25**, 2543–2550.
15. Raimondi, E., Balzaretti, M., Moralli, D., Vagnarelli, P., Tredici, F., Bensi, M., De Carli, L. (1996) Gene targeting to the centromeric DNA of a human minichromosome Hum *Gene Ther.* 7, 1103–9.
16. Moralli, D., Vagnarelli, P., Bensi, M., De Carli, L., Raimondi, E. (2001) Insertion of a loxP site in a size-reduced human accessory chromosome *Cytogenet Cell Genet.* **94**, 113–20.
17. Freeman, S. M., Whartenby, K. A., Freeman, J. L., Abboud, C. N, Marrogi, A. J. (1996) In situ use of suicide genes for cancer therapy *Semin Oncol.* **23**, 31–45.

Chapter 5

Insertional Engineering of Chromosomes with *Sleeping Beauty* Transposition: An Overview

Ivana Grabundzija, Zsuzsanna Izsvák, and Zoltán Ivics

Abstract

Novel genetic tools and mutagenesis strategies based on the *Sleeping Beauty* (*SB*) transposable element are currently under development with a vision to link primary DNA sequence information to gene functions in vertebrate models. By virtue of its inherent capacity to insert into DNA, the *SB* transposon can be developed into powerful tools for chromosomal manipulations. Mutagenesis screens based on *SB* have numerous advantages including high throughput and easy identification of mutated alleles. Forward genetic approaches based on insertional mutagenesis by engineered *SB* transposons have the advantage of providing insight into genetic networks and pathways based on phenotype. Indeed, the *SB* transposon has become a highly instrumental tool to induce tumors in experimental animals in a tissue-specific manner with the aim of uncovering the genetic basis of diverse cancers. Here, we describe a battery of mutagenic cassettes that can be applied in conjunction with *SB* transposon vectors to mutagenize genes, and highlight versatile experimental strategies for the generation of engineered chromosomes for loss-of-function as well as gain-of-function mutagenesis for functional gene annotation in vertebrate models.

Key words: Transposon, Transgenesis, Insertional mutagenesis, Gene trap, Cre/loxP, Cancer, Embryonic stem cells

1. DNA Transposons and Transposition

Class II transposable elements (TEs) or DNA transposons are discrete pieces of DNA that possess the ability to change their position within the genome via a "cut and paste" mechanism called transposition. In this process, the transposase enzyme mediates the excision of the element from its donor locus, which is then followed by reintegration of the transposon into another DNA sequence environment (Fig. 1). In nature, these elements exist as single units containing the transposase gene flanked by inverted

Gyula Hadlaczky (ed.), *Mammalian Chromosome Engineering: Methods and Protocols*, Methods in Molecular Biology, vol. 738, DOI 10.1007/978-1-61779-099-7_5, © Springer Science+Business Media, LLC 2011

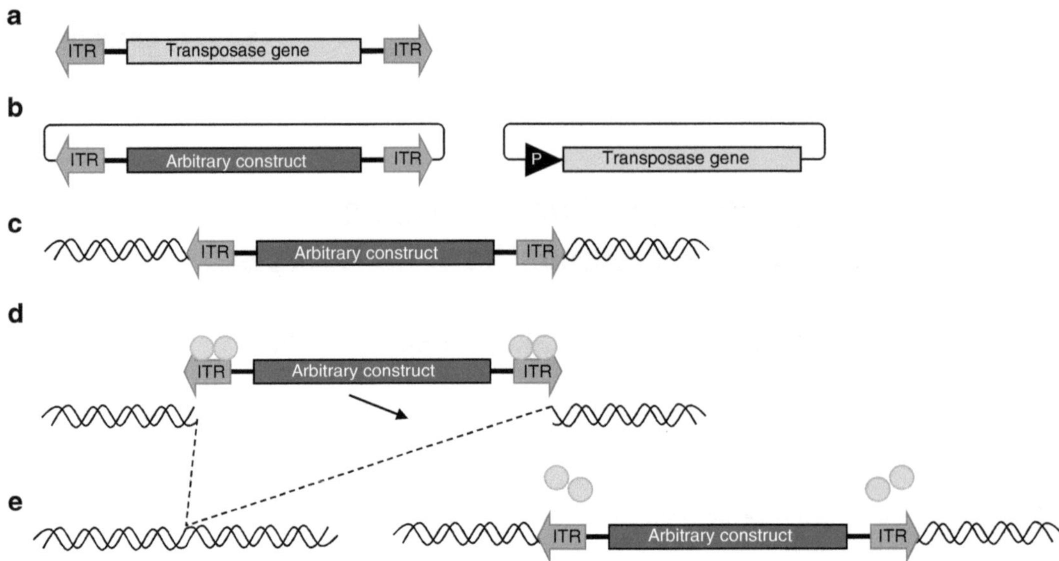

Fig. 1. General organization of class II transposable elements and mechanism of transposition. (**a**) Autonomous transposable elements consist of inverted terminal repeats (ITR) that flank the transposase gene. (**b**) Bi-component transposon vector system for delivering transgenes that are maintained in plasmids. One component contains a gene of interest (arbitrary construct) between the transposon ITRs carried by a plasmid vector, while the other component is a transposase expression plasmid. *Black triangle* represents the promoter element (P), while the transposase gene is depicted by a *rectangle*. (**c**–**e**) Schematic outline of the transposition reaction, when initiated from a chromosomal context. (**c**) The transposon carrying a DNA of interest is integrated at a chromosomal donor site. (**d**) Transposase proteins (*circular elements*) bind to the transposon ITRs and facilitate the excision of the transposon from the donor site. (**e**) Transposase mediates the integration of the excised transposon into a new donor locus, whereas the double-stranded gap at the excision site is repaired by the host repair machinery.

terminal repeats (ITRs) that carry transposase binding sites (Fig. 1a). However, in the laboratory, it is possible to use them as nonautonomous, controllable, bi-component systems, in which virtually any DNA sequence of interest can be placed between the transposon ITRs and mobilized by *trans*-supplementing the transposase in form of an expression plasmid (Fig. 1b) or mRNA synthesized *in vitro*. This feature makes transposons natural and easily controllable DNA delivery vehicles that can be used as tools for versatile applications ranging from somatic and germline transgenesis to functional genomics and gene therapy.

2. The *Sleeping Beauty* Transposon System

Even though DNA transposons have been extensively harnessed as tools for genome manipulation in invertebrates (1–3), there was no known transposon that was active enough to be used for such purposes in vertebrates. In 1997, Ivics et al. succeeded to engineer the *Sleeping Beauty* (*SB*) transposon system by molecular reconstruction of an ancient, inactive Tc1/*mariner*-type transposon found in several fish genomes (4). This newly reactivated element allowed highly efficient transposition-mediated gene transfer in the major vertebrate model organisms without the potential risk of cross mobilization of endogenous transposon copies in host genomes. Indeed, *SB* has been successfully used as a tool for genetic modifications of a wide variety of vertebrate cell lines and species including humans (reviewed in (5–7)), and became the reference system for all the transposon tools applied today in vertebrates.

However, although the newly resurrected *SB* element was active enough to be mobilized in vertebrate cells, its transpositional activity still presented a major bottleneck for some applications. Requirements for transfection of primary cells and other hard-to-transfect cell types or for re-mobilization of transposons from chromosomally resident, single-copy donor sites demanded an enzyme with more robust activity. In the past years significant efforts have been put into enhancing *SB*'s transpositional efficiency and engineering hyperactive versions by mutagenizing and modifying the transposon ITRs and the transposase coding region (8–15). These endeavors yielded a novel hyperactive *SB* transposase (referred to as *SB100X*) (13) that is up to 100-times more active than the originally reconstructed *SB* enzyme with its efficiency in transgene delivery reaching those of viral vectors.

In addition to the overall activity of the transposase in a cell type of interest, there are several other factors that can significantly influence *SB* transposition. One of these features is overproduction inhibition (8, 9), a phenomenon that results in the inhibition of transposition by excess cellular concentrations of the transposase (reviewed in (16)). Luckily, this can easily be circumvented by using the *SB* transposon as a two-component system and by carefully dosing the transposase component. Another practical consideration when using the *SB* system as a gene delivery vector is its cargo capacity. As all other transposons studied to date, the *SB* transposon is sensitive to the size of transgene to be mobilized (17). Namely, transpositional efficiency decreases proportionally to the increase in the size of the cargo component placed between its ITR's, which becomes pronounced with transgenes over 10 kb. One way to solve this problem is to use a specifically arranged vector termed "sandwich" transposon (two complete transposons flanking the cargo) which allows for transposon cargo size to exceed 10 kb (8).

In addition to the considerations described above, target site selection and the mutagenic potential of the *SB* transposon is of paramount importance for chromosomal engineering. The insertional preferences of *SB* were extensively analyzed, and revealed no or weak preference for integration into transcription units and a predominantly intronic localization of those insertions that occur in genes (18, 19). When transposition is initiated from plasmids, insertion site distribution on the genomic level is fairly random; apparently all chromosomes can serve as targets, and no chromosomal clustering of transposon insertions is evident (18, 19). However, the integrational profile of *SB* changes dramatically when chromosomally resident elements are mobilized. In such an arrangement, *SB* exhibits a phenomenon known as "local hopping", which results in 30–50% of all new transposition events occuring in the vicinity of the original donor site. This trait of the *SB* transposon system, first reported in mouse embryonic stem (ES cells) (20) was confirmed in other *in vivo* studies in mice (21–23). Notably, Horie et al. (24) have found that 25% of the transposition events occurred in a region approximately 200 kb around the donor sites, whereas the majority of transposition events occurred within a 3 Mb region. Similar results were obtained with germline mutagenesis screens in mice, where 30–80% of reinsertions occurred locally in the vicinity of transposon donor locus (reviewed in (6)).

Finally, a parameter of transposition that should be considered before utilizing transposons as gene delivery tools is transposon insertion frequency in a given cell type. This is especially important in light of the different demands of different applications with respect to average transposon copy number per cell. Namely, transgenesis typically requires low transposon copy number, whereas phenotype-driven forward mutagenesis screens can capitalize on multiple transposon insertions in the same cell. Importantly, in a bi-component system, it is possible to titrate the transposon and transposase components that are delivered to the cells in order to influence insertion frequencies. Use of the *SB100X* enzyme presents a significant advantage in this respect, as its robust activity allows for a considerable decrease of the transposon substrate component in order to generate low insertion frequencies in large numbers of cells. Conversely, delivery of high transposon concentrations to cells promotes the generation of high transposon copy numbers, which was not possible with earlier generations of *SB* vectors (13).

3. Transposons and Functional Genomics

The postgenomic era presented the scientific community with the new challenge of functional annotation of every gene and identification of elaborate genetic networks. Diverse methods have been employed to address this task, including mutational analysis that proved to be one of the most direct ways to decipher gene functions. There are versatile strategies for creating mutations. For example, reverse genetics approaches rely on targeting and disrupting genes of interest by homologous recombination, preferably in ES cells, thereby creating engineered alleles that allow for the functional dissection of genes of interest on the cellular as well as on whole-animal levels. Even though this approach is in itself very useful, frequently the phenotype caused by disruption of a given gene cannot be guessed solely from its sequence, which is often considered as a guideline in the choice of the mutation of interest. A shorter route to address this problem is designing genetic screens, in which phenotypes of interest are identified first and then followed by the identification of causative gene mutations. Indeed, forward genetic approaches aim to obtain mutant phenotypes by introducing loss-of-function or gain-of-function mutations in genomes of model organisms in a random and genome-wide fashion. In that manner, large-scale insertional mutagenesis can be applied not only to decipher the functions of individual mutated genes, but also to provide insight into genetic pathways and networks.

3.1. Forward Genetics with Mutagenic Sleeping Beauty Transposons

3.1.1. Mutagenic Cassettes

Mutating genes by insertion of discrete pieces of foreign DNA has the advantage that the inserted DNA fragment can serve as a molecular tag that allows rapid, usually PCR-based, identification of the mutated allele. Since the function of the gene in which the insertion has occurred is often disturbed, such loss-of-function insertional mutagenesis is frequently followed by functional analysis of mutant phenotypes. In many instances, retroviral vectors were utilized to introduce mutagenic cassettes into genomes, but their chromosomal insertion bias does not allow full coverage of genes (25). The random integration pattern of the *SB* transposon combined with its ability to efficiently integrate versatile transgene cassettes into chromosomes established this system as a highly useful tool for insertional mutagenesis in both ES cells (26, 27) as well as in somatic (28, 29) and germline tissues (21–23, 30–35) in animal models.

Most intronic integrations of foreign DNA (also applicable to *SB*, as discussed above) are expected to end up spliced out without having a mutagenic effect on endogenous gene expression (Fig. 2a). Thus, considerable effort has been put into improving mutagenicity as well as reporting capabilities of the internal components of mutagenic vectors (36, 37). There are several types of mutagenic cassettes that can be efficiently combined with *SB*-based gene delivery for insertional mutagenesis (Fig. 2). 5′ gene-trap cassettes include splice acceptors and polyadenylation sequences so that transcription of genes can be disrupted upon vector insertion into introns (Fig. 2b) (25). Frequently, such cassettes are also equipped with a reporter gene (usually, a fluorescent protein, β-galactosidase, or antibiotic resistance) whose expression is dependent on the correct splicing between exons of the trapped gene and a splice acceptor (SA) site carried by the transposon vector (24, 37). Commonly, such gene-traps have been

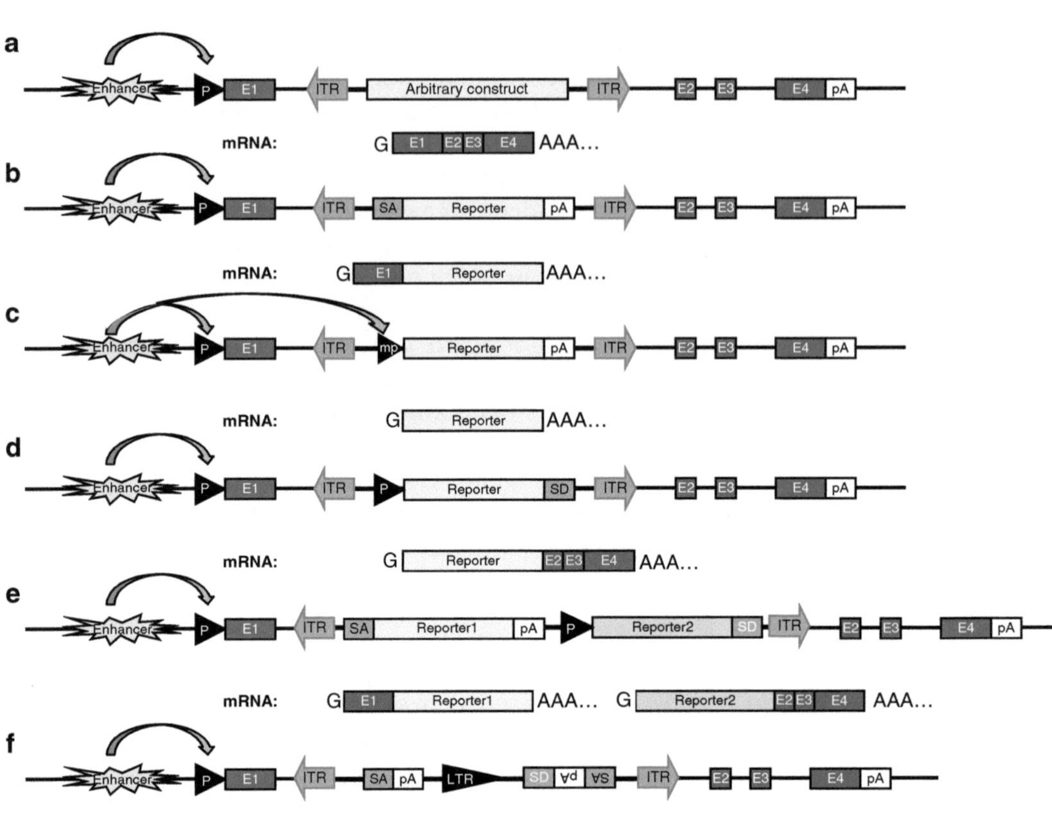

Fig. 2. Transposon gene-trapping strategies. (**a**) Non-mutagenic transposon insertion within an intron of a Pol II transcription unit. Transposon sequence is, in this case, most likely to be spliced out of the primary RNA transcript together with the intron sequence. Structure of a putative endogenous target gene is depicted with upstream regulatory element/ enhancer (Enhancer), a promoter sequence (P), 4 exons (E1, E2, E3, and E4) and a polyadenylation sequence (pA). G represents 5′-guanine cap, while AAA… represents the poly-A tail. Transposon inverted terminal repeats (ITRs) are indicated by *gray arrows*. (**b**) A gene-trap transposon contains a splice acceptor sequence (SA) followed by a reporter gene (Reporter) and a poly-A signal. The reporter gene is only expressed when its transcription is initiated from the promoter of a disrupted, actively transcribed endogenous transcription unit. Therefore, the expression pattern of the transposon reporter reflects that of the endogenous gene harboring the transposon insertion. (**c**) An enhancer-trap transposon includes a reporter gene with a minimal promoter (mP) and a poly-A signal. The minimal promoter can drive expression of the reporter gene only when it is affected by the endogenous enhancer element. (**d**) A poly-A-trap transposon includes a promoter sequence followed by a reporter gene and a splice donor sequence (SD). This trapping cassette does not contain a poly-A signal; thus, the expression of the reporter gene is dependent on successful splicing to the downstream exons of an endogenous Pol II transcription unit that includes a poly-A signal. (**e**) Dual-tagging transposon vectors contain both gene-trap and poly-A-trap cassettes. (**f**) Oncogene-trap transposons include SA and poly-A signals in both orientations in order to disrupt transcription of endogenous genes as well as a strong viral promoter/enhancer sequence (LTR) that can drive transcription toward the outside of the transposon and thereby overexpress a product of a gene the transposon has inserted into.

used for ES cell mutagenesis (38, 39), and were also employed in combination with the *SB* delivery system in several studies for mutagenizing ES cells (27) as well as the germlines of experimental animals (21–23, 30–35). In addition, the efficiency of gene-trapping can be further improved by inserting an internal ribosome entry site (IRES) sequence in front of the reporter gene, which allows for the expression of the reporter cassette irrespective of the reading frame of the disrupted gene. This principle was demonstrated in a study performed by Bonaldo and coworkers (40), where insertion of the IRES sequence between the splice acceptor site and a β-geo selection cassette not only led up to a 15-fold increase in the number of detectable gene-trap events as compared to a conventional vector, but also promoted trapping of genes expressed at very low levels in ES cells. Trapping and discovery of low expressing genes can be further facilitated by using transcriptional transactivation systems, in which an initial, low level of the gene-trap reporter signal is amplified by transactivation of a second reporter and hence made detectable. This strategy was applied by Clark et al. in a study that described an *SB*-based IRES-gene-trap vector conditionally expressing the tTA (tetracycline controllable) transcriptional activator that could amplify the trap signal by activating transcription of a reporter in zebrafish embryos (41). Advantage of this system for identifying expression patterns was further confirmed by Geurts et al. (31) in mice. In another study, gene trapping was elegantly combined with a coat color marker in order to provide a method for fast and noninvasive identification of new transposition events and homozygous animals (35).

Another type of cassette that can be incorporated into *SB* transposons is called an "enhancer-trap". This contains a reporter gene supplied with a minimal promoter that drives the expression of the reporter gene only when affected by an endogenous enhancer element (Fig. 2c). Such *SB*-based enhancer-trap approach was established by two groups in order to characterize novel gene expression patterns (reviewed in (42)) in medaka (43) and zebrafish (44). Since both gene- and enhancer-trap vectors depend on the transcriptional activities of endogenous promoters/enhancers to drive the expression of their reporter cassettes, they can only report insertions into or near (in case of the enhancer-traps) the genes that are actively transcribed in the tissue of interest in which mutagenesis is done. Poly-A-trap cassettes are, on the other hand, equipped with an internal promoter, a reporter cassette, and a splice donor (SD) site, but lack a poly-A-signal (Fig. 2d). If such cassette lands in a transcription unit in the right orientation, the RNA transcript initiated by the internal promoter splices to the endogenous, downstream exons and is processed and polyadenylated. Therefore, poly-A-traps can be used for trapping genes regardless of their transcriptional activity, but are only expected to be sufficiently mutagenic when combined with gene-breaking trap cassettes (reviewed in (6)). Such *SB*-based dual vectors (Fig. 2e) containing a *lacZ* reporter gene in a mutagenic gene-trap cassette as well as a GFP reporter in the poly-A-trap part of the transposon vector was used for germline mutagenesis in mice (24) and rats (34). In both studies, the *lacZ* reporter cassette was employed to visualize endogenous, locus-specific expression patterns of trapped genes, while the fluorescent protein reporter cassette allowed rapid and noninvasive selection of animals which harbored insertions of the transposon vector in transcription units. As for the gene-traps, efficiency of poly-A-trapping vectors can be improved by incorporating the IRES sequence, which suppresses nonsense mediated RNA decay (NMD) of chimeric transcripts, between the reporter cassette and the splice donor site, thereby removing the bias in preferentially detecting those vector integrations that occurred in the last introns of the trapped genes (45). In a recent paper, Tsakiridis and coworkers demonstrated that the activity of human β-actin promoter in mouse ES cells facilitates NMD-independent poly-A-trapping (46). The poly-A-trap vectors used in this study were also equipped with a *cis*-acting mRNA destabilizing AU-rich element (ARE) in order to improve the performance of the SD sequence by reducing the incidence of background SD read-through events, so that only true splicing events with the downstream SA are reported. Further verification of this system was provided by showing that this method can be used to target and trap genes with low expression levels (46). A specific combination of gene- and poly-A-trap features was developed to discover proto-oncogenes as well as novel tumor suppressor genes in mice. This "oncogene-trap" contains splice acceptor signals and poly-A-signals in both orientations in order to disrupt transcription of endogenous genes (Fig. 2f and Fig. 3e, f). In addition, oncogene-trap cassettes also include strong viral enhancers/promoters that can drive transcription outwards from the vector, thereby leading to overexpression of a full-length or truncated protein product of the trapped gene (Fig. 2f and Fig. 3c, d) (reviewed in (6)).

3.1.2. Engineered Chromosomes for Cancer Gene Discovery

SB vectors harboring oncogene-traps have been successfully applied for large-scale cancer gene discovery screens in experimental animals (Fig. 3) (reviewed in (7)). In these studies, *SB* transposons were somatically mobilized from donor chromosomal concatemers (Fig. 3a, b), which contained either low (25) (28) or high (150–350) numbers (29) of the oncogene-trap transposon. Here, dominant mutations in somatic tissues of double transgenic mice carrying a transposase source and the mutagenic transposons resulted in the generation of experimental tumors in cancer-predisposed (28) as well as wild-type (29) animals. In a follow-up study, Collier et al. demonstrated that combination of low-copy oncogene-trap lines with the *SB11* transposase (an early generation hyperactive *SB* variant) expressed from the *Rosa26* locus can achieve whole-body transposon mobilization at rates that are sufficient to promote penetrant tumorigenesis without the complications of embryonic lethality or genomic instability (47). Thus, this approach can be successfully employed not only

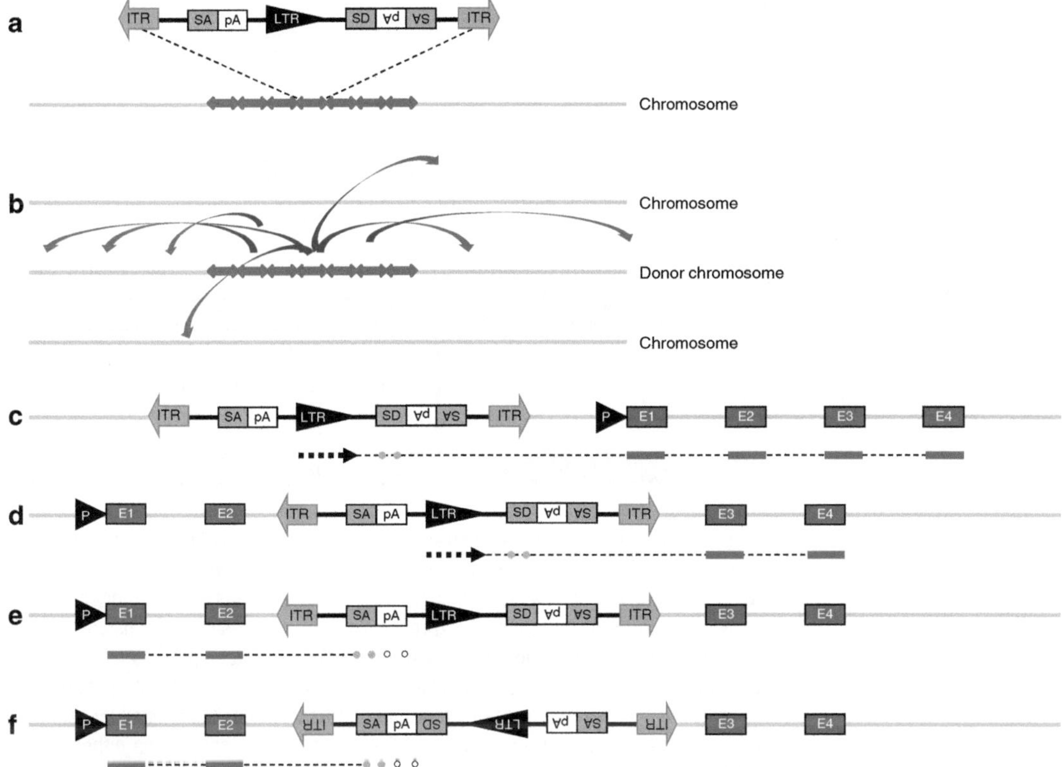

Fig. 3. Schematic outline of transposon mutagenesis strategies for cancer gene discovery screens. (**a**) Concatemeric array of *Sleeping Beauty* (*SB*) oncogene-trap transposons (*bidirectional arrows*) on a donor chromosome (*gray line*). *Arrows* represent transposon inverted repeats (ITRs). The transposon contains splice acceptor (SA), splice donor (SD), and poly-A (pA) signals as well as a viral promoter/enhancer element (LTR). (**b**) Transposition of mutagenic *SB* transposons out of the donor locus upon expression of the transposase enzyme. Excised transposons reinsert either within the donor chromosome in the vicinity of the donor locus due to the local hopping phenomenon or they integrate into different genomic loci on other chromosomes. Chromosomes are represented by *gray lines*. (**c**, **d**) Reinsertion of the oncogene-trap transposons within or close to oncogenes can induce gain-of-function mutations. Structure of the endogenous gene is represented by a promoter sequence (P) and four exons (E1, E2, E3, and E4). (**c**) Overexpression of the full-length gene product occurs upon transposon insertion into the promoter region of the oncogene. (**d**) Overexpression of a truncated gene product by transposon insertion within the transcription unit of an oncogene. In both cases (**c** and **d**), overexpression of the whole or partial gene product is driven by the promoter/splice donor sequences incorporated within the transposon. (**e**, **f**) Integration of the mutagenic transposon within a tumor suppressor transcription unit can induce loss-of-function mutations. Transcription of the tumor suppressor gene can be interrupted by the splice acceptor and poly-A sequences present in the transposon in both orientations (**e** and **f**). The predicted splicing events between the transposon-contained elements (*dots*) and endogenous gene elements (*solid lines*) as well as the predicted endogenous gene splicing patterns are represented with *dashed lines* below each figure.

to identify novel cancer genes, but also combinations of cancer genes that act together to transform a cell.

Current efforts are concentrating on customized, tissue-specific screens for cancer development studies. The strategies employed to achieve this goal focus on establishing mouse lines that either conditionally express the transposase from tissue-specific promoters, or rely on generation of Cre recombinase-inducible

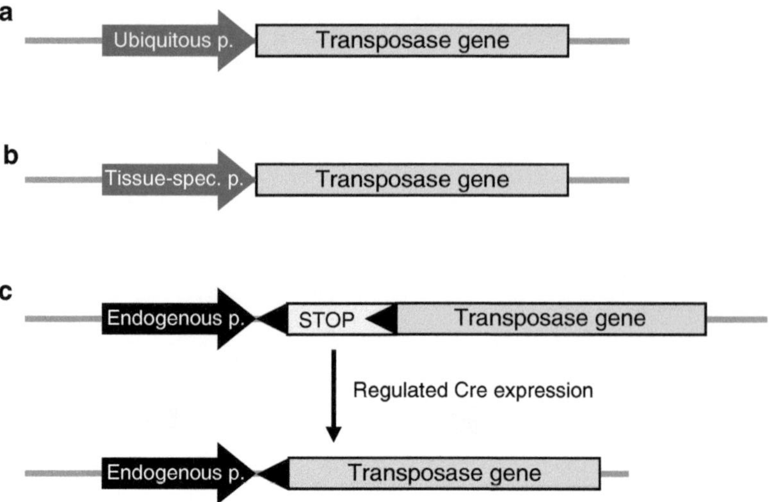

Fig. 4. *SB* transposase expression strategies in transposon-mediated cancer gene discovery screens. Transposase can be expressed in all tissues from a ubiquitous promoter (e.g. *Rosa26*) (**a**) or in a tissue-specific manner (**b** and **c**). Transposon mutagenesis can be restricted to certain tissues by expressing the transposase from a tissue-specific promoter (**b**) or by expressing the transposase from a ubiquitous promoter, but disrupting its expression with a "floxed stop cassette" (**c**). This cassette efficiently prevents the expression of downstream genes by terminating transcription. It is flanked by *loxP* sites and can be removed upon expression of Cre recombinase that can be either tissue-specific, ligand-mediated, or both. Using this approach, many different Cre-expressing mouse strains can be utilized for inducing cell- or organ-specific mutagenesis.

transposase alleles, which can be used in conjunction with mice that express Cre in a tissue-specific manner (Fig. 4) (48–50). For example, this approach was addressed by Dupuy and coworkers (51), who were able to experimentally modify the spectrum of tumors by creating a Cre-inducible *SB* transposase allele (RosaSBase^{LsL}). With this strategy, they managed to overcome the obstacle associated with high embryonic lethality associated with ubiquitous *SB* transposase expression in the presence of the pT2/Onc2 oncogene-trap (48, 52) and to generate a model of germinal center B-cell lymphoma. They achieved this by activating *SB* transposase expression with an AidCre allele that drove Cre-mediated recombination in germinal center B-cells. In another approach, ubiquitous expression of the *SB* transposase was combined with the novel T2/Onc3 oncogene-trap transposon vector. Here, the MSCV (mouse stem cell virus) 5′ LTR that was previously used to drive oncogene expression was replaced by the ubiquitously active CAGGS promoter that resulted in removing the bias from inducing mostly lymphomas and in reducing embryonic lethality. This strategy powerfully underscores that the change in the design of the mutagenic transposon (e.g. promoter choice) can have profound effects on tumor types induced by transposition.

Notably, this approach resulted in a production of nearly 200 independent tumors of more than 20 types and identification of novel, candidate cancer genes, suggesting that the combination of tissue-specific promoters and inducible transposase alleles could provide a fine mechanism of control in tumorigenesis studies.

3.1.3. Loss-of-Function Mutagenesis in Germline-Competent Cells

Insertional mutagenesis can be applied in cultured ES cells. One advantage of this approach lies in the possibility to perform preselection of modified ES cell clones before going for creation of mutant animals, as well as in the possibility to differentiate selected clones into many different tissue types *in vitro*. This principle was demonstrated in mouse ES cells by Bonaldo et al. who preselected gene-trapping events in developmentally regulated genes by selecting for activation or repression of the gene-trap reporter *LacZ* cassette in the presence of specific growth/differentiation factors (40). It is possible to perform large-scale, *SB*-based, insertional mutagenesis screens in ES cells by simply transfecting or electroporating transposon donor and transposase expression plasmids in to the cells. Especially advantageous here is the fact that the amounts of delivered plasmids can be adjusted to obtain the desired insertion frequencies per cell (see above). In addition, *SB* transposons can also be remobilized from chromosomally resident loci and reintegrated somewhere else in the genome by providing the transposase source; such excision–reintegration events can be monitored by using double selection systems, in which excision results in activation of the first and reintegration in activation of the second selection marker (20).

Following this direction, another method in which TEs are utilized for insertional mutagenesis in animal models was derived. This strategy employs a "jumpstarter and mutator" scheme (21, 23, 33). In this arrangement mutator transgenic lines carry *SB* transposon-based gene-trapping vectors, while a jumpstarter line expresses the transposase preferably in the male germline (22, 24). Crossing of the two lines results in transposition in the germline of the F1 double-transgenic males, which are then repeatedly crossed with wild-type females to segregate the transposition events that occurred in their sperm cells to separate F2 animals. In the mouse system a single sperm cell of an F1 male contains, on average, two transposon insertions (21), and up to 90% of the F2 progeny can carry transposon insertions (33). The applicability of this approach has been demonstrated by the identification of mouse genes with either ubiquitous or tissue-specific expression patterns (23, 24, 53, 54). Recently, a similar system for *SB* insertional mutagenesis was also established in rats (34, 35).

3.1.4. Local Chromosomal Mutagenesis by Sleeping Beauty

As mentioned above, transposition of *SB* out of chromosomal donor loci is subject to the "local hopping" phenomenon. For example, in a study performed in mouse ES cells, Liang et al.

mobilized a mutagenic *SB* transposon from an intronic site in the *Hprt* locus (26). Excision of the mutagenic vector from *Hprt* restored its function, and cells in which excision took place were selected in medium containing HAT (hypoxanthine, aminopterin, and thymidine). Subsequent analysis of the reinsertion sites in HAT-resistant clones in which reintegration of transposon took place (determined by Southern blotting) revealed clustering of *SB* reinsertion sites within 4 Mb around the *Hprt-SB* donor site on the mouse X chromosome. Local hopping dramatically increases mutagenicity in a restricted region of the genome, allowing for effective, region-specific mutagenic screens (24, 32). Indeed, a study by Keng et al. demonstrated that all genes within 4-Mb regions around donor loci can be mutated by local hopping of *SB* (32). These results revealed the power of this system for functional genomics screens using saturation mutagenesis within a defined chromosomal region. Such screens could be particularly useful for characterizing contiguous gene syndromes, identifying putative tumor suppressor genes mapped to particular intervals, or saturating specific chromosomal region with mutations. However, the efficiency of local hopping can vary substantially (20–23), which is probably related to the different chromatic contexts of the transposon donor sites. Thus, in order to saturate larger portions of the genome with transposon insertions, many transposon launch pads could be created throughout the genome.

3.2. Reverse Genetics and Chromosome Engineering with Sleeping Beauty

Recently emerging evidence combined with advances in whole genome sequencing indicate that individual genes should not be viewed as separate units, but placed within a larger genomic context that also includes neighboring genes and *cis*-regulatory sequences. Genomic regions involved in such elaborate gene regulatory networks sometimes span over hundreds of kilobases around a specific locus of interest, and are a general feature of vertebrate genomes. The local hopping trait of the *SB* transposon can be combined with chromosome engineering strategies and utilized as an effective tool for studying such regulatory architecture in vertebrate animal models. A classical chromosome engineering approach relies on targeting components of the Cre/*loxP* site-specific recombination system in ES cells. In general, two *loxP* sites are inserted sequentially by gene targeting into two loci in the ES cell genome. Transient expression of the Cre recombinase in these cells induces recombination between the two targeted *loxP* sites in order to generate chromosomal rearrangements (55). Kokubu et al. developed a Local Hopping Enhancer Detector (LHED) system, a strategy that elegantly combined the local-hopping feature of *SB* with site-specific chromosomal engineering with Cre in mouse ES cells (Fig. 5) (27). This system allows monitoring of enhancer activities along the targeted

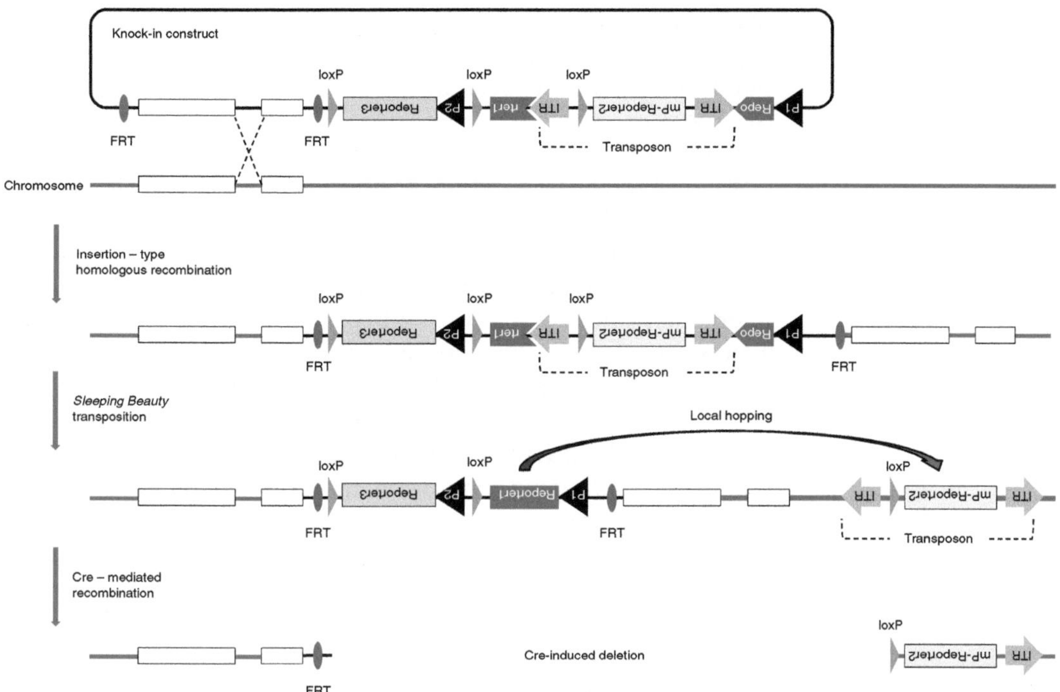

Fig. 5. Transposon-based system for chromosomal engineering. P1 and P2 represent promoter sequences. *Rectangles* labeled "Reporter1" and "Reporter2" represent two different reporter genes, while the third reporter gene is fused to the minimal promoter and represented by a *rectangle* labeled "mP-reporter2." *Arrows* depict transposon inverted repeats. *Triangles* and *oval* shapes stand for the *loxP* and *FRT* sites, respectively. An insertion-type targeting vector can be knocked into a chromosomal donor site. The transposon is mobilized by the transposase and reinserted predominantly into closely linked sites. Cre recombinase can then introduce chromosomal deletions between the fixed and the transposed *loxP* sites. After transposition it is also possible to excise the remaining vector sequence from the knock-in locus by Flp/*FRT* recombination.

genomic region and generation of nested series of deletions for the purpose of studying loss-of-function effects of the genomic neighborhood. In this strategy the LHED transposon vector was knocked into the 5′-flank of the *Pax1* transcription factor gene locus deletion interval, which was located within a region of evolutionary conserved synteny along with other key developmental genes. The knock-in cassette contained an *SB* transposon carrying an enhancer-trap cassette equipped with one *loxP* site. The transposon vector was breaking a *puro* reporter cassette, which allowed for puromycin selection of the cells in which excision took place, whereas a *neo* reporter cassette flanked by two *loxP* sites that was also included in the knock-in construct provided possibility for the selection of successful knock-in events. After the transposon was mobilized by the *SB* transposase and excised from the donor locus, it predominantly inserted into closely linked sites situated along a 1 Mb region of interest. The three *loxP* sites, one within the transposon and two outside, at a fixed position in the knock-in donor locus, were then employed for chromosomal engineering by introducing a nested series of deletion/inversion mutations by the Cre recombinase. Additional *FRT* sites flanking the knock-in cassette could be used, in combination with the Flp recombinase, to excise the remaining knock-in vector after transposition. Preselection of transposition-positive clones allowed the generation of an ES cell-based transposon insertion library, in which nested, Cre-mediated chromosomal deletions could be produced, and their phenotypic effects followed *in vivo* by subsequent derivation of mouse embryos and comparative reporter expression analysis. This strategy resulted in the discovery that *Pax1* enhancers act over a large region along the chromosome with their influence decreasing with increasing distance and in mapping of the distant genomic regions of the enhancer activity that drive *Pax1* gene expression in ventral parts or the entire mouse sclerotome.

A similar approach would be possible by utilizing two different TEs for marker delivery and local mutagenesis (6). In this organization the outer transposon, e.g. *piggyBac* or *Tol2*, would be employed as a vector to carry the other transposon, e.g. *SB*, within its inverted repeats. Both transposons would be equipped with *loxP* sites and selection markers, which could report the excision and integration events of the respective transposon units. Transposition of the outer transposable element would then be used to create sets of different launch pads for the inner transposon. Mobilization and reintegration of the inner element could be followed by Cre-mediated recombination of the *loxP* sites situated within the outer transposon residing at the original donor site and those within the inner transposon integrated in a new chromosomal location. Recombination between these *loxP* sites could produce deletions, inversions, and translocations depending on the position and orientation of the inner transposon.

References

1. Thibault, S. T., Singer, M. A., Miyazaki, W. Y., Milash, B., Dompe, N. A., Singh, C. M., Buchholz, R., Demsky, M., Fawcett, R., Francis-Lang, H. L., Ryner, L., Cheung, L. M., Chong, A., Erickson, C., Fisher, W. W., Greer, K., Hartouni, S. R., Howie, E., Jakkula, L., Joo, D., Killpack, K., Laufer, A., Mazzotta, J., Smith, R. D., Stevens, L. M., Stuber, C., Tan, L. R., Ventura, R., Woo, A., Zakrajsek, I., Zhao, L., Chen, F., Swimmer, C., Kopczynski, C., Duyk, G., Winberg, M. L. and Margolis, J. (2004) A complementary transposon tool kit for Drosophila melanogaster using P and piggyBac. *Nat Genet.* **36**, 283–7.
2. Zwaal, R. R., Broeks, A., van Meurs, J., Groenen, J. T. and Plasterk, R. H. (1993) Target-selected gene inactivation in Caenorhabditis elegans by using a frozen transposon insertion mutant bank. *Proc Natl Acad Sci USA.* **90**, 7431–5.
3. Cooley, L., Kelley, R. and Spradling, A. (1988) Insertional mutagenesis of the Drosophila genome with single P elements. *Science.* **239**, 1121–8.
4. Ivics, Z., Hackett, P. B., Plasterk, R. H. and Izsvak, Z. (1997) Molecular reconstruction of Sleeping Beauty, a Tc1-like transposon from fish, and its transposition in human cells. *Cell.* **91**, 501–10.
5. Miskey, C., Izsvak, Z., Kawakami, K. and Ivics, Z. (2005) DNA transposons in

vertebrate functional genomics. *Cell Mol Life Sci.* **62**, 629–41.

6. Mates, L., Izsvak, Z. and Ivics, Z. (2007) Technology transfer from worms and flies to vertebrates: transposition-based genome manipulations and their future perspectives. *Genome Biol.* **8 Suppl 1**, S1.
7. Ivics, Z., Li, M. A., Mates, L., Boeke, J. D., Nagy, A., Bradley, A. and Izsvak, Z. (2009) Transposon-mediated genome manipulation in vertebrates. *Nat Methods.* **6**, 415–22.
8. Zayed, H., Izsvak, Z., Walisko, O. and Ivics, Z. (2004) Development of hyperactive sleeping beauty transposon vectors by mutational analysis. *Mol Ther.* **9**, 292–304.
9. Geurts, A. M., Yang, Y., Clark, K. J., Liu, G., Cui, Z., Dupuy, A. J., Bell, J. B., Largaespada, D. A. and Hackett, P. B. (2003) Gene transfer into genomes of human cells by the sleeping beauty transposon system. *Mol Ther.* **8**, 108–17.
10. Yant, S. R., Park, J., Huang, Y., Mikkelsen, J. G. and Kay, M. A. (2004) Mutational analysis of the N-terminal DNA-binding domain of sleeping beauty transposase: critical residues for DNA binding and hyperactivity in mammalian cells. *Mol Cell Biol.* **24**, 9239–47.
11. Baus, J., Liu, L., Heggestad, A. D., Sanz, S. and Fletcher, B. S. (2005) Hyperactive transposase mutants of the Sleeping Beauty transposon. *Mol Ther.* **12**, 1148–56.
12. Wilson, M. H., Kaminski, J. M. and George, A. L., Jr. (2005) Functional zinc finger/sleeping beauty transposase chimeras exhibit attenuated overproduction inhibition. *FEBS Lett.* **579**, 6205–9.
13. Mates, L., Chuah, M. K., Belay, E., Jerchow, B., Manoj, N., Acosta-Sanchez, A., Grzela, D. P., Schmitt, A., Becker, K., Matrai, J., Ma, L., Samara-Kuko, E., Gysemans, C., Pryputniewicz, D., Miskey, C., Fletcher, B., Vandendriessche, T., Ivics, Z. and Izsvak, Z. (2009) Molecular evolution of a novel hyperactive Sleeping Beauty transposase enables robust stable gene transfer in vertebrates. *Nat Genet.* **41**, 753–61.
14. Izsvák, Z., Khare, D., Behlke, J., Heinemann, U., Plasterk, R. H. and Ivics, Z. (2002) Involvement of a bifunctional, paired-like DNA-binding domain and a transpositional enhancer in *Sleeping Beauty* transposition. *J. Biol. Chem.* **277**, 34581–34588.
15. Cui, Z., Geurts, A. M., Liu, G., Kaufman, C. D. and Hackett, P. B. (2002) Structure-function analysis of the inverted terminal repeats of the *Sleeping Beauty* transposon. *J. Mol. Biol.* **318**, 1221–1235.
16. Izsvak, Z. and Ivics, Z. (2004) Sleeping beauty transposition: biology and applications for molecular therapy. *Mol Ther.* **9**, 147–56.
17. Izsvak, Z., Ivics, Z. and Plasterk, R. H. (2000) Sleeping Beauty, a wide host-range transposon vector for genetic transformation in vertebrates. *J Mol Biol.* **302**, 93–102.
18. Vigdal, T. J., Kaufman, C. D., Izsvak, Z., Voytas, D. F. and Ivics, Z. (2002) Common physical properties of DNA affecting target site selection of sleeping beauty and other Tc1/mariner transposable elements. *J Mol Biol.* **323**, 441–52.
19. Yant, S. R., Wu, X., Huang, Y., Garrison, B., Burgess, S. M. and Kay, M. A. (2005) High-resolution genome-wide mapping of transposon integration in mammals. *Mol Cell Biol.* **25**, 2085–94.
20. Luo, G., Ivics, Z., Izsvak, Z. and Bradley, A. (1998) Chromosomal transposition of a Tc1/mariner-like element in mouse embryonic stem cells. *Proc Natl Acad Sci USA.* **95**, 10769–73.
21. Dupuy, A. J., Fritz, S. and Largaespada, D. A. (2001) Transposition and gene disruption in the male germline of the mouse. *Genesis.* **30**, 82–8.
22. Fischer, S. E., Wienholds, E. and Plasterk, R. H. (2001) Regulated transposition of a fish transposon in the mouse germ line. *Proc Natl Acad Sci USA.* **98**, 6759–64.
23. Carlson, C. M., Dupuy, A. J., Fritz, S., Roberg-Perez, K. J., Fletcher, C. F. and Largaespada, D. A. (2003) Transposon mutagenesis of the mouse germline. *Genetics.* **165**, 243–56.
24. Horie, K., Yusa, K., Yae, K., Odajima, J., Fischer, S. E., Keng, V. W., Hayakawa, T., Mizuno, S., Kondoh, G., Ijiri, T., Matsuda, Y., Plasterk, R. H. and Takeda, J. (2003) Characterization of Sleeping Beauty transposition and its application to genetic screening in mice. *Mol Cell Biol.* **23**, 9189–207.
25. Hansen, G. M., Markesich, D. C., Burnett, M. B., Zhu, Q., Dionne, K. M., Richter, L. J., Finnell, R. H., Sands, A. T., Zambrowicz, B. P. and Abuin, A. (2008) Large-scale gene trapping in C57BL/6N mouse embryonic stem cells. *Genome Res.* **18**, 1670–9.
26. Liang, Q., Kong, J., Stalker, J. and Bradley, A. (2009) Chromosomal mobilization and reintegration of Sleeping Beauty and PiggyBac transposons. *Genesis.* **47**, 404–8.
27. Kokubu, C., Horie, K., Abe, K., Ikeda, R., Mizuno, S., Uno, Y., Ogiwara, S., Ohtsuka, M., Isotani, A., Okabe, M., Imai, K. and Takeda, J. (2009) A transposon-based

chromosomal engineering method to survey a large cis-regulatory landscape in mice. *Nat Genet.* **41**, 946–52.

28. Collier, L. S., Carlson, C. M., Ravimohan, S., Dupuy, A. J. and Largaespada, D. A. (2005) Cancer gene discovery in solid tumours using transposon-based somatic mutagenesis in the mouse. *Nature.* **436**, 272–6.
29. Dupuy, A. J., Akagi, K., Largaespada, D. A., Copeland, N. G. and Jenkins, N. A. (2005) Mammalian mutagenesis using a highly mobile somatic Sleeping Beauty transposon system. *Nature.* **436**, 221–6.
30. Roberg-Perez, K., Carlson, C. M. and Largaespada, D. A. (2003) MTID: a database of Sleeping Beauty transposon insertions in mice. *Nucleic Acids Res.* **31**, 78–81.
31. Geurts, A. M., Wilber, A., Carlson, C. M., Lobitz, P. D., Clark, K. J., Hackett, P. B., McIvor, R. S. and Largaespada, D. A. (2006) Conditional gene expression in the mouse using a Sleeping Beauty gene-trap transposon. *BMC Biotechnol.* **6**, 30.
32. Keng, V. W., Yae, K., Hayakawa, T., Mizuno, S., Uno, Y., Yusa, K., Kokubu, C., Kinoshita, T., Akagi, K., Jenkins, N. A., Copeland, N. G., Horie, K. and Takeda, J. (2005) Region-specific saturation germline mutagenesis in mice using the Sleeping Beauty transposon system. *Nat Methods.* **2**, 763–9.
33. Horie, K., Kuroiwa, A., Ikawa, M., Okabe, M., Kondoh, G., Matsuda, Y. and Takeda, J. (2001) Efficient chromosomal transposition of a Tc1/mariner- like transposon Sleeping Beauty in mice. *Proc Natl Acad Sci USA.* **98**, 9191–6.
34. Kitada, K., Ishishita, S., Tosaka, K., Takahashi, R. I., Ueda, M., Keng, V. W., Horie, K. and Takeda, J. (2007) Transposon-tagged mutagenesis in the rat. *Nat Methods.*
35. Lu, B., Geurts, A. M., Poirier, C., Petit, D. C., Harrison, W., Overbeek, P. A. and Bishop, C. E. (2007) Generation of rat mutants using a coat color-tagged Sleeping Beauty transposon system. *Mamm Genome.* **18**, 338–46.
36. Durick, K., Mendlein, J. and Xanthopoulos, K. G. (1999) Hunting with traps: genome-wide strategies for gene discovery and functional analysis. *Genome Res.* **9**, 1019–25.
37. Stanford, W. L., Cohn, J. B. and Cordes, S. P. (2001) Gene-trap mutagenesis: past, present and beyond. *Nat Rev Genet.* **2**, 756–68.
38. Joyner, A. L., Auerbach, A. and Skarnes, W. C. (1992) The gene trap approach in embryonic stem cells: the potential for genetic screens in mice. *Ciba Found Symp.* **165**, 277-88; discussion 288–97.
39. Skarnes, W. C., Auerbach, B. A. and Joyner, A. L. (1992) A gene trap approach in mouse embryonic stem cells: the lacZ reported is activated by splicing, reflects endogenous gene expression, and is mutagenic in mice. *Genes Dev.* **6**, 903–18.
40. Bonaldo, P., Chowdhury, K., Stoykova, A., Torres, M. and Gruss, P. (1998) Efficient gene trap screening for novel developmental genes using IRES beta geo vector and in vitro preselection. *Exp Cell Res.* **244**, 125–36.
41. Clark, K. J., Geurts, A. M., Bell, J. B. and Hackett, P. B. (2004) Transposon vectors for gene-trap insertional mutagenesis in vertebrates. *Genesis.* **39**, 225–33.
42. Balciunas, D. and Ekker, S. C. (2005) Trapping fish genes with transposons. *Zebrafish.* **1**, 335–41.
43. Grabher, C., Henrich, T., Sasado, T., Arenz, A., Wittbrodt, J. and Furutani-Seiki, M. (2003) Transposon-mediated enhancer trapping in medaka. *Gene.* **322**, 57–66.
44. Balciunas, D., Davidson, A. E., Sivasubbu, S., Hermanson, S. B., Welle, Z. and Ekker, S. C. (2004) Enhancer trapping in zebrafish using the Sleeping Beauty transposon. *BMC Genomics.* **5**, 62.
45. Shigeoka, T., Kawaichi, M. and Ishida, Y. (2005) Suppression of nonsense-mediated mRNA decay permits unbiased gene trapping in mouse embryonic stem cells. *Nucleic Acids Res.* **33**, e20.
46. Tsakiridis, A., Tzouanacou, E., Rahman, A., Colby, D., Axton, R., Chambers, I., Wilson, V., Forrester, L. and Brickman, J. M. (2009) Expression-independent gene trap vectors for random and targeted mutagenesis in embryonic stem cells. *Nucleic Acids Res.* **37**, e129.
47. Collier, L. S., Adams, D. J., Hackett, C. S., Bendzick, L. E., Akagi, K., Davies, M. N., Diers, M. D., Rodriguez, F. J., Bender, A. M., Tieu, C., Matise, I., Dupuy, A. J., Copeland, N. G., Jenkins, N. A., Hodgson, J. G., Weiss, W. A., Jenkins, R. B. and Largaespada, D. A. (2009) Whole-body sleeping beauty mutagenesis can cause penetrant leukemia/lymphoma and rare high-grade glioma without associated embryonic lethality. *Cancer Res.* **69**, 8429–37.
48. Dupuy, A. J., Jenkins, N. A. and Copeland, N. G. (2006) Sleeping beauty: a novel cancer gene discovery tool. *Hum Mol Genet.* **15 Spec No 1**, R75–9.
49. Keng, V. W., Villanueva, A., Chiang, D. Y., Dupuy, A. J., Ryan, B. J., Matise, I., Silverstein, K. A., Sarver, A., Starr, T. K., Akagi, K., Tessarollo, L., Collier, L. S., Powers, S., Lowe,

S. W., Jenkins, N. A., Copeland, N. G., Llovet, J. M. and Largaespada, D. A. (2009) A conditional transposon-based insertional mutagenesis screen for genes associated with mouse hepatocellular carcinoma. *Nat Biotechnol.*

50. Starr, T. K., Allaei, R., Silverstein, K. A., Staggs, R. A., Sarver, A. L., Bergemann, T. L., Gupta, M., O'Sullivan, M. G., Matise, I., Dupuy, A. J., Collier, L. S., Powers, S., Oberg, A. L., Asmann, Y. W., Thibodeau, S. N., Tessarollo, L., Copeland, N. G., Jenkins, N. A., Cormier, R. T. and Largaespada, D. A. (2009) A Transposon-Based Genetic Screen in Mice Identifies Genes Altered in Colorectal Cancer. *Science.*
51. Dupuy, A. J., Rogers, L. M., Kim, J., Nannapaneni, K., Starr, T. K., Liu, P., Largaespada, D. A., Scheetz, T. E., Jenkins, N. A. and Copeland, N. G. (2009) A modified sleeping beauty transposon system that can be used to model a wide variety of human cancers in mice. *Cancer Res.* **69**, 8150–6.
52. Collier, L. S. and Largaespada, D. A. (2005) Hopping around the tumor genome: transposons for cancer gene discovery. *Cancer Res.* **65**, 9607–10.
53. Yae, K., Keng, V. W., Koike, M., Yusa, K., Kouno, M., Uno, Y., Kondoh, G., Gotow, T., Uchiyama, Y., Horie, K. and Takeda, J. (2006) Sleeping beauty transposon-based phenotypic analysis of mice: lack of Arpc3 results in defective trophoblast outgrowth. *Mol Cell Biol.* **26**, 6185–96.
54. Geurts, A. M., Collier, L. S., Geurts, J. L., Oseth, L. L., Bell, M. L., Mu, D., Lucito, R., Godbout, S. A., Green, L. E., Lowe, S. W., Hirsch, B. A., Leinwand, L. A. and Largaespada, D. A. (2006) Gene mutations and genomic rearrangements in the mouse as a result of transposon mobilization from chromosomal concatemers. *PLoS Genet.* **2**, e156.
55. Yu, Y. and Bradley, A. (2001) Engineering chromosomal rearrangements in mice. *Nat Rev Genet.* **2**, 780–90.

Chapter 6

Rodent Transgenesis Mediated by a Novel Hyperactive Sleeping Beauty Transposon System

Lajos Mátés

Abstract

DNA-based transposons are natural gene delivery vehicles. Similarly to retroviruses, these elements integrate into the chromosomes of host cells, but their life-cycle does not involve reverse transcription and they are not infectious. Transposon-based gene delivery has several advantageous features compared to viral methods; however, its efficacy has been the bottleneck of transposon utilization. Recently, using a novel strategy for in vitro evolution, we created a new hyperactive version (SB100X) of the vertebrate-specific *Sleeping Beauty* (SB) transposase. SB100X, when coupled with enhanced inverted terminal repeat structure T2 type SB transposons, is over 100-fold more active in mammalian cells than the prototype. We established protocol for SB100X mediated rodent transgenesis resulting on the average 35% transgenic founders with a low average number (1–2) of transgene insertions per founder. Due to these characteristics the SB100X based protocol opens the possibility of designing SB based transgenes also for in vivo knockdown experiments. By the same token, single copy transgene units introduced by the SB transposon system, more than being less prone to transgene silencing, also allow better control of transgene expression levels and patterns.

Key words: Sleeping Beauty, SB100X, Transgenesis, Transgene silencing, In vivo knockdown

1. Introduction

1.1. Class II (or DNA) Transposons

Class II transposons (also called DNA transposons) that move in the host genome via a "cut-and-paste" mechanism are particularly useful tools for genome manipulations due to their easy laboratory handling and controllable nature. Schematic outlines of the structure and the transposition process of a class II transposon are presented in Fig. 1a. Class II transposons are simply organized; they consist of a transposase-coding gene flanked by the inverted terminal repeats (ITRs). The ITRs contain the transposase binding sites necessary for transposition. The process of transposition

Gyula Hadlaczky (ed.), *Mammalian Chromosome Engineering: Methods and Protocols*, Methods in Molecular Biology, vol. 738, DOI 10.1007/978-1-61779-099-7_6, © Springer Science+Business Media, LLC 2011

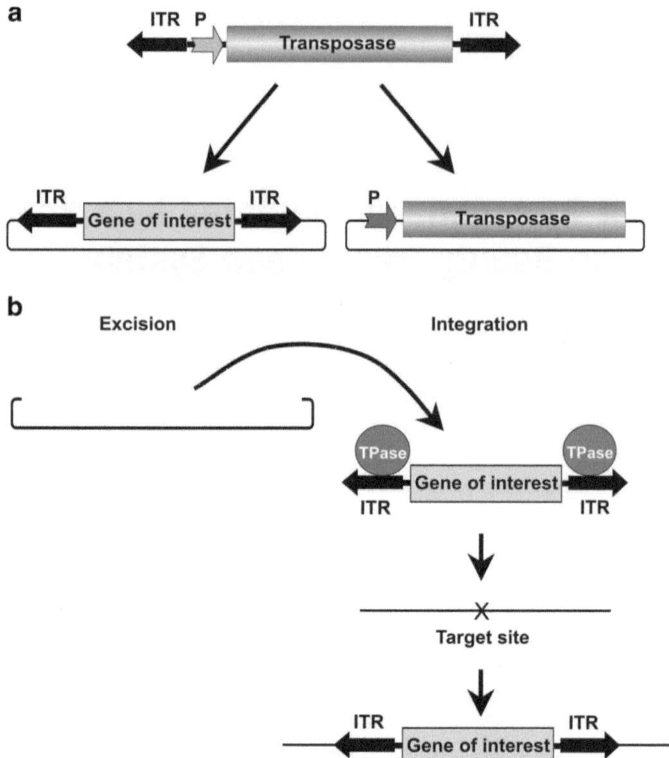

Fig. 1. *Cut and paste DNA transposition.* (**a**) Scheme of a class II (or DNA) transposon, and that of a binary transposition system created by dissecting the transposase source from the transposon. (**b**) Outline of the mechanism of the "cut and paste" transposition.

can easily be controlled by separating the transposase source from the transposable DNA harboring the ITRs, thereby creating a nonautonomous transposon (Fig. 1a). In such a two-component system, the transposon can only be moved by supplementing the transposase *in trans*. In practice, any sequence of interest can be positioned between the ITR elements according to the experimental needs. Transposition will result in excision of the element from the donor vector and subsequent integration into a new sequence environment (Fig. 1b).

1.2. Transposon-Mediated Transgenesis

Classic ways to induce expression of foreign genes in vertebrates rely on microinjection of nucleic acids into oocytes or fertilized eggs. Two main limitations of these approaches are low rates of genomic integration and that the injected DNA generally integrates as a multicopy transgene concatemer. Both drawbacks can be circumvented by utilizing transposition mediated gene delivery because it can increase the efficiency of chromosomal integration and results in single-copy insertion events. Single-unit expression cassettes are less prone to transgene silencing than are the concatemeric insertions created using classical methods. Another particular problem concerning transgenesis is that founders that develop from the injected oocytes or eggs are predominantly mosaic for the transgene because integration generally occurs relatively late during embryonic development. For example, in zebrafish transgenesis, the mosaicism extent is high, due to the integration of the transgene into the chromosome at later cleavage stages of the embryos (1, 2). Therefore, many transgenic founders transmit the transgene to only a few percent of their offspring or do not transmit at all. In contrast, mosaicism does not seem to be a routine problem in rodents, where the same integration events predominantly happen at the one-cell stage in injected embryos. Transposon-mediated transgenesis catalyzed by delivery of DNA encoding the transposase *in trans* faces the same transgene mosaicism problem because the oocytes are in a transcriptionally quiescent state and the embryonic genome activation (EGA) starts species-specifically at different stages after the fertilization. In mice, the major onset of transcription, EGA, begins during the 2-cell stage (3); it begins during the 4-cell stage in rats (4), and during the 8-cell to 16-cell stage in cattle (5). However, co-injection of engineered transposons with in vitro transcribed transposase messenger RNA (mRNA) helps to overcome this limitation because only translation of the synthesized mRNA is necessary to produce the transposase protein. By this approach it is possible to shift the window of transposon-mediated integration events to early stages in order to promote lower mosaicism and successful transmission of the transgene to the next generation in spite of the actual transcriptional quiescent state of the zygote. Currently the Sleeping Beauty (SB), piggyBac (PB), and Tol2 transposon systems are predominantly harnessed for transgenic purposes in vertebrates due to their sufficient activity in the vertebrate hosts. Lately, the mRNA co-injection method, where the in vitro synthesized mRNA of the transposase is co-injected with the transposon DNA, became the procedure predominantly used for transposon-mediated transgenesis. This method has been employed to generate transgenic zebrafish with Tol2 (6) and SB (7); transgenic Xenopus with SB (8) and Tol2 (9); and transgenic mice with SB (10) and more recently with a novel improved version of the SB transposase, SB100X (11).

In case of animal transgenesis, a single copy transgene insertion not disturbing endogenous gene functions is desirable. The insertional spectrum of the SB transposon system satisfies this criterion the best, because it integrates randomly at the genome level, and do not exhibit pronounced bias for integration into genes or 5′ regulatory regions (12, 13) . However, PB and Tol2 tolerate bigger cargo sizes (14, 15), which can be important for certain transgenes. Therefore the transposon system should be selected carefully based on the actual experimental design.

Table 1
Transposon-mediated transgenesis in mice

Transposon name	Transposon source (cc injected)	Transposase source (cc injected)	Transgenic efficiency (transpositional only)	Average transgene copy number	Source
PB	Circular plasmid DNA (1.33 ng/μl)	Circular plasmid DNA (0.66 ng/μl)	35–65% (depending on the construct used)	5–10	Ref. (14)
SB	Linear DNA (4 ng/μl)	mRNA (10 ng/μl)	14%	3	Ref. (10)
SB100X	Circular plasmid DNA (0.4 ng/μl)	mRNA (5 ng/μl)	37%	1–2	Ref. (11)

Table 1 summarizes the reported transgenic studies in mice using the PB and SB transposon systems. In the case of the PB mediated transgenic study, plasmid DNA was used as a transposase source. However, integration and continued expression of a gene encoding the transposase could be problematic if it led to transposon remobilization and reintegration. Therefore, and considering later onset of EGA in some species, the use of in vitro synthesized transposase mRNA is recommended. The PB transposon system produced transgenic founders with high efficiency; however, the generated transgene copy number per founder is also high (Table 1). The SB system generates lower transgene copy number per founder and using the novel hyperactive SB transposase version SB100X (11) the efficiency is also in the high range, as reported with the PB transposon system (Table 1).

The SB100X mediated transgenesis has a particular advantage as compared to other systems, by efficiently mediating transposition events to the genome of the host at low amounts of transposon template DNA available. On the one hand applying low amount of DNA decreases the toxicity of the injected material, and therefore enhances the survival rate of the embryos, as an apparent increasing toxic effect was observed by increasing the transposon DNA amounts injected to mouse zygotes ((11), and unpublished observations). On the other hand this also allows setting up microinjection protocols where low amount of circular plasmid is used as a transposon donor to avoid concatemer transgene integrations and harvest on the average 1–2 exclusively transposition mediated insertions in positive founders at stably high overall transgenic rates.

Therefore, further on I will focus on the rodent transgenesis mediated by the novel hyperactive SB100X transposase (11) coupled with the enhanced ITR structure T2 SB transposon (16).

2. Materials

2.1. Equipment

1. Refrigerated centrifuge capable of high speed >10,000 g.
2. FHS/LS-1B macro-visualization equipment (BLS, Ltd).
3. Water bath, 37°C for REN digestion.
4. NanoDrop® ND-1000 Spectrophotometer (Peqlab) or similar.
5. Milli-Q Water Purification System (Millipore) or similar.
6. Agarose gel running apparatus.
7. Eppendorf tubes (RNase-free).

2.2. Buffers and Reagents

1. mMessage mMachine® T7 kit, (Ambion).
2. Plasmid DNA preparation kit (Qiagen).
3. Transposon donor plasmid (see below).
4. Transposase source plasmid (see below).
5. Agarose.
6. Ethidium bromide, 1% solution in water (Merck) or similar.
7. *Cla* I REN.
8. Taq DNA polymerase.
9. Injection buffer, EmbryoMax®, (Millipore) or similar.
10. Sodium acetate 3 M at pH 5.5 (RNase free), Ambion, or similar.
11. Sodium hydroxide 0.2 N.
12. Ethanol (RNase free) (ROTH), or similar.
13. 70% Ethanol (RNase free).
14. Water (RNase free) (SIGMA), or similar.
15. Phenol/chloroform/isoamyl alcohol, Roti®-Phenol/C/I, (ROTH), or similar.
16. Chloroform/isoamyl alcohol, Roti®-C/I, (ROTH), or similar.
17. 1× TBE; 89 mM Tris, 89 mM Borate, and 2 mM EDTA (pH 8.0).
18. 1× TE; 10 mM Tris–HCl (pH 7.5) and 1 mM EDTA 9 (pH 8.0).
19. Transposon donor plasmids: pT2/BH, pT2/HB, or pT2/SVNeo. The plasmids are available from Dr. Perry Hackett, Ph.D. (University of Minnesota, Minneapolis, MN, USA, http://www.cbs.umn.edu/labs/perry).
20. Transposase source plasmid: pcGlobin2-SB100X. The plasmid is available from Dr. Zsuzsanna Izsvak, Ph.D. Max Delbrück-Centrum für Molekulare Medizin, Berlin, Germany, http://www.mdc-berlin.de/en/research/research_teams/mobile_dna/index.html.

3. Methods

The transposon-mediated transgenesis protocol is similar to the classical transgenesis. For detailed technical information about all aspects of the standard transgenesis protocol like pronuclear injection of the mouse zygote, the reader is referred to the manual initially published by Nagy et al. (17). Here I will focus on the difference as compared to the standard technology that lies in the material injected into the fertilized oocytes. The material prepared for microinjection contains two components; the transposon donor plasmid and the in vitro synthesized transposase mRNA (see Note 1). Both components have to be prepared RNase free in injection buffer, and the appropriate volumes have to be mixed to have the final injection mixture. Due to the mRNA content of the injected material it is necessary to handle the injection needles (capillaries) with gloves, to avoid RNase contamination and degradation of the transposase mRNA.

3.1. Preparation of the Transposon Donor Plasmid

1. Clone your gene of interest between the ITRs of a T2 type SB transposon donor plasmid. The SB100X transposase (11) has been created using the T2 type enhanced SB ITRs (16); this type of ITRs provide the best gene transfer efficiencies with the novel transposase. Suitable vectors (pT2/BH, pT2/HB, or pT2/SVNeo) are listed above (see Subheading 2.2). Preferably your transgene size should not exceed 5 kB (18) (see Note 2).
2. After cloning your gene of interest prepare the transposon donor plasmid using the plasmid DNA preparation kit. Follow the instructions of the manufacturer.
3. Make the plasmid DNA RNase free by phenol/chloroform extraction. All centrifugations during the following phenol/chloroform extraction procedure are done at 12,000 × *g* (top speed of a tabletop centrifuge) at room temperature unless otherwise noted.
   (a) Increase the volume of the plasmid prepared to 400 μl with TE buffer in a 1.5-ml eppendorf tube.
   (b) Add 400 μl of phenol/chloroform/isoamyl alcohol to the tube.
   (c) Vortex for 15 s and leave on the table for 2 min. Repeat this step three times to completely inactivate the residual RNase.
   (d) Centrifuge for 5 min.
   (e) Transfer the top layer to a new RNase-free microcentrifuge tube and add 400 ml of chloroform/isoamyl alcohol.

(f) Vortex for 15 s and centrifuge for 5 min.

(g) Transfer the top layer to a new RNase free eppendorf tube, add 1/10 volume sodium acetate 3 M at pH 5.5 (RNase free) and 2.5 volume of 100% ethanol (RNase free), and let the DNA precipitate for 30 min at −20°C.

(h) Spin down for 15 min in a refrigerated centrifuge at 4°C and discard the supernatant.

(i) Wash the pellet in cold 70% ethanol (RNase free). Keep the ethanol on the pellet for 10 min and then remove it (centrifuge if necessary). Repeat this step once more to completely wash away any residual chemicals which may not be tolerated by the embryos.

(j) Air-dry the pellet for 5–10 min and resuspend the pellet in 100 μL of RNase free injection buffer (EmbryoMax®, Millipore) (see Note 3).

(k) Measure the concentration of the plasmid using a NanoDrop® or other spectrophotometer.

3.2. Preparation of the Transposase mRNA

The SB100X CDS is cloned into the pcGlobin2 (19) vector supporting in vitro mRNA synthesis (11). This vector allows the in vitro synthesis of an SB100X mRNA with Zebrafish β-globin 5′ and 3′ UTRs and a 30 mer synthetic poly (A) sequence, from a T7 promoter (19).

1. Linearize at least 2 μg of pcGlobin2-SB100X plasmid with *Cla* I. digestion, according to the instructions of the enzyme supplier. 1 μg of linearized plasmid will be necessary for one round of mRNA synthesis. Check the complete linearization on 1% agarose gel.
2. Make the fully digested plasmid RNase free, by phenol/chloroform extraction as described above with the modification that the volumes of the digested plasmid DNA, the phenol/chloroform/isoamyl alcohol, and the chloroform/isoamyl alcohol are set to 100 μl each.
3. Synthesize the mRNA using a mMessage mMachine® T7 kit. Follow the manufacturer's instructions. After synthesis, use the Turbo DNAse treatment and phenol/chloroform extraction options suggested in the kit manual, with the modification that after the isopropanol precipitation following the phenol/chloroform extraction, wash the pellet twice in cold 70% ethanol (RNase free). Air-dry the pellet. Resuspend the mRNA in 20 μl of RNase free water.
4. Check your synthesized mRNA on an RNase free 1% agarose gel.

   (a) Wash the running chamber, gel tray, comb and flask for gel preparation with 70% ethanol.

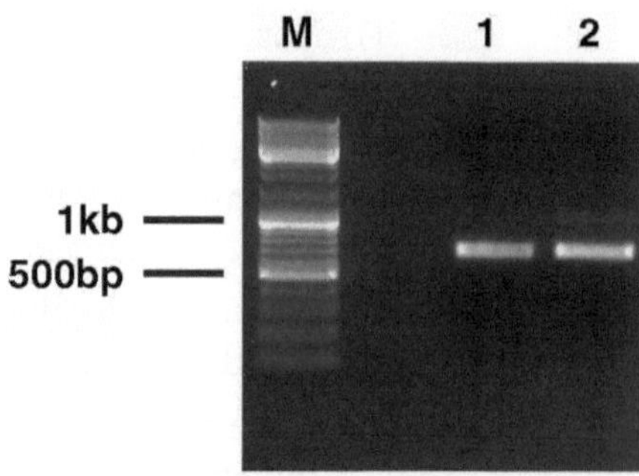

Fig. 2. *Result of the in vitro mRNA synthesis and test of the microinjection buffer.* Samples run on 1% RNase free nondenaturing agarose gel. Lanes: M, DNA size marker, and GeneRuler DNA ladder Mix (Fermentas); 1, 1 μl of in vitro synthesized mRNA; 2, 1 μl of in vitro synthesized mRNA incubated (1 h, 37°C) in 10 μl of injection buffer.

(b) Incubate the running chamber, gel tray, comb and flask for gel preparation in 0.2 N NaOH for 1 h.

(c) Rinse the running chamber, gel tray, comb and flask for gel preparation with sterile Milli-Q water (see Subheading 2.1).

(d) Prepare 1× TBE gel running buffer using sterile 10× TBE buffer and sterile Milli-Q water.

(e) Prepare the 1% agarose gel using sterile 1× TBE buffer, sterile Milli-Q water, and Agarose powder.

(f) Load 1 μl of the final in vitro synthesized mRNA in RNA loading buffer and a DNA size marker and run the gel. The RNA loading buffer is supplied in the mMessage mMachine® T7 kit.

A typical result is shown in Fig. 2. The SB100X mRNA prepared using the T7 promoter on the *Cla I*-digested pcGlobin2-SB100X, runs on a normal agarose gel between 700 and 800 bp corresponding to dsDNA size marker (Fig. 2, lane 1). Typically the in vitro synthesized mRNA runs as one band on nondenaturing gel (Fig. 2, lane 1). Alternatively you may see two bands (Fig. 2, lane 2) due to secondary structures (see Note 4).

5. Measure the concentration of the in vitro synthesized mRNA using a NanoDrop® or similar spectrophotometer. The typical yield is around 1 ug/μl and a total volume of 20 μl.

6. Make diluted aliquots for later use in RNase-free injection buffer (see Note 3) (see Subheading 3.3 below). Store the concentrated and diluted aliquots at –80°C.

3.3. Preparation of the Microinjection Mixture

In order to create the final microinjection mixture specific amounts of SB100X mRNA and transposon donor plasmid have to be mixed to reach the desired final concentrations in the injection mixture.

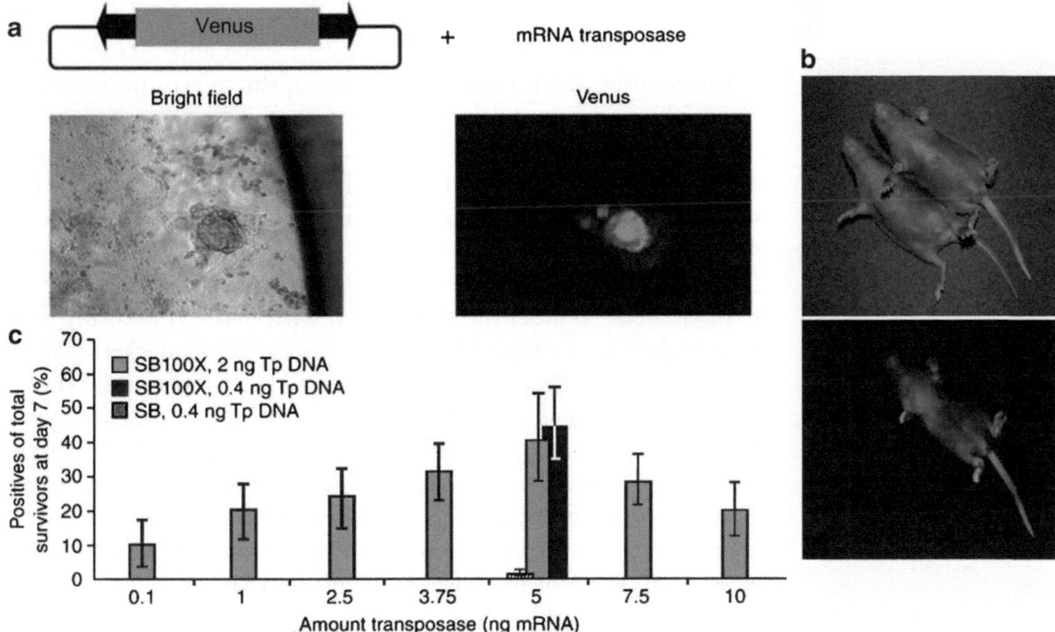

Fig. 3. *Optimization of the microinjection mixture in cultured embryos and detection of the Venus expression in mice.* (**a**) Circular pT2/CAGGS-Venus plasmid DNA was injected together with SB and SB100X transposase mRNAs into mouse zygotes. The embryos were cultured ex vivo, and transferred onto feeder cells for culture at the blastocyst stage. Reporter gene expression was scored at day 7 postinjection, when the embryos were already hatched and the background fluorescence was negligible. (**b**) Macro-detection of the Venus transgene in 3-week-old founder mice. The picture on the upper panel was taken at normal light without filter. The picture on the lower panel was taken using FHS/LS-1B light source and YFP filter (BLS Ltd, Hungary). (**c**) Optimization of the transposon donor plasmid and transposase mRNA concentrations injected into mouse zygotes. The data are collected from two experiments (30–40 embryos per experiment), except injections with 5 ng transposase mRNA that were done six, three, or two times at the different transposon DNA concentrations (*gray*, *black*, and *striped*, respectively). Under these conditions, the 80–90% survival rate of the embryos was similar to the noninjected controls. Error bars, s.d. (Reproduced from Ref. 11 by permission from NPG).

In our hand the best working injection mixture contains 5 ng/μl of SB100X mRNA and 0.4 ng/μl of transposon donor plasmid (Fig. 3c) (11). Prepare 2× concentration stocks of transposon donor plasmid (0.8 ng/μl) (see Note 2) and transposase mRNA (10 ng/μl) in RNase-free injection buffer (EmbryoMax®, Millipore). These stock solutions may be stored at −80°C or mixed directly at a ratio of 1:1 to create the final microinjection mixture. Prepare 10 μl aliquots of the final microinjection mixture and keep them frozen at −80°C until use. Follow the standard microinjection protocol as described by Nagy et al. (17). Pay attention to avoid RNase contamination of the injection mixture (see Note 5).

3.4. Screening Founder Animals for Transpositional Transgenesis

We found that the efficiency of the SB100X mediated rodent transgenesis is varying between 10 and 75% with an average around 35%; the transgene mosaicism occurs at a low extent and most of the transgenes are transmitted to the next generation (unpublished observations). The average number of transgenes in positive founders is between 1 and 2 in mice and rats using the current conditions for the preparation of the microinjection mixture ((11), and unpublished observations). The expression of the Venus transgene driven by the CAGGS promoter was always uniformly detected across the body in F1 and later generations without visible signs of transgene silencing ((11), and unpublished observations).

3.4.1. Detection of the Transgene by Marker Gene Expression

It is beneficial to label the transposon with markers allowing phenotypic detection of the transgene, like coat color markers (20) or fluorescent proteins (11). One copy of a CAGGS promoter driven Venus (21) marker allows the stable and easy detection of the presence of the transgene in the genome (11). Figure 3b shows the detection of Venus fluorescence in founder mice using a macro-visualization equipment (BLS, Budapest, Hungary). Animals were exposed to a light source with a wavelength of $\lambda = 460–495$ nm to detect expression of the fluorescent protein Venus.

3.4.2. Detection of the Transgene by PCR

Setting up a genotyping PCR sensitive enough for the detection of one copy of the transgene in a diploid genome may also be necessary, especially if the use of marker genes is not feasible, for example, due to transposon size limitations. To reach this goal, one should optimize the genotyping PCR to stably detect 50 fg of transposon donor plasmid mixed with 500 ng of wild type genomic DNA. One example of optimized genotyping PCR detecting the Venus transgene is shown in Fig. 4a. This sensitivity guarantees that single copy transgenes are always detected using 500 ng genomic DNA as PCR template (see Note 6). Genotyping PCR of a cohort of F1 generation mice, descendants of pT2/Venus (11) transgenic founders, is shown in Fig. 4b as an example.

4. Notes

1. Decreased viability of the injected embryos may be due to residual harmful chemicals in the microinjection mixture. Therefore, the careful washing of precipitated plasmid DNA and mRNA with 70% ethanol is important. Diethylpyrocarbonate (DEPC), a chemical widely used for the preparation of RNase free solutions is not tolerated by the embryos. Therefore the RNase free solutions of the protocol are prepared without using DEPC.
2. Using SB100X, transgenes significantly bigger than 5 kB may still transpose with an acceptable efficiency. Increasing the amount of transposon donor plasmid, preferably not over the

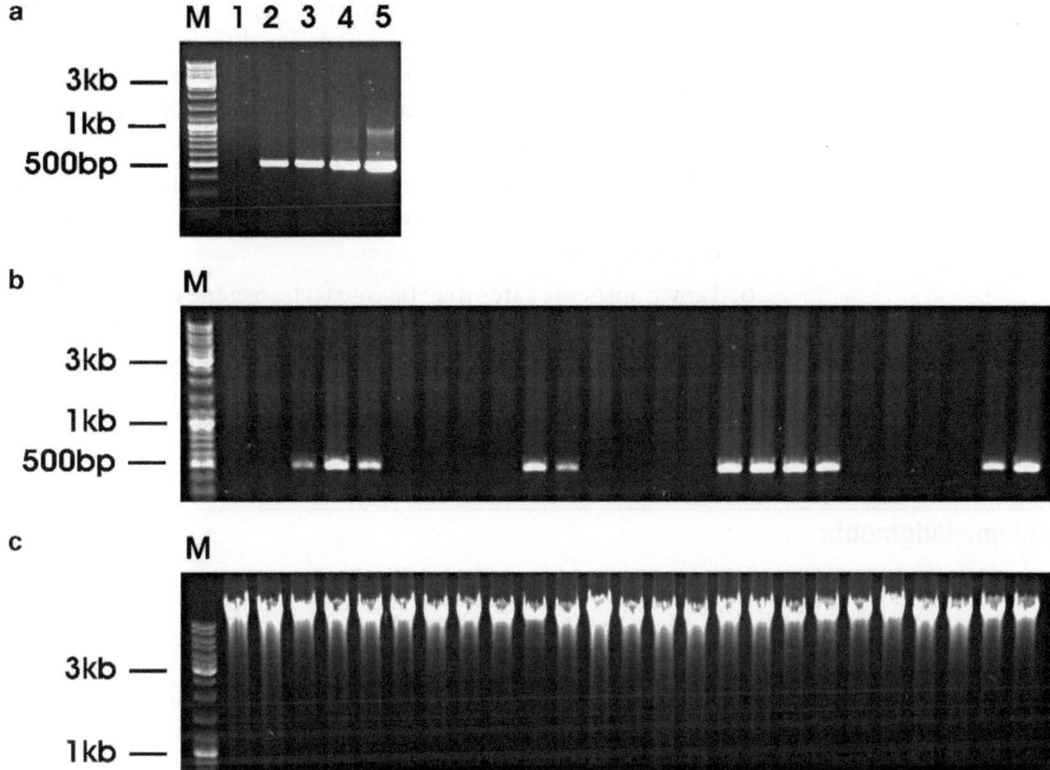

Fig. 4. *Genotyping PCR for the detection of the Venus transgene.* (**a**) Different amounts of pT2/Venus plasmid have been mixed with 500 ng of wt mouse DNA per reaction to evaluate the sensitivity of the PCR. Lanes: M, DNA size marker, and GeneRuler DNA ladder Mix (Fermentas);1, no pT2/Venus template was added into the reaction; 2–4, 50, 100, and 500 fg of pT2/Venus template was added into the reaction, respectively; 5, 10 ng of pT2/Venus plasmid alone was used as the template. (**b**) Venus-specific PCRs for genotyping the F1 descendants of a Venus positive founder. 500 ng of genomic DNA was used as the template. (**c**) Genomic DNA samples, 500 ng each.

2 ng/μl value, in the final injection mixture may help to increase the efficiency in case of larger transgenes. Plasmid concentrations higher than 2 ng/μl has been found to decrease the viability of the injected zygotes (Mates, 2009, and unpublished observations).

3. The new batches of the injection buffer, EmbryoMax® (Millipore) or equivalent must be tested for the presence of RNase as the manufacturers do not guarantee they are RNase free. Incubate 1 μl of in vitro synthesized mRNA in 10 μl of injection buffer for 1 h at 37°C and run on RNase free 1% agarose gel as described in Subheading 3.2. RNase is typically not detected. Figure 2, lane 2 shows the result of an injection buffer test without visible sign of mRNA degradation.

4. It is not necessary to run a Northern gel to test the result of in vitro mRNA synthesis. However, if more than two bands

are detected on the nondenaturing gel, a Northern gel may help to identify whether the bands are different length products or the mRNA runs aberrantly due to secondary structures (also see the troubleshooting instructions of the mMessage mMachine® T7 kit).

5. After microinjection it is recommended to discard the used aliquot of injection mixture due to the increased risk of RNase contamination.
6. Low transgenic rates may be due to transgene detection problems. Even the optimized genotyping PCR may fail to detect a single copy transgene if the genomic DNA template is degraded. Good quality genomic DNA runs on agarose gel as a dominant high molecular weight band (Fig. 4c).

Acknowledgments

The author would like to thank Prof. Zsuzsanna Izsvak for supporting the project, Dr. Boris Jerchow and Katja Becker for managing the pronuclear microinjections in mice, and for Prof. Michal Pravenec and Dr. Vladimir Landa for their essential contribution to produce SB mediated transgenic rats.

References

1. Stuart, G.W., J.V. McMurray, and M. Westerfield, *Replication, integration and stable germ-line transmission of foreign sequences injected into early zebrafish embryos.* Development, 1988. **103**(2): p. 403–12.
2. Stuart, G.W., et al., *Stable lines of transgenic zebrafish exhibit reproducible patterns of transgene expression.* Development, 1990. **109**(3): p. 577–84.
3. Hamatani, T., et al., *Dynamics of global gene expression changes during mouse preimplantation development.* Dev Cell, 2004. **6**(1): p. 117–31.
4. Harrouk, W., B. Robaire, and B.F. Hales, *Paternal exposure to cyclophosphamide alters cell-cell contacts and activation of embryonic transcription in the preimplantation rat embryo.* Biol Reprod, 2000. **63**(1): p. 74–81.
5. Misirlioglu, M., et al., *Dynamics of global transcriptome in bovine matured oocytes and preimplantation embryos.* Proc Natl Acad Sci USA, 2006. **103**(50): p. 18905–10.
6. Kawakami, K., A. Shima, and N. Kawakami, *Identification of a functional transposase of the Tol2 element, an Ac-like element from the Japanese medaka fish, and its transposition in the zebrafish germ lineage.* Proc Natl Acad Sci USA, 2000. **97**(21): p. 11403–8.
7. Davidson, A.E., et al., *Efficient gene delivery and gene expression in zebrafish using the Sleeping Beauty transposon.* Dev Biol, 2003. **263**(2): p. 191–202.
8. Sinzelle, L., et al., *Generation of trangenic Xenopus laevis using the Sleeping Beauty transposon system.* Transgenic Res, 2006. **15**(6): p. 751–60.
9. Hamlet, M.R., et al., *Tol2 transposon-mediated transgenesis in Xenopus tropicalis.* Genesis, 2006. **44**(9): p. 438–45.
10. Dupuy, A.J., et al., *Mammalian germ-line transgenesis by transposition.* Proc Natl Acad Sci USA, 2002. **99**(7): p. 4495–9.
11. Mates, L., et al., *Molecular evolution of a novel hyperactive Sleeping Beauty transposase enables robust stable gene transfer in vertebrates.* Nat Genet, 2009. **41**(6): p. 753–61.
12. Vigdal, T.J., et al., *Common physical properties of DNA affecting target site selection of sleeping beauty and other Tc1/mariner transposable elements.* J Mol Biol, 2002. **323**(3): p. 441–52.

13. Carlson, C.M., et al., *Transposon mutagenesis of the mouse germline*. Genetics, 2003. **165**(1): p. 243–56.

14. Ding, S., et al., *Efficient transposition of the piggyBac (PB) transposon in mammalian cells and mice*. Cell, 2005. **122**(3): p. 473–83.

15. Urasaki, A., G. Morvan, and K. Kawakami, *Functional dissection of the Tol2 transposable element identified the minimal cis-sequence and a highly repetitive sequence in the subterminal region essential for transposition*. Genetics, 2006. **174**(2): p. 639–49.

16. Cui, Z., et al., *Structure-function analysis of the inverted terminal repeats of the sleeping beauty transposon*. J Mol Biol, 2002. **318**(5): p. 1221–35.

17. Nagy, A., et al., *Manipulating the Mouse Embryo – A laboratory Manual*. 3 rd ed. 2003, New York: Cold Spring Harbor Press, Cold Spring Harbor.

18. Izsvak, Z., Z. Ivics, and R.H. Plasterk, *Sleeping Beauty, a wide host-range transposon vector for genetic transformation in vertebrates*. J Mol Biol, 2000. **302**(1): p. 93–102.

19. Ro, H., et al., *Novel vector systems optimized for injecting in vitro-synthesized mRNA into zebrafish embryos*. Mol Cells, 2004. **17**(2): p. 373–6.

20. Lu, B., et al., *Generation of rat mutants using a coat color-tagged Sleeping Beauty transposon system*. Mamm Genome, 2007. **18**(5): p. 338–46.

21. Nagai, T., et al., *A variant of yellow fluorescent protein with fast and efficient maturation for cell-biological applications*. Nat Biotechnol, 2002. **20**(1): p. 87–90.

Chapter 7

Construction and Use of a Bottom-Up HAC Vector for Transgene Expression

Masashi Ikeno and Nobutaka Suzuki

Abstract

Recent technological advances have enabled visualization of the organization and dynamics of local chromatin structures; however, the global mechanisms by which chromatin organization modulates gene regulation are poorly understood. We designed and constructed a human artificial chromosome (HAC) vector that allows regulation of transgene expression and delivery of a gene expression platform into many vertebrate cell lines. This technology for manipulating a transgene using a HAC vector could be used in applied biology.

Key words: Alphoid DNA, BAC, HAC, MMCT, Cre/lox recombination, Transgene

1. Introduction

A human artificial chromosome (HAC) is a mini-chromosome that is constructed artificially in human cells (1, 2). Using its own self-replicating and segregating systems, a HAC can behave as a stable chromosome that is independent from the chromosomes of host cells. HACs were constructed using a bottom-up strategy based on the transfection of cloned or synthetic centromeric alphoid DNA precursors with CENP-B boxes into a cultured human cell line, HT1080 (3–5). The HACs were built up to megabase size (1–10 Mb) by multimerization of alphoid precursors. This "bottom-up construction" strategy involves the de novo construction of HACs by introducing DNA elements necessary for the maintenance of chromosome function into cells. By contrast, "top-down construction" refers to the truncation of natural chromosomes into smaller sizes by using targeting vectors containing telomeric sequences (6, 7).

Gyula Hadlaczky (ed.), *Mammalian Chromosome Engineering: Methods and Protocols*, Methods in Molecular Biology, vol. 738, DOI 10.1007/978-1-61779-099-7_7,

We produced a HAC that carries a site-directed insertion system (HAC vector) (8). The expression of transgenes (cDNA or genomic DNA) integrated into chromosomes in cultured cells and in transgenic mice is often subject to position effects. However, transgenes can be inserted at a certain position in the HAC vector, and the transgene in the HAC can be expressed in mammalian cells in a promoter-dependent manner under the desired stable control (8). This HAC vector provides several potential advantages over viral and integrating vectors for evaluating gene expression, including long-term stability, low toxicity, and accommodation of a huge size of inducible DNA.

Microcell-mediated chromosome transfer (MMCT) has been used to deliver large-sized chromosomal material (9). At present, HACs have been transferred successfully into many vertebrate cell lines by MMCT and are stably transferred during mitosis (8, 10). HACs can be transferred into mouse embryonic stem cells by MMCT for straightforward development of a transgenic mouse containing exogenous genes (10). The establishment of a reliable method to create a transgenic animal will enable use of the HAC vector for gene therapy and regenerative medicine.

2. Materials

2.1. Cell Lines and Culture

1. HT1080 (ATCC: CCL-121): Dulbecco's Modified Eagle's Medium (DMEM) (Sigma) supplemented with 10% fetal bovine serum (FBS).
2. A9 (ATCC: CCL-1.4): DMEM (Sigma) supplemented with 10% FBS.
3. CHO-K1 (ATCC: CCL-61): Ham's F-12 nutrient mixture (Sigma) supplemented with 10% FBS.
4. HeLa (ATCC: CCL-2): DMEM (Sigma) supplemented with 10% FBS.
5. Indian Muntjac (ATCC: CCL-157): Ham's F-10 nutrient mixture (Sigma) supplemented with 10% FBS.

2.2. Precursors of the HAC Vector

1. BAC vector: pBelo-BAC11 (New England BioLabs).
2. Alphoid DNA: Human chromosome 21 alphoid DNA (GenBank D29750.1).
3. Insulator (11–13): Human β-globin 5′HS5 (3.4-kb *Eco*RI fragment; 4,818–8,173 in GenBank NG 000007) and 3′-HS1 (5.6-kb *Sph*I–*Sac*I fragment; 8,255–13,891 in GenBank AC104389) cloned from the YAC clone A201F4.3.
4. CAG promoter: The sequence derived from pCAGGS (14).

5. Lox71 sequence (14): 5′TACCGTTCGTATAGCATACATTA TACGAAGTTAT3′.
6. Neo: The coding sequence derived from pSV-neo.
7. QIAGEN Large Construct Kit (QIAGEN).

2.3. Cre–Lox Recombination

1. Puro: The coding sequence derived from pGK-puro.
2. Lox66 sequence (14): 5′ATAACTTCGTATAGCATACAT TATACGAACGGTA3′.
3. Cre expression plasmid: CAG-Cre (14).

2.4. DNA Transfection

1. Lipofectamine™ 2000 (Invitrogen).
2. FuGENE® HD (Roche).
3. G418 (Sigma).
4. Puromycin dihydrochloride (Sigma).

2.5. Red-ET Recombination

1. Quick and easy BAC modification kit (Gene Bridges).
2. The target sequences for homologous recombination set in position (6,915–7,114) and position (51–2,509) in Belo-BAC11.

2.6. Cell Fusion and Microcell-Mediated Chromosome Transfer

1. PEG (1:1.4) (5 g autoclaved PEG1000, 1 ml DMSO, and 6 ml serum-free DMEM).
2. PEG (1:3) (3 g autoclaved PEG1000, and 9 ml serum-free DMEM).
3. 6-Thioguanine (Sigma-Aldrich).
4. Ouabain octahydrate (Sigma-Aldrich).
5. Colcemid (1 mg/ml).
6. Cytochalasin B (Sigma-Aldrich) (10 mg/ml in DMSO).
7. Percoll (GE Healthcare).
8. 50% PEG1500 (Roche).
9. HAT Media Supplement (50×) Hybri-Max™ (Sigma-Aldrich).

2.7. Fluorescence In Situ Hybridization

1. 11-4 alphoid DNA (15).
2. Digoxigenin-11-dUTP (Roche).
3. Anti-digoxigenin rhodamine Fab fragment (Roche).
4. Biotin-16-dUTP (Roche).
5. Alexa Fluor® 488-conjugated streptavidin (Invitrogen).

2.8. PCR

1. Primer for CAG promoter: 5′CTCTGCTAACCATGTTC ATG3′.
2. Primer for Puro: 5′CTTGTACTCGGTCATGGTAAGC3′.

3. Methods

The HAC vector is a useful tool for transgene expression when the cell line carrying the HAC vector has already been established. It takes a good amount of time to construct a unique HAC vector using a bottom-up strategy because the vector precursors are required as materials. The method of introducing the HAC vector into target cells is dependent on MMCT. It is important to plan the scale of cell culture because the introduction efficiency of the HAC vector using MMCT is about 5×10^{-6}. Insertion of the gene of interest into the HAC vector is achieved using a general molecular biology technique, DNA transfection.

3.1. Construction of a Bottom-Up HAC Vector in Human Cell Line HT1080

3.1.1. Preparation of Precursors (Alphoid-BAC and Lox-BAC) of the HAC Vector

1. Clone human alpha satellite (alphoid) DNA (see Note 1) longer than 50 kb into the BAC vector (see Note 2) from human genomic DNA, or screen for ~50-kb alphoid DNA from BAC, PAC, or cosmid clone libraries to construct a first precursor (alphoid-BAC).
2. Construct a DNA cassette consisting of a promoter/lox71/drug resistance gene (in the example here, CAG promoter/lox71/Neo) by standard techniques (see Note 3).
3. Clone insulator fragments, for example, human β-globin 5′HS5 and 3′-HS1 from genomic DNA, or BAC, or YAC clone libraries (see Note 4).
4. Then, insert the CAG/lox71/neo, I-*Sce*I recognition sequence, 5′HS5, and 3′HS1, into pBelo-BAC (without loxP) to construct a second precursor (Lox-BAC) that contains 5′HS5, I-*Sce*I, CAG/lox71/Neo, and 3′HS1, in that order (Fig. 1).

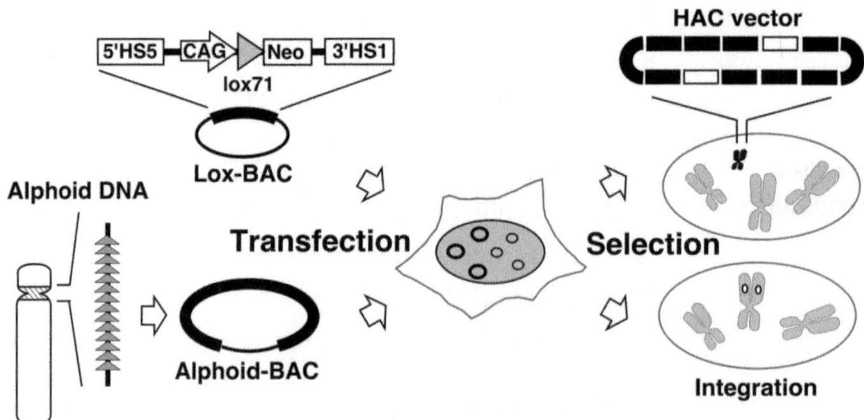

Fig. 1. Construction scheme of the HAC vector. Alphoid-BAC contains 50 kb of alphoid DNA, and Lox-BAC contains the CAG/lox71/neo cassette with insulator elements (5′HS5 and 3′HS1). The HAC vector is constructed by co-transfection of alphoid-BAC and Lox-BAC and by subsequent selection using drug resistance (Neo) and FISH analysis.

3.1.2. DNA Transfection into HT1080 Cells to Generate a HAC Vector

1. Co-transfect 1.0 μg of the first precursor DNA (alphoid-BAC) and 0.5 μg of the second precursor DNA (Lox-BAC) into HT1080 cells (5×10^5) using 5 μl of Lipofectamine™ 2000 according to the manufacturer's instructions.
2. Plate the transfected cells in ten 10-cm dishes, followed by the addition of 400 μg/ml G418. Pick colonies after 10–14 days to establish drug-resistant cell lines (see Note 5).

3.1.3. Selection of an HT1080 Cell Line Carrying a HAC by Fluorescence In Situ Hybridization

1. Prepare metaphase chromosomes of drug-resistant cell lines on glass slides after methanol/acetate (3:1) fixation.
2. Perform FISH using BAC vector DNA and 11-4 alphoid DNA (15) as probes for detecting HAC vectors according to a standard procedure.
3. Visualize the digoxigenin-labeled BAC sequence with TRITC-conjugated anti-digoxigenin, and biotin-labeled 11-4 alphoid DNA with Alexa Fluor® 488-conjugated streptavidin.
4. Make photographs using a CCD camera mounted on a Zeiss microscope AxioPlan2. Images are processed using AxioVision.

3.1.4. Confirmation of the Number of Expression Cassettes in the HAC Vector

1. Prepare the genomic DNA from HT1080 cell lines carrying the HAC vector in a 1% low-melting agarose block.
2. Digest the DNA in the agarose block with the I-*Sce*I restriction enzyme for 4 h and size-separate the results in a 1% agarose gel using the CHEF mapper system. The running condition depends on the auto algorithm from 10 to 1,000 kb.
3. Transfer the digested DNA to a nylon membrane and hybridize with ^{32}P-labeled DNA probes prepared from the Belo-BAC vector for detection of the number and size of fragments derived from the HAC vector (see Note 6).

3.2. Transfer of HAC Vectors into Vertebrate Cells

The scheme of MMCT is shown in Fig. 2.

3.2.1. Creation of an HT1080-A9 Hybrid Cell Line Harboring HAC by Cell Fusion

1. Mix HT1080 cells carrying the HAC vector (4×10^5) and A9 cells (4×10^5) and seed onto a 6-cm dish. Culture cells for 24 h (see Note 7).
2. Wash cells twice in 3 ml of PBS and treat in 3 ml of PEG (1:1.4) at room temperature for 60 s.
3. Remove the solution and treat cells in 3 ml of PEG (1:3) at room temperature for 60 s.
4. Wash cells three times in 3 ml of serum-free DMEM and incubate for 24 h in DMEM supplemented with 10% FBS.
5. Harvest fused cells by trypsin treatment and plate onto eight 10-cm dishes.

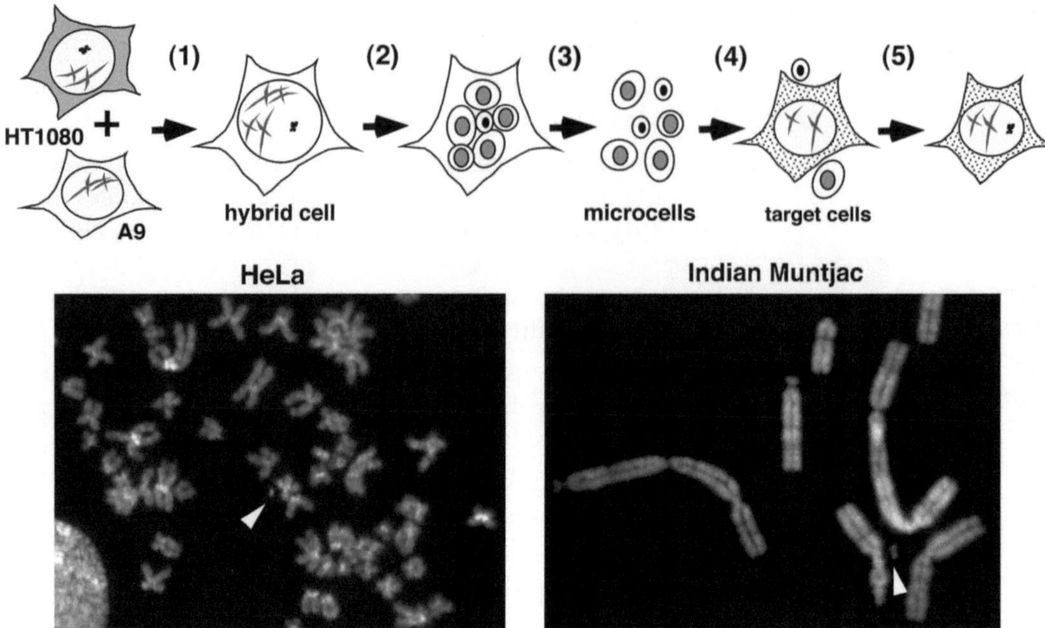

Fig. 2. Strategy of microcell-mediated chromosome transfer. (1) Whole cell fusion of HT1080 cells carrying the HAC vector and A9 cells. (2) Formation of microcells by colcemid treatment. (3) Purification of microcells by Percoll centrifugation after cytochalasin B treatment. (4) Fusion of microcells and target cells. (5) Selection of cell lines carrying the HAC vector. The *small black shapes* show HAC vectors and *gray shapes* show nuclei. Transferred HAC was confirmed by FISH using metaphase cells of the HeLa and Indian Muntjac lines. *Arrowheads* show HAC vectors.

6. Select for the HT1080-A9 hybrid cell line carrying the HAC vector using 600 μg/ml G418, 10 μM ouabain, and 30 μM 6-thioguanine.
7. Confirm the presence of the HAC vector in drug-resistant cell lines using FISH (Subheading 3.1.3).

3.2.2. Microcell-Mediated Chromosome Transfer

1. Grow HT1080-A9 hybrid cells carrying HAC on 40 10-cm dishes to 70–80% confluence and then culture for 72 h in DMEM containing colcemid (0.05 μg/ml).
2. Harvest cells by trypsin treatment and suspend in 60 ml serum-free DMEM.
3. Incubate collected cells for 5 min at 37°C in serum-free DMEM containing cytochalasin B (20 μg/ml), then add an equal volume of Percoll prewarmed at 37°C. Centrifuge the suspension in a Hitachi R20A2 rotor at 27,000 × *g* for 90 min at 37°C. Three bands should be visible in the tubes.
4. Collect the upper two bands in four 15-ml tubes. Dilute the collected microcells in serum-free DMEM up to 14 ml and centrifuge at 800 × *g* for 5 min.
5. Wash the collected microcells in serum-free DMEM twice and finally suspend in two 15-ml tubes.

6. Mix the microcells with recipient cells and centrifuge at 800×*g* for 5 min.
7. Suspend the resulting pellet in 1 ml of 50% PEG1500 and incubate at room temperature for 90 s (see Note 8).
8. Add 5 ml of serum-free DMEM and centrifuge the mixture at 200×*g* for 5 min.
9. Wash the fused cells twice in serum-free DMEM.
10. Plate the cells in suitable medium containing serum and 1× HAT media supplement and culture for 24 h.
11. Isolate resistant colonies in medium at 800 μg/ml (K562), 400 μg/ml (HT1080, HeLa, Indian Muntjac, CHO), and 1× HAT media supplement.
12. Assay for the presence of HAC vector by FISH with a BAC vector sequence (see Notes 9 and 10).

3.2.3. Mitotic Stability of HAC Vectors in Transferred Cell Lines

1. At 10-day intervals, during culture for about 100 divisions, under non-selective growth conditions, prepare spreads of metaphase chromosomes and determine the presence of HAC by FISH.
2. Measure the percentage of metaphase cells with HACs using more than 50 cells.

Calculate the loss rate (R) of HAC using the formula as follows: $N_n = N_0 \times (1-R)^n$, where N_0 is the number of metaphase spreads containing a HAC vector under selective conditions and N_n is the number of metaphase spreads containing a HAC after n days of culture under non-selective conditions.

3.3. Insertion of the Transgene into the HAC Vector

3.3.1. Construction of a Donor Plasmid Containing a Small Size of Transgene

1. Fuse the synthetic lox66 sequence with a promoter-less puromycin-resistant gene to construct a lox66/puro cassette by standard techniques.
2. Add the genes of interest consisting of a promoter, coding sequence, and poly A signal to the lox66/puro cassette to construct the donor plasmid (Fig. 3).

3.3.2. Construction of a Donor Plasmid Containing a Large Size of Transgene (BAC Clones)

1. To replace the loxP site in BAC clones with the lox66/puro cassette, prepare the DNA fragment composed of lox66/puro cassette with a more than 50-mer homologous sequence to the BAC vector (see Subheading 2) at both ends by PCR, using homologous sequences as primers.
2. Modify the BAC clone to contain the lox66/puro cassette using Red–ET recombination in *Escherichia coli* according to the manufacturer's instructions (Gene Bridges).
3. Purify the modified BAC DNA using the QIAGEN Large Construct Kit.

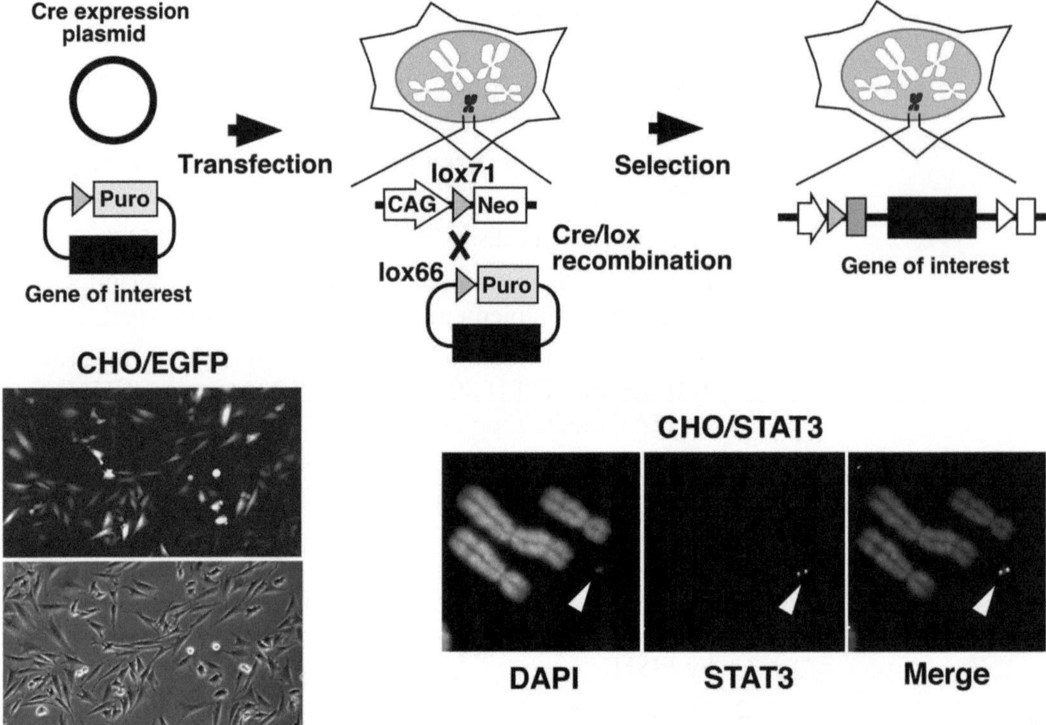

Fig. 3. Gene insertion into the HAC vector. The genes of interest were inserted into the HAC vector using Cre–lox recombination in cells. CMV-EGFP and human STAT3 genomic DNA (217 kb) were inserted into the HAC vector in CHO cells. Fluorescence of EGFP and FISH signal of STAT3 are shown.

3.3.3. DNA Transfection for Insertion of Transgenes

1. Co-transfect 0.5 μg of transgene fused with lox66/puro and 0.5 μg of CAG-Cre plasmid into HT1080 cells (5×10^5) carrying the HAC vector using FuGENE® HD according to the manufacturer's instructions.
2. Select the cells with puromycin (HT1080: 0.25 μg/ml, CHO: 6.0 μg/ml) and pick colonies after about 10 days to establish drug-resistant cell lines (see Note 11).
3. Prepare genomic DNA from drug-resistant cell lines and confirm successful insertion by PCR with CAG promoter and puro sequences as primers (see Note 12). Finally, confirm the integrity of the HAC by FISH with the BAC sequence as a probe. Examples of insertion of genes into HAC vector in CHO cells are shown in Fig. 3: fluorescence of CMV-EGFP and FISH analysis of STAT3 genomic DNA.

4. Notes

1. Alphoid DNAs from human chromosome 21 or chromosome 17 that contain a CENP-B box frequently are good materials for first precursors.

2. The loxP sequence should be removed from the cloning vector pBelo-BAC.
3. The cassette is used during drug selection for the construction of a HAC and for gene insertion.
4. Human β-globin 5′ and 3′ DNase I hypersensitive sites (5′HS5 and 3′HS1) are reported to possess chromatin boundary function in addition to enhancer-blocking function.
5. In this period, the HAC is generated by in vivo multimerization of the precursor DNA (alphoid-BAC and Lox-BAC).
6. The number of expression cassettes in HAC vectors varies among cell lines.
7. Donor cell lines suitable for microcell delivery are limited to a few cell lines that include CHO and mouse A9 cells. The HT1080 cell line that is the host cell for production of a HAC vector is not suitable. Thus, a new cell line that carries the HAC vector and is suitable for production of microcells is required for the MMCT method by whole cell fusion of HT1080 carrying the HAC vector and A9.
8. The minimum volume of PEG solution should be 1 ml.
9. The elements responsible for centromere and replication origins in human cells remain functional in cells derived from many different species.
10. The HAC vector has been transferred successfully into many vertebrate cell lines, including mouse embryonic stem cells, by MMCT.
11. The puromycin-resistance shows successful insertion of the gene of interest into the HAC vector by Cre–lox recombination of the lox66 site at the promoterless cassette and the lox71 site in the gene expression cassette of the HAC vector.
12. The major event in puromycin-resistant cell lines is the insertion of the gene of interest into the expression cassette.

Acknowledgments

We would like to thank Dr. D. Engel for the YAC clone A201F4.3 and Dr. K. Yamamura for the plasmid pCAGGS. This work was supported by a Grant-in-Aid for Scientific Research from The New Energy and Industrial Technology Development Organization (NEDO) and from The Ministry of Education, Culture, Sports, Science, and Technology (MEXT) of Japan.

References

1. Larin, Z. and Mejia, J.E. (2002) Advances in human artificial chromosome technology. *Trends in genetics* **18**, 313–319.
2. Grimes, B.R. and Monaco, Z.L. (2005) Artificial and engineered chromosomes: developments and prospects for gene therapy. *Chromosoma* **114**, 230–241.
3. Harrington, J.J., Van Bokkelen, G., Mays, R.W., Gustashaw, K. and Willard, H.F. (1997) Formation of de novo centromere and construction of first-generation human artificial microchromosomes. *Nat. Genet.* **4**, 345–355.
4. Ikeno, M., Grimes, B., Okazaki, T., Nakano, M., Saitoh, K., Hoshino, H., McGill, N. I., Cooke, H., and Masumoto,H. (1998) Construction of YAC-based mammalian artificial chromosomes. *Nat. Biotech.* **16**, 431–439.
5. Ohzeki, J., Nakano, M., Okada, T. and Masumoto, H. (2002) CENP-B is required for de novo centromere chromatin assembly on human alphoid DNA. *J. Cell Biol.* **159**, 765–775.
6. Tomizuka, K., Yoshida, H., Uejima, H., Kugoh, H., Sato, K., Ohguma, A., Hayasaka, M., Hanaoka, K., Oshimura, M., and Ishida, I. (1997) Functional expression and germline transmission of a human chromosome fragment in chimaeric mice. *Nat. genet.* **16**, 133–143.
7. Kuroiwa,Y., Tomizuka,K., Shinohara,T., Kazuki,Y., Yoshida,H., Ohguma,A., Yamamoto,T., Tanaka,S., Oshimura,M. and Ishida,I. (2000) Manipulation of human minichromosomes to carry greater than megabase-sized chromosome inserts. *Nat. Biotech.* **18**, 1086–1090.
8. Ikeno, M., Suzuki, N., Hasegawa., and T. Okazaki. (2009) Manipulating transgenes using a chromosome vector. *Nucleic Acids Res.*, **37**, e44.
9. Doherty, A.M.O. and Fisher, E.M.C. (2003) Microcell-mediated chromosome transfer (MMCT): small cells with huge potential. *Mamm. Genome* **14**, 583–592.
10. Suzuki, N., Nishii, K., Okazaki, T. and Ikeno, M. (2006) Human artificial chromosomes constructed using the bottom-up strategy are stably maintained in mitosis and efficiently transmissible to progeny mice. *J. Biol. Chem.* **281**, 26615–26623.
11. Gaszner, M., Felsenfeld, G. (2006) Insulator: exploiting transcriptional and epigenetic mechanism. *Nat. Rev. Genet.* 7, 703–713.
12. Chung, J.H., Whiteley, M. and Felsenfeld, G. (1993) A 5' element of the chicken beta-globin domain serves as an insulator in human erythroid cells and protects against position effect in Drosophila. *Cell* **74**, 505–514.
13. Lindenbaum,M., Perkins,E., Csonka,E., Fleming,E., Garcia,L., Greene,A., Gung,L., Hadlaczky,G.,Lee,E.,Leung,J.,MacDonald,N., Maxwell,A., Mills,K., Monteith,D., Perez,C.F., Shellard,J., Stewart,S., Stodola,T., Vandenborre,D., Vanderbyl,S., Ledebur,H.C. Jr. (2004) A mammalian artificial chromosome engineering system (ACE System) applicable to biopharmaceutical protein production, transgenesis and gene-based cell therapy. *Nucleic Acids Res.* **32**, e172.
14. Araki, K., Araki, M. and Yamamura, K. (1997) Targeted integration of DNA using mutant lox sites in embryonic stem cells. *Nucleic Acids Res.* **25**, 868–872.
15. Ikeno, M., Masumoto, H. & Okazaki, T. (1994) Distribution of CENP-B boxes reflected in CREST centromere antigenic sites on long-range alpha-satellite DNA arrays of human chromosome 21. *Hum. Mol. Genet.* **3**, 1245–1257.

Chapter 8

De Novo Generation of Satellite DNA-Based Artificial Chromosomes by Induced Large-Scale Amplification

Erika Csonka

Abstract

Mammalian artificial chromosomes (MACs) are engineered chromosomes with defined genetic content that can function as non-integrating vectors with large carrying capacity and stability. The large carrying capacity allows the engineering of MACs with multiple copies of the same transgene, gene complexes, and to include regulatory elements necessary for the regulated expression of transgene(s). In recent years, different approaches have been explored to generate MACs (Vos Curr Opin Genet Dev 8:351–359, 1998; Danielle et al. Trends Biotech 23:573–583, 2005; Duncan and Hadlaczky Curr Opin Biotech 18:420–424, 2007): (1) the de novo formation by centromere seeding, the "bottom-up" approach, (2) the truncation of natural chromosomes or the modification of naturally occurring minichromosomes, the "top-down" approach, and (3) the in vivo "inductive" approach. Satellite DNA-based artificial chromosomes (SATACs) generated by the in vivo "inductive" method have the potential to become an efficient tool in diverse gene technology applications such as cellular protein manufacturing (Kennard et al. BioPharm Int 20:52–59, 2007; Kennard et al. Biotechnol Bioeng 104:526–539, 2009; Kennard et al. Biotechnol Bioeng 104:540–553, 2009), transgenic animal production (Telenius et al. Chromosome Res 7:3–7, 1999; Co et al. Chromosome Res 8:183–191, 2000; Monteith et al. Methods Mol Biol 240:227–242, 2003), and ultimately a safe vector for gene therapy (Vanderbyl et al. Stem Cells 22:324–333, 2004; Vanderbyl et al. Exp Hematol 33:1470–1476, 2005; Katona et al. Cell. Mol. Life Sci 65:3830–3838, 2008). A detailed protocol for the de novo generation of satellite DNA-based artificial chromosomes (SATACs) via induced large-scale amplification is presented.

Key words: Artificial chromosomes, Gene therapy, Mammalian artificial chromosomes, Satellite DNA, Satellite DNA-based artificial chromosomes

1. Introduction

The in vivo "inductive" approach is a reproducible method to generate mammalian artificial chromosomes in cell lines of different species. It is based on an inducible intrinsic cellular mechanism, which can facilitate large-scale amplifications and formation

Gyula Hadlaczky (ed.), *Mammalian Chromosome Engineering: Methods and Protocols*, Methods in Molecular Biology, vol. 738, DOI 10.1007/978-1-61779-099-7_8, © Springer Science+Business Media, LLC 2011

of de novo centromeres and chromosomes in mammalian cells (13, 14), upon the targeted integration of exogenous DNA sequences into the specific region of the chromosomes. Induced amplification of the pericentric/centromeric region of mammalian chromosomes leads to the formation of new chromosomes. SATACs are composed of co-amplified satellite DNAs and exogenous DNA sequences. They can be regarded as artificially generated accessory chromosomes with predictable DNA sequences, and contain defined genetic information (15).

There are several lines of direct and indirect evidence suggesting that the ribosomal-DNA (rDNA) containing chromosomal sites can be responsible for the large-scale amplification events, and the rDNA itself may have importance in the de novo chromosome formations (Fodor, K. et al. unpublished). Integration of exogenous DNA (Fig. 1b) into close proximity of a so-called megareplicator (16) could lead to a replication error, which initiates large-scale amplification (Fig. 1c, d) of surrounding sequences and brings about the formation of large inverted repeats (amplicons). These amplicons are composed of co-amplified endogenous and exogenous DNA sequences, and they are the building blocks of the new chromosomal segments ("sausage" chromosome) (Fig. 1c, d). Amplified centromeric regions can eventually form active centromere that leads to the formation of a dicentric chromosome (Fig. 1c). The presence of two active centromeres on the same chromosome causes specific breakage between the two centromeres (Fig. 1e, f), which ultimately brings about the existence of a new chromosome (Fig. 1g, h). This newly formed independent chromosome consists of multiple copies of the "foreign" DNA and "neutral" endogenous rDNA, and noncoding satellite sequences only. SATACs can be generated by induced de novo chromosome formation in cells of different mammalian species. During the recent years, a number of mouse and human SATACs have been generated by this technology (13–17, 21). In this chapter, a detailed protocol for de novo generation of SATACs is provided.

2. Materials

2.1. Cell Lines, Cell Culturing, DNAs

1. To generate a SATAC, preferably choose a cell line having chromosomes where the rDNA and centromeric region are in close proximity to each other (see Subheading 1) (see Note 1). All SATACs have been generated in adherent cells, so far. There is no report about successful SATAC formation in cells cultured in suspension.
2. Cell culture facilities: CO_2 thermostat, sterile fume hood, and cell storage.

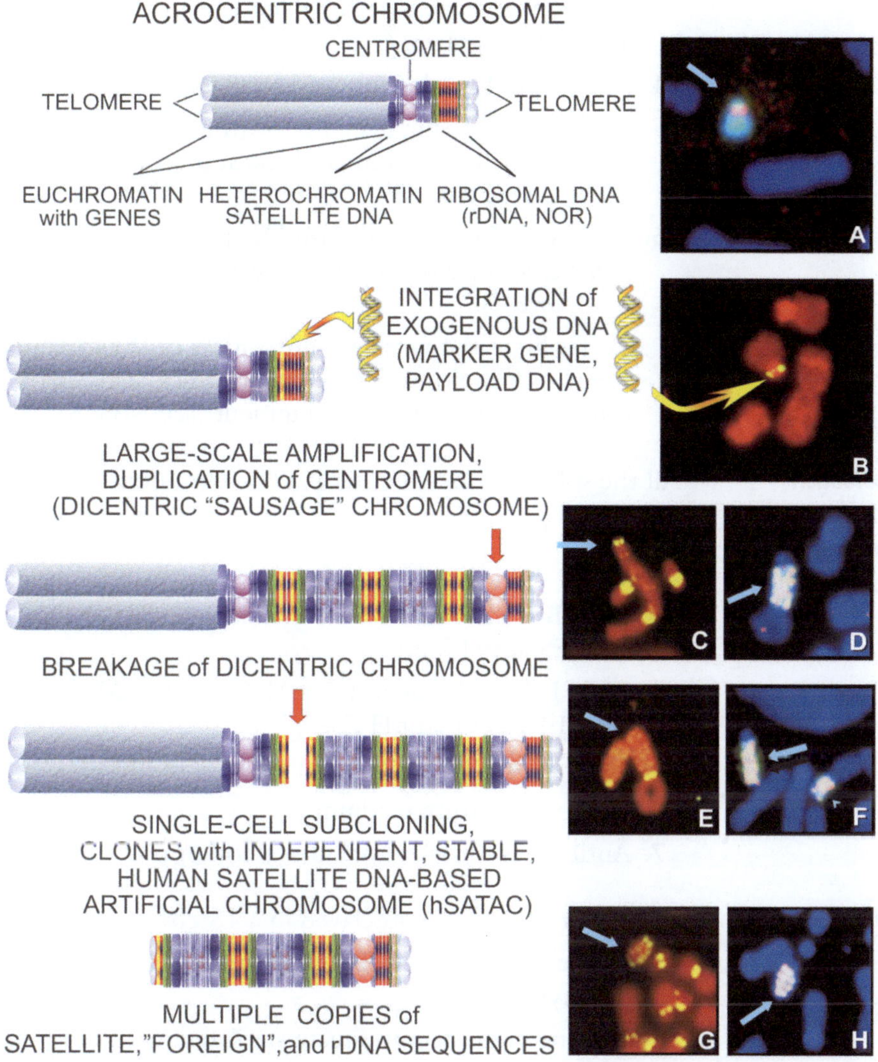

Fig. 1. *Generation of MAC via in vivo inductive approach.* Subsequent steps of induced de novo chromosome formations are shown on (**b–h**). Chromosomes on (**a**, **d**, **f**, **h**) are counterstained with DAPI (*blue*), overlapping *green* and *red* signals of the probes appear as *pink–white*. Chromosomes on (**b**, **c**, **e**, **g**) are counterstained with propidium iodide (*red*). (**a**) A mammalian NOR chromosome. In situ hybridization on human chromosome #15 (*arrow*) in a human/hamster monochromosomal hybrid cell (94-3) with human chromosome specific painting probe (*green*) and human chromosome #15 centromere-specific alpha satellite DNA (*red*) probes. (**b**) Integration of exogenous DNA into close proximity to a megareplicator. Yellow in situ hybridization signal with plasmid probe (*arrow*) demonstrates the integration of "foreign" DNA into the centromeric/short arm region of chromosome #15. (**c**) Large-scale amplification of the centromeric/short arm region of chromosome #15 result in the formation of a new chromosome arm with a de novo centromere. *Arrow* points to the newly formed centromere of the dicentric "sausage" chromosome. *Yellow* signals correspond to the centromeres visualized by indirect immunofluorescence with LU851 human anti-centromere antibody. (**d**) Two color in situ hybridization on the "sausage" chromosome with chromosome #15 alpha satellite (D15Z1) (*red*) and exogenous DNA (*green*) probes. (**e–f**) Breakage of the dicentric chromosome (*arrows*). The *arrowhead* on (**f**) indicates the formerly dicentric chromosome. (**g–h**) Independent de novo SATACs (*arrows*) with anti centromere staining (**g**), and with in situ hybridization. Probes are the same as on (**c**, **e**) and (**d**, **f**) (note the faint double signals on the amplified chromosome arms (**c**, **e**) and on the SATAC (**g**) that correspond to multiple inactive centromeres).

3. Solution of trypsin (0.25%) and ethylene diamine tetraacetic acid (EDTA) (1 mM).
4. Antibiotics for selection (see Note 2).
5. Culture media supplemented with fetal bovine serum (FBS): the type of the medium and the supplements depend on the cell line (see Note 3).
6. A "foreign" DNA, any plasmid, or other construct (e.g. PAC (18)) carrying a mammalian selectable marker gene is suitable (see Note 4).
7. To facilitate the targeted integration of the foreign DNA to the rDNA region of the chromosomes, use carrier DNA; it can be any kind of rDNA fragment (see Note 5).

2.2. Transfection of Mammalian Cells by Calcium Phosphate DNA Coprecipitation Method

All the solutions must be sterilized by filtration.

1. 2.5 M $CaCl_2$.
2. Sterile double distilled water.
3. 2× HBS: 280 mM NaCl, 1.5 mM Na_2HPO_4, and 50 mM HEPES at pH 7.2.
4. PBS: 137 mM NaCl, 2.7 mM KCl, 10 mM Na_2HPO_4, and 2 mM KH_2PO_4 at pH 7.4.
5. Glycerol–PBS: 25% (v/v) of glycerol in PBS.
6. Complete culture medium.
7. Antibiotics for selective culture conditions.

2.3. Establishment of Stable Clones

1. A pair of forceps autoclaved in a 15-cm glass Petri dish.
2. Sterile cloning cylinders (glass, stainless steel, or polystyrene).
3. Silicone high vacuum grease sterilized in a Petri dish.
4. Solution of trypsin (0.25%), and EDTA (1 mM).
5. Culture medium.
6. 6-well plates or 3-cm Petri dishes.
7. Micropipette (100 μl).

2.4. Fixation of Mammalian Cells

1. 1 mg/ml of colchicine was dissolved in sterile distilled water.
2. Hypotonic solution: 75 mM KCl was freshly diluted from the 750 mM stock solution with double distilled water. The stock solution is sterilized by filtration and kept at 4°C.
3. Fixative: freshly prepared methanol–glacial acetic acid in 3:1 ratio. Keep on ice.
4. 10-ml screw capped Falcon tubes.
5. Plastic Pasteur pipette.
6. Waterbath at 37°C.

2.5. Freezing of Mammalian Cells

1. Solution of trypsin (0.25%) and EDTA (1 mM).
2. Freezing solution: 80% complete culture (selective) medium, 10% Fetal Bovine Serum (FBS), 10% dimethylsulfoxide (DMSO). Always prepare freshly in a sterile Falcon tube. Mix well, because the DMSO is susceptible to collect at the bottom of the tube. Keep on ice.
3. Freezing vial (2 ml).
   - Deep freezer (–80°C), liquid nitrogen.

2.6. Cytological Analysis of Primary Clones by FISH

2.6.1. Slide Preparation

1. Microscope slides.
2. Slide washing solution: 99 ml of absolute ethanol + 1 ml of 37% HCl.
3. Clean cotton tissue.
4. Ice.
5. Freshly prepared ice-cold fixative (see Subheading 2.4).
6. 80–100-ml staining jar.
7. Phase contrast microscope.
8. Forceps.

2.6.2. Labeling of Probes

1. Biotin-Nick Translation Mix (Roche, #1 745 824).
2. Digoxigenin-Nick Translation Mix (Roche, #1 745 816).
3. Sterile double distilled water.
4. Waterbath at 15 and 65°C.
5. 0.5 M EDTA, adjust the pH to 8.0 with NaOH (pellets), sterilize by filtration.
6. Nick-Spin Column (Pharmacia, #17-0862-02) or equivalent e.g. QIAquick Removal Kit (Qiagen, #28304).

2.7. First Day

2.7.1. RNase Treatment

1. Ribonuclease A (RNase A).
2. Double distilled water (to avoid DNase contamination during the whole procedure double distilled water is recommended to use).
3. 20× SSC: 3 M NaCl, 0.3 M tri-Na-citrate, adjust the pH to 6.5–7.0 with 1 M HCl and sterilize by filtration. Make at least 500 ml.
4. 2× SSC: make a 10× dilution of 20× SSC with double distilled water (100 ml 2× SSC: 10 ml 20× SSC + 90 ml double distilled water). Always make freshly, no sterilization is needed.
5. 100-ml Erlenmeyer flask.
6. Staining jar (100-ml).
7. 37°C waterbath or thermostat.

8. Boiling water.
9. 70%, 90%, and absolute ethanol.
10. Vacuum or fume hood to dry the preparations.

2.7.2. Denaturation

1. 100-ml beaker.
2. Another 100-ml staining jar for ice-cold 70% ethanol.
3. Denaturation solution: 70% high quality formamide + 30% 2× SSC. Make 100 ml: 70 ml formamide + 3 ml 20× SSC + 27 ml double distilled water. Always prepare freshly, no sterilization is needed.
4. Waterbath at 70–75°C.
5. Ice-cold 70%, 90%, and absolute ethanol.
6. Ice.
7. Vacuum or fume hood to dry the preparations.

2.7.3. Preparation of the Hybridization Mixture and Hybridization

1. HybriSol VII (Qbiogene, #RIST 1390) or equivalent (Kreatech, #KBI-FHB).
2. Waterbath 70–75°C.
3. Ice.
4. Coverslips, 24 × 32 mm.
5. Humid chamber: pad Petri dishes with 15 cm (or bigger) diameter using a piece of wet filter paper.
6. Rubber cement.
7. 37°C thermostat.
8. Adhesive tape.

2.8. Second Day

2.8.1. Immunodetection and Microscopy

1. Two 100-ml staining jars.
2. 100 ml 2× SSC: prepare freshly, do not sterilize.
3. Buffer-A: 50% formamide – 50% 2× SSC. Make 400 ml: 200 ml formamide + 20 ml 20× SSC and double distilled water up to 400 ml. Always make freshly, no sterilization is needed.
4. Buffer-B: 2× SSC. Make 400 ml: 40 ml 20× SSC and double distilled water up to 400 ml. Always prepare freshly, no sterilization is needed.
5. Buffer-C: 4× SSC-0.05% Triton X-100. Make 1,000 ml: 200 ml 20× SSC + 5 ml 10% Triton X-100, and double distilled water up to 1,000 ml. Always prepare freshly, no sterilization is needed.
6. 10% Triton X-100: Make 100 ml (10 ml Triton X-100 + 90 ml double distilled water). It is difficult to handle because the Triton X-100 is rather viscous. It is better to pour 10 ml into a measuring cylinder instead of pipetting. Dissolve by gentle shaking. Sterilize by filtration. Store at 4°C.

7. Waterbath at 42–45°C.
8. For biotin detection: FITC–Avidin (Vector, A-2011), Biotinylated anti-Avidin (Vector, BA-0300).
9. For digoxigenin detection: Mouse Monoclonal anti-DIG (Sigma, D-8156), Sheep anti-Mouse Ig-DIG (Chemicon, AQ300D), anti-DIG Rhodamine (Roche, 1 207 750).
10. Coverslips, 24 mm × 32 mm, or Hybri-Slips (SIGMA) (see Note 13).
11. Humid chamber.
12. 37°C thermostat.
13. Mounting medium: VectaShield (Vector, H-1000).
14. DAPI. (Make a 0.25 mg/ml stock in sterile double distilled water. For counter staining use 2–8 μl of DAPI stock solution into 1 ml VectaShield mounting medium).
15. Nonfluorescent nail polish.
16. Fluorescence microscope.
17. Fluorescence-free immersion oil.

2.9. Single Cell Subcloning

1. Solution of trypsin (0.25%), and ethylene diamine tetraacetic acid (EDTA) (1 mM).
2. Culture medium.
3. Bürker counting chamber or any kind of cell counter.
4. 24-well plates.

3. Methods

3.1. Preparation of Mammalian Cells and DNAs for Transfection

1. The day before transfection split approximately 5×10^6 adherent cells into 10-cm Petri dish(es). The cells should be in the exponential growth phase. Work in sterile hood.
2. Purify the plasmids (or other DNA) as you usually do; high quality of the DNA is crucial. Measure the DNA concentration by a DNA meter.
3. Choose a restriction endonuclease, which cuts only once in your constructs without destroying any genes or elements (promoter, polyA site) necessary for the expression of e.g. mammalian selectable marker gene or other transgene (see Note 6). Digest that amount of your DNA you intend to use in transfection.
4. After linearization, remove the salts and enzyme by phenol–chloroform extraction.
5. Precipitate the DNA and dissolve in sterile double distilled water.

3.2. Transfection of Mammalian Cells by Calcium Phosphate DNA Coprecipitation Method (19)

1. Feed cells 3–5 h before transfection. During the transfection work in sterile hood.
2. Prepare DNA: use 1:10 ratio of your DNA and carrier DNA in a total amount of 20–100 μg per 10-cm semiconfluent Petri dish (see Note 7). Make up the DNA mixture volume up to 450 μl with sterile double distilled water. Add 50 μl of 2.5 M $CaCl_2$ to the 450-μl DNA mix in an Eppendorf tube, and finally add 500 μl of 2× HBS dropwise, so that the final volume is 1 ml. Bubble with a pipet and leave for 30 min at room temperature to form precipitates. When they form, the transfection solution gets opaque.
3. Wash cells with sterile PBS, then add the 1-ml DNA precipitate dropwise to the cells; rock the plate gently and allow the plate to stand for 30 min at room temperature. It is important that the transfection mixture be spread evenly on the surface of the plate.
4. Add 9 ml of culture medium, gently rock the plate, and incubate the cell for 3.5 h at 37°C in CO_2 thermostat.
5. Glycerol shock: remove the medium and add 2 ml of glycerol–PBS dropwise for 1–2 min and then dilute out immediately with 10 ml of prewarmed (37°C) medium. Maintain the incubation time, because glycerol is toxic to mammalian cells. If cells are becoming rounded – while observing them under an inverted microscope – it is time to add fresh medium.
6. Change the medium containing glycerol for complete medium and grow cells for 48–72 h.
7. Start selection for stable transformants by adding antibiotics and change the selective medium every 3 days.

3.3. Establishment of Stable Clones

1. Pick the antibiotic resistant clones when each colony is at least 2–3 mm in diameter. The colonies should be well separated from each other to avoid contamination. Prepare 6-well plates or 3-cm Petri dishes containing 2 ml of culture medium. Work in sterile hood.
2. Remove the medium from the cells.
3. Pick up a cloning cylinder with forceps and dip it in vacuum grease. The grease must be distributed in the periphery, make sure that it doesn't fill the ring.
4. Place the cloning cylinder on the colony and make sure the cylinder sticks well to the dish. Repeat this on not more than five colonies at a time. Wet the rest of the plate with a few (approximately 1 ml) drops of culture medium thereby protecting the other colonies from drying out. Add into each cylinders 50–100 μl of trypsin and carefully remove it immediately using a micropipette.

5. Wait until the cells detach and fill up the cylinder with 50–100 µl of culture medium.
6. Remove the cells from each cloning cylinder by pipetting up and down and transfer to 6-well plates or 3-cm Petri dishes. Be quick to avoid "overtrypsinization."
7. Remove the 1-ml medium and then the cloning cylinders. Choose another five clones to pick up and perform the same procedure as described above.
8. Culture them separately, passage several times, and make a fixation (see Subheading 3.4) from every primary clone for further cytological analysis (see Subheading 3.6).
9. Freeze (see Subheading 3.5) at least two 6-cm Petri dish of cells from each clone.

3.4. Fixation of Mammalian Cells (20)

1. Culture the cells in the presence of 5 µg/ml colchicine for 5 h (to 10-cm Petri dish containing 10 ml of medium add 50 µl from 1 mg/ml colchicine stock).
2. Collect the cells blocked in mitosis in a 10-ml Falcon tube by gently washing down the rounded mitotic cells from the surface of the cell layer using a plastic Pasteur pipette. Centrifuge at room temperature at $100 \times g$ for 10 min.
3. Remove and discard the supernatant and resuspend the cell pellet by adding 10 ml of 75 mM KCl, which has been prewarmed to 37°C and draw the cells gently in and out of a Pasteur pipette.
4. Incubate at 37°C for 10 min and spin as above. This gives a total time of 20 min in the hypotonic solution.
5. Remove the supernatant, leaving about 0.5 ml of fluid above the cell pellet. Resuspend the cell pellet in this fluid by tapping at first, and then by gently sucking the cells up and down a Pasteur pipette until no large cell clumps remain.
6. Draw this fine cell suspension into the Pasteur pipette and fill the tube with 10 ml of freshly prepared ice-cold fixative. Drop back the suspension slowly into the fixative in the tube and mix gently using the Pasteur pipette. Leave the cell suspension for 10 min and spin down as above.
7. Remove the supernatant and add 10 ml of fresh fixative. Resuspend the cell pellet and spin down as previously. Repeat this step once more.
8. After the last fixation remove the supernatant and resuspend the cells in a small volume of fixative, e.g. 0.5 ml. Store in a screw capped Falcon tube at −20°C.

3.5. Freezing of Mammalian Cells

1. Work in sterile hood.
2. Trypsin treatment: add 1 ml of trypsin solution to the cells in a 6-cm Petri dish, remove immediately and wait for a while.

Check cells under an inverted microscope, when they start to become round proceed to the next step.

3. Detach the cell layer with 1 ml of precooled freezing solution by gentle pipetting. Collect the cell suspension into a freezing vial.
4. Chill on ice for a while.
5. Wrap the vials in multiple layers of paper towels and put into a small plastic bag. Store in a paper box at –80°C for at least 24 h.
6. For long-term storage, transfer the vials into a cell storage unit containing liquid nitrogen (see Note 8).

3.6. Cytological Analysis of Primary Clones

Fluorescent in situ hybridisation (FISH) is the most reliable technique to select the appropriate clones. Applying the DNA used in transfection as probe amplifications or SATAC formations can be detected by FISH (Fig. 1). As a parallel probe satellite DNA can be used in double colored FISH.

3.6.1. Slide Preparation

Use a quality microscopic slide, wash it, and wipe it with a clean cotton tissue. During the slide preparation, the fixative and fixed cell samples must be kept on ice. Drop fixed cells onto the slide. After the initial drop of cell suspension has been spread out on the slide and the edges of the drop have begun to dry, flood the slide with two or three drops of ice-cold fixative and gently drain off before allowing the slide to air dry at room temperature. Control the quality of slides under a phase contrast microscope. Use only well spread, clean metaphases. Leave the preparations to age in a staining jar in dry conditions at room temperature overnight.

3.6.2. Labeling of Probes

Use the appropriate labeling kit (see Note 9). Labeled probes can be used directly; removal of unincorporated nucleotides is optional, but not necessary. For further probe purification use a Nick-Spin Column or an equivalent (see Note 10).

3.7. First Day

3.7.1. RNase Treatment

Perform RNase A treatment at 37°C for 60 min at 100 μg/ml of RNase A concentration. Prepare 100 ml of 2× SSC by diluting from 20× SSC and prewarm in an Erlenmeyer flask at 37°C. Weigh 10 mg crystalline RNase A into an Eppendorf tube and dissolve in 1 ml of 2× SSC. To make it DNase free, immerse into boiling water for 10 min, and mix with 100 ml of 2× SSC and apply to slides in the staining jar (see Note 11). After the 60-min incubation time change the RNase A buffer to 70% ethanol. Dehydrate the preparations in 70%, 90%, and absolute ethanol at room temperature by changing the ethanols every 5 min. Dry under vacuum, if it is not possible a working fume hood is suitable as well.

3.7.2. Denaturation

Denaturation: is performed in freshly prepared and prewarmed denaturation solution at 70–75°C for 2 min in a waterbath.

During the RNase treatment, prepare 100 ml of denaturation solution, pour into a 100-ml beaker, cover with a piece of aluminium foil, and place it into a waterbath set at 70–75°C. During the denaturation step there are only two slides simultaneously in the beaker, because more slides would cool the denaturation buffer down, and the temperature of the buffer would be lower than the required 70–75°C. At lower temperature denaturation does not occur. After the 2-min denaturation step transfer the two slides immediately to ice-cold 70% ethanol in another staining jar. Perform the same procedure with every two slides. Dehydrate with ice-cold ethanol series for every 5 min and dry as above.

3.7.3. Preparation of the Hybridization Mixture and Hybridization

Prepare the hybridization mixture: 0.5–2.0 μl of biotin labeled probe, 0.5–2.0 μl of digoxigenin labeled probe, and HybriSol up to 30 μl. After vortexing denature the probe mix at 70–75°C for 10 min in a waterbath and keep on ice till usage. Apply 30 μl of probe per slide, cover carefully (avoid air bubbles) with 24 × 32 mm coverslip, and seal the edges of the coverslip with rubber cement. After the rubber cement bond place the slides into the bottom of the humid chamber. Place the lid on and seal around with an adhesive tape. Incubate at 37°C over night in a thermostat.

3.8. Second Day

3.8.1. Immunodetection and Microscopy

1. Carefully remove the rubber cement using the forceps and soak off coverslips in 2× SSC in a staining jar. Transfer the slides into another staining jar containing prewarmed buffer-A and incubate at 42–45°C in a waterbath. Change the buffer every 3 min; four times in total. Change the last portion of buffer-A to prewarmed buffer-B. Incubate the slides as mentioned above at 42–45°C in a waterbath for 4 × 3 min. Finally rinse the slides in buffer-C for 5 min at room temperature. All subsequent washes and antibody dilutions are performed in buffer-C. It is very important to remember that during the washes and immunodetection steps do not allow the slides to dry.
2. Immunodetection 1.

   In the same Eppendorf tube make a 1,000× dilution of FITC–Avidin for biotin detection and 1,000× dilution of Mouse Monoclonal anti-DIG for digoxigenin detection; dilute with buffer-C and mix well. Apply 60 μl of the antibody mixture per slide taking care to see that the slides do not dry (see Note 12). Cover with 24 × 32 mm coverslip, place the slides in humid a chamber, and incubate at 37°C for 30–45 min in a thermostat (if you are in a hurry, 30 min is sufficient) (see Note 13).
3. Immunodetection 2.

   Shake off the coverslips and wash the slides in prewarmed solution-C at 42–45°C in waterbath for 2 × 3 min. Apply 60 μl of the secondary antibodies, which is a mixture of

500× diluted Biotinylated anti-Avidin and 500× diluted anti-Mouse Ig-DIG diluted in buffer-C. Cover with coverslips and incubate as above.

4. Immunodetection 3.

   Shake off the coverslips and wash slides as in the previous step. Apply tertiary antibodies, 1,000× diluted FITC-Avidin and 200× diluted anti-DIG Rhodamine diluted in buffer-C, in the same manner as above. Incubate as before.

5. DNA staining and mounting.

   Shake off the coverslips and wash slides. Mount the preparations with 20–25 μl of mounting medium containing 2–0.5 ng/μl of DAPI. Cover with 24 × 32 mm coverslip, gently remove the excess fluid with filter paper and eliminate the air bubbles. Seal the edge of the coverslip with nonfluorescent nail polish. Preparations are now ready for analysis, and can be stored for weeks at −20°C (see Note 14).

6. Microscopy.

   Regular fluorescence microscope with green, red, and DAPI filters can be used. DAPI imparts a blue counter stain; the biotinylated probe (FITC-Avidin) gives green signal and the digoxigenin labeled probe (anti-DIG Rhodamine) appears red (see Note 15).

3.9. Single Cell Subcloning

Usually the primary clone is heterogenic; SATACs may vary in size and structure in different cells. In this case, single cell subcloning is required to establish a uniform SATAC carrying cell line.

1. After a trypsin treatment, collect the cells with 5–10 ml of the culture medium. Work in sterile hood.
2. Determine the cell number by counting the cells in Bürker chamber or with any other cell counter.
3. Make a dilution of 0.5 cells/ml with the culture medium.
4. Add 1 ml of diluted cells into wells of a 24-well plate. If required, perform the same procedure with any number of 24-well plates.
5. Select those lines which originate from one single cell.
6. Analyze the cell lines by FISH.

4. Notes

1. Chinese Hamster Ovary (CHO) cell line, for example, is not the best choice because the rDNA region on the CHO chromosomes has telomeric/subtelomeric localization. Previously,

the LMTK⁻ mouse cell line was used to generate mouse SATACs, and the 94-3 cell line was used to generate human SATACs (13–17, 21). The 94-3 is a monochromosomal hybrid cell line, which carries human acrocentric chromosomes including chromosome #15 on a Chinese-hamster background (Fig. 1a). In humans, the acrocentric chromosomes are the NOR chromosomes bearing the rDNA region on the p arm (NOR = Nucleolus Organiser Region).

2. To establish the appropriate concentration of an antibiotic for a given cell line, make a killing curve using different concentration of the antibiotic. E.g. the appropriate puromycin concentration for 94-3 cell line is 10 μg/ml.
3. 94-3 cell line can be propagated in Dulbecco's Modified Eagle's Medium (high glucose, GibcoBRL) supplemented with 10% heat inactivated fetal bovine serum (GibcoBRL) (DMEM-10FBS).
4. Platform SATAC can be built by using *pattB* constructs as the "foreign" DNA (3, 21).
5. A cloned ~8 kb fragment of coding sequence of rDNA (accession number AY390526) was used as carrier DNA to generate SATACs (17, 21).
6. Linearized construction is recommended. If it is not possible circular form of DNA can be used as well, though the efficiency of stable transfection is lower with circular form than the linearized form. However, in the case of a bigger, e.g. PAC sized DNA, the circular form is preferable, because the linearization of large DNA molecule makes it fragile.
7. Too much DNA is toxic for mammalian cells.
8. Thawing: put the vial containing the frozen cells under running hot tap water and wait till it becomes fluid again. Split immediately into a 6-cm Petri dish containing prewarmed complete culturing medium and after the cells have attached, change the medium.
9. The labeling efficiency can be monitored by applying a portion of the labeling mixture in a standard gel electrophoresis assay. The optimal probe size is 200–500 bp.
10. Certain probes (e.g. human chromosome specific alpha satellite probes) are commercially available in the labeled form. Using these probes the FISH procedure should be modified.
11. Do not insert the upper (frosted) part of the slides to the liquids; avoid contamination of DNase, bacteria, etc. Use forceps, not your fingers.
12. Take one slide out of the staining jar containing buffer-C and place it into the bottom of the humid chamber. Apply 60 μl of the antibody mixture and cover with a coverslip. Take another slide and repeat the previous steps.

13. During the immunodetection steps Hybri-Slips can be used instead of normal coverslips. These special coverslips are more adhesive and move less on the surface of the liquid layer thus preventing the preparations from drying out.
14. Protect the slides from direct light.
15. Any image analysis system is suitable to analyze and record the chromosomal images.

References

1. Vos J-MH (1998) Mammalian artificial chromosomes as tools for gene therapy. *Curr Opin Genet Dev* **8**, 351–359.
2. Danielle V. Irvine, Margaret L. Shaw, K. H. Andy Choo and Richard Saffery (2005). Engineering chromosomes for delivery of therapeutic genes. *Trends Biotech.* **23**, (12) 573–583.
3. Duncan, A., Hadlaczky, Gy. (2007). Chromosomal Engineering. *Curr Opin Biotech* **18**, 420–424.
4. Kennard M.L, Goosney D.G, Ledebur H.C. 2007. Generating stable, high–expressing cell lines for recombinant protein manufacture. *BioPharm Int* **20**, (3):52–59.
5. Kennard, M.L., Goosney, D.L., Monteith, D., Roe, S., Fischer, D., Mott, J. (2009). Auditioning of CHO host cell lines using the Artificial Chromosome Expression (ACE) technology. *Biotechnology and Bioengineering* **104**(3), 526–539.
6. Kennard, M.L., Goosney, D.L., Monteith, D., Zhang, L., Moffat, M., Fischer, D., Mott, J. (2009). The generation of stable, high MAb expressing CHO cell lines based on the Artificial Chromosome Expression (ACE) technology. *Biotechnology and Bioengineering* **104**(3), 540–553.
7. Telenius, H., Szeles, A., Kereső, J., Csonka, E., Praznovszky, T., Imreh, S., Maxwell, A., Perez, C.F., Drayer, J.I., Hadlaczky, Gy. (1999). Stability of a functional murine satellite DNA-based artificial chromosome across mammalian species. *Chromosome Res.* 7, 3–7.
8. Co, D.O., Borowski, A.H., Leung, J.D., van der Kaa, J., Hengst, S., Platenburg, G., Pieper, F.R., Perez, C.F., Jirik, F.R., Drayer, J.I. (2000). Generation of transgenic mice and germline transmission of a mammalian artificial chromosome introduced into embryos by pronuclear microinjection. *Chromosome Res.* **8**, 183–191.
9. Monteith, D.P., Leung, J.D., Borowski, A.H., Co, D.O., Praznovski, T., Jiric, F.R., Hadlaczky, Gy., and Perez, C.F. (2003). Pronuclear microinjection of purified artificial chromosomes for generation of transgenic mice: Pick-and-Inject Technique. In Mammalian Artificial Chromosomes: *Methods and Protocols.* Eds.: Sgaramella, V. and Eridani, S. *Methods in Mol. Biol.* **240**, 227–242
10. S. Vanderbyl, G. N. MacDonald, S. Sidhu, L. Gung, A. Telenius, C. Perez, E. Perkins (2004). Transfer and stable transgene expression of a mammalian artificial chromosome into bone marrow-derived human mesenchymal stem cells. *Stem Cells.* **22**, 324–333.
11. Vanderbyl SL, Sullenbarger B, White N, Perez CF, MacDonald GN, Stodola T, Bunnell BA, Ledebur HC Jr, Lasky LC. (2005). Transgene expression after stable transfer of a mammalian artificial chromosome into human hematopoietic cells. *Exp Hematol.* **33**(12), 1470–1476.
12. R. L. Katona, I. Sinkó, G. Holló, K. Székely Szűcs, T. Praznovszky, J. Kereső, E. Csonka, K. Fodor, I. Cserpán, B. Szakál, P. Blazsó, A. Udvardy and G. Hadlaczky (2008). A combined artificial chromosome-stem cell therapy method in a model experiment aimed at the treatment of Krabbe`s disease in the Twitcher mouse. *Cell. Mol. Life Sci.* **65**, 3830–3838.
13. Hadlaczky, Gy., Praznovszky, T., Cserpán, I., Kereső, J., Péterfy, M., Kelemen, I., Atalay, E., Szeles, A., Szelei, J., Tubak, V. (1991). Centromere formation in mouse cells cotransformed with human DNA and a dominant marker gene. *Proc. Natl. Acad. Sci. USA.* **88**, 8106–8110.
14. Praznovszky, T. Kereső, J. Tubak, V., Cserpán, I., Fátyol, K., Hadlaczky, Gy. (1991). *De novo* chromosome formation in rodent cells. *Proc. Natl. Acad. Sci. USA.* **88**, 11042–11046.
15. Kereső, J., Praznovszky, T., Cserpán, I., Fodor, K., Katona, R., Csonka, E., Fátyol, K., Holló, Gy., Szeles, A., Ross, A.R., Sumner, A.T., Szalay, A.A., Hadlaczky, Gy. (1996). *De novo* chromosome formations by large-scale amplification of the centromeric region of mouse chromosomes. *Chromosome Res.* **4**, 226–239.
16. Holló, Gy., Kereső, J., Praznovszky, T., Cserpán, I., Fodor, K., Katona, R., Csonka, E.,

Fátyol, K., Szeles, A., Szalay, A.A., Hadlaczky, Gy. (1996). Evidence for a megareplicon covering megabases of centromeric chromosome segments. *Chromosome Res.* **4**, 240–247.

17. Csonka, E., Cserpán, I., Fodor, K., Holló, Gy., Katona, R., Kereső, J., Praznovszky, T., Szakál, B., Telenius, A., deJong, G., Udvardy, A., Hadlaczky, Gy. (2000). Novel Generation of Human Satellite DNA-based Artificial Chromosomes in Mammalian Cells. *J Cell Sci.* **113**, 3207–3216.
18. Hadlaczky, Gy. (2001) Satellite DNA-based artificial chromosomes for use in gene therapy. *Current Opinion in Molecular Therapeutics* **3**(2), 125–132.
19. Chen, C, Okayama, H. (1987). High-efficiency transformation of mammalian cells by plasmid DNA. *Mol. Cell. Biol.* 7, 2745–2752.
20. Macgregor, H., Varley, J. (1983) Mitotic chromosomes. Working with animal chromosomes. John Wiley & Sons Ltd.
21. Lindenbaum, M., Perkins, E., Csonka, E., Greene, A., Fleming, E., Hadlaczky, Gy., MacDonald, N., Maxwell, A., Perez, C. and Ledebur, H.C. Jr. (2004). The ACE System: engineering artificial chromosomes to rapidly generate high-expressing cell lines for manufacture of recombinant proteins. *Nucl. Acids Res.* **32**(21), e172.

Chapter 9

Downstream Bioengineering of ACE Chromosomes for Incorporation of Site-Specific Recombination Cassettes

Amy L. Greene and Edward L. Perkins

Abstract

Advances in mammalian artificial chromosome technology have made chromosome-based vector technology amenable to a variety of biotechnology applications including cellular protein production, genomics, and animal transgenesis. A pivotal aspect of this technology is the ability to generate artificial chromosomes *de novo*, transfer them to a variety of cells, and perform downstream engineering of artificial chromosomes in a tractable and rational manner. Previously, we have described an alternative artificial chromosome technology termed the ACE chromosome system, where the ACE platform chromosome contains a multitude of site-specific, recombination sites incorporated during the creation of the ACE platform chromosome. In this chapter we review a variant of the ACE chromosome technology whereby site-specific, recombination sites can be integrated into the ACE chromosome following its *de novo* synthesis. This variation allows insertion of user-defined, site-specific, recombination systems into an existing ACE platform chromosome. These bioengineered ACE platform chromosomes, containing user-defined recombination sites, represent an ideal circuit board to which an array of genetic factors can be plugged-in and expressed for various research and therapeutic applications.

Key words: Mammalian artificial chromosome, ACE chromosome, *LoxP*, DsRed

1. Introduction

Mammalian artificial chromosomes (MACs) provide an alternative means to introduce large payloads of genetic information into the cell as an autonomously replicating, non-integrating chromosome-based vector system. For example, MACs potentially allow for the engineering of large segments of genomic DNA, such as fragments containing long-range genetic elements required for appropriate regulation of gene expression, developmentally regulated multi-gene loci, or multiple copies of two or more genes in fixed stoichiometry. Advances in MAC technology and methodology

Gyula Hadlaczky (ed.), *Mammalian Chromosome Engineering: Methods and Protocols*, Methods in Molecular Biology, vol. 738, DOI 10.1007/978-1-61779-099-7_9,

for the generation of these non-integrating vectors for gene therapy have previously been described (1–4). Artificial and engineered chromosomes can be generated by several means including: (1) Co-transfection of a permissive cell line with defined chromosomal elements (*i.e.*, telomere elements, centromeric alpha satellite DNA multimers, and mammalian replication origins) and a drug-selectable marker, which then self-assemble into an artificial chromosome, (2) Reduction of individual host cell chromosomes to mini-chromosomes consisting of minimal functional centromere regions or neocentromeres by a process of targeted telomere integration and excision, (3) Modification of stably maintained centric fragments or small accessory chromosomes to accept foreign genes, (4) Creation of a yeast–human hybrid artificial chromosome by transformation-associated recombination (TAR) cloning of human centromere sequences into an existing YAC, and (5) Generation of satellite-DNA amplified chromosomes (SATACs) by targeted amplification and fragmentation of pericentromeric sequences from acrocentric chromosomes into stably maintained chromosome vectors (Fig. 1). The overriding principle common to all of these methods is the recapitulation of functional mammalian centromeres and telomeres in a form suitable for

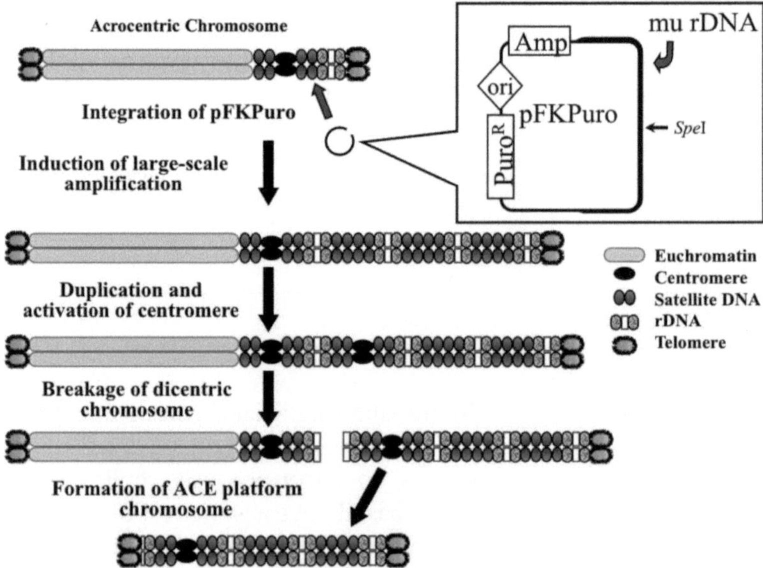

Fig. 1. Production of the ACE platform chromosome. Plasmid pFKPuro is digested with *SpeI* and transfected into mouse LTMK– cell line. Plasmid pFKPuro, a derivative of plasmid pFK161 (17), contains a portion of the murine 45S pre-rRNA gene and the puromycin resistance cassette in the vector backbone (13). Upon integration of pFKPuro near the pericentric heterochromatin of an acrocentric chromosome, a large-scale amplification occurs resulting in duplication of the centromere. A new artificial chromosome (termed ACE chromosome) is formed upon cleavage and resolution of the formed dicentric chromosome. The ACE chromosome is composed primarily of satellite repeat sequences, interspersed with co-amplified pFKPuro and endogeneous rDNA.

downstream engineering. Previously, a modification of the SATACs construction methodology was used to generate engineered artificial chromosomes (termed artificial chromosome expression system or ACE chromosome) containing a multitude of site-specific recombination sites (5). In this version, the ACE system consists of a platform chromosome (ACE chromosome) containing >50 site-specific, recombination acceptor sites (attP), that can carry single or multiple copies of genes of interest using specially designed ACE targeting vectors (ATV) and a site-specific integrase (ACE Integrase). The ACE Integrase is a derivative of the bacteriophage lambda integrase (λINT) engineered to direct site-specific recombination in mammalian cells *in lieu* of bacterial encoded, host integration accessory factors. The ACE bioengineering system is, therefore, applicable to a variety of cell-based applications including cellular protein production, gene therapy, and animal transgenesis (5–11 and reviewed in ref. 2).

The ACE system can be reproducibly generated *de novo* in cell lines of different species and readily purified from the host cells' chromosomes by flow cytometry and chromosome sorting. In turn, purified mammalian ACE chromosomes can be readily introduced into a variety of cell lines (for example, primary cells, transformed cell lines, murine embryonic stem cells, human mesenchymal, and hematopoietic stem cells) by transfection and maintained for extended periods without enforced genetic selection (5, 8, 12–16). In addition to the ACE chromosome containing a multitude of attP recombination sites, a variation of the ACE technology enables the retrofitting of ACE chromosomes to carry other site-specific recombination systems (13). As such, ACE chromosomes can be tailor-made to end-user specific vector/recombination preferences. As an example of this approach, we will outline the techniques and methodology used to retrofit an ACE chromosome to contain the Cre–*lox*P site-specific, recombination technology.

2. Materials

Unless otherwise indicated, all chemicals and antibiotics were obtained from Sigma-Aldrich, Inc. and cell culture media and transfection reagents are from Invitrogen, Inc. Oligonucleotides were synthesized by Invitrogen, Inc.

2.1. Generation of the ACE Platform Chromosome

1. Mouse fibroblast cell line LMTK– (American Type Cell Culture, Inc., Note 1).
2. Cell culture media (Dulbecco's Modified Eagle's Medium (D-MEM) containing high glucose, sodium pyruvate and GlutaMax™) supplemented with fetal bovine serum to a final concentration of 10% (Thermo Scientific Hyclone, Inc.).

3. Large-scale plasmid isolation kit (Qiagen EndoFree Plasmid Maxi Kit, Qiagen, Inc.).
4. Calcium Phosphate Transfection Kit.
5. Puromycin stock solution (10 mg/ml in water).
6. Restriction enzyme *Spe*I (New England Biolabs, Inc.).
7. Plasmid pFKPuro (Fig. 1; (13)).

2.2. Validation of ACE Chromosome Construction by Fluorescent In Situ Hybridization

1. Colchicine stock solution (50 mg/ml in water).
2. Biotin-nick-Translation Mix (Roche Applied Science).
3. Digoxigenin-11-dUTP solution (1 mM DIG-dUTP, Roche Applied Science).
4. Biotin-16-dUTP solution (1 mM Biotin-dUTP, Roche Applied Science).
5. Taq PCR Master Mix Kit (Qiagen, Inc.).
6. Mounting Medium (Vectashield Mounting Medium with DAPI, Vector Laboratories, Inc.).
7. Primers for FISH probes (5′-mmajor: 5′-ATACTCTTCAGG ACCTGGAATATGGCGAG-3′, 3′-mmajor: 5′-ATACTCTT CGTCCTTCAGTGTGCATTTCTCATTTTTC-3′, 5′-mminor: 5′-GGAAAATGATAAAAACCTAC-3′, 3′-mminor: 5′-ATGTTT-CTAATTGTAACTCAT-3′).

2.3. Validation of ACE Chromosome Construction by Flow Cytometry

1. Hypotonic swelling solution (75 mM KCl).
2. Polyamine Buffer (80 mM KCl, 70 mM NaCl, 0.1% β-mercaptoethanol, 15 mM Tris–HCl, 2 mM EDTA, 0.5 mM EGTA, 0.2 M spermine, 0.5 M spermidine and 0.25% Triton X-100 with the solution adjusted to pH 7.2).
3. Hexylene glycol/glycine buffer: 2% hexylene glycol with 200 mM glycine prepared in water.
4. Chromosome/DNA staining stock solutions: Propidium iodide (1.0 mg/ml solution in water), Hoechst 33258 (10 mg/ml in water), and Chromomycin A3 (10 mg/ml in ethanol, Enzo Life Sciences, Inc.).
5. 10× Na citrate/Na sulfite solution (100 mM sodium citrate, 250 mM sodium sulfite, pH 10.0).

2.4. Retrofitting of the ACE Chromosome

1. Lipofectamine transfection reagent (Invitrogen, Inc.) with Plus Reagent (Invitrogen, Inc.).
2. Qiaex II Gel Extraction Kit (Qiagen, Inc.).
3. Restriction enzymes *Pml*I and *Eco*RV (New England Biolabs, Inc.).
4. Plasmids pDsRed-N1 (Clontech, Inc.) and pBSFKLoxD-sRedLox (Fig. 3).

3. Methods

For *de novo* production of ACE chromosomes, exogenous DNA sequences are introduced into cells and, upon integration into the pericentric heterochromatic regions of acrocentric chromosomes, a large-scale amplification of the short arms of the acrocentric chromosome (rDNA/centromere region) is triggered (Fig. 1; (2, 5, 13, 17–22)). During the amplification event, the centromere is duplicated resulting in a dicentric chromosome with two active centromeres. Subsequent mitotic events results in cleavage and resolution of the dicentric chromosome leading to a break-off chromosome typically 40–80 Mb in size and composed predominantly of satellite repeat sequences with interfused subdomains of coamplified transfected transgene that may also contain amplified copies of rDNA. The newly generated chromosome (ACE chromosome) is validated by fluorescent *in situ* hybridization (FISH) for the presences of centromeric heterochromatin, telomeric sequences and, if needed, additional transfected DNA sequences (*e.g.* drug selection marker).

Both murine and human ACE chromosomes can be reliably produced (5, 13, 17, 19–22) and validated by flow cytometry due to their size and unique composition (Fig. 2). This unique property permits ACE chromosomes to be purified by chromosome sorting. After isolation, ACE chromosomes can readily be introduced into a variety of cell types by transfection (5, 8, 12–16, 23), as well as utilized for transgenic animal production (6, 8, 24, 25). For downstream bioengineering applications such as knocking-in gene(s) or genomic fragments onto the ACE chromosomes, the addition of site-specific recombination sequences can be incorporated during ACE chromosome construction. As discussed below, we outline a method by which site-specific recombination sequences can be targeted onto the ACE chromosome after its initial construction independent of enforced drug selection.

3.1. Generation of the ACE Platform Chromosome

1. Prepare a large, endotoxin-free prep of plasmid pFKPuro using the Qiagen EndoFree Plasmid Maxi Kit according the manufacturers' protocol. Assess the purity and quantity of the plasmid preparation by standard OD 260/280 spectrophotometry or with the use of a NanoDrop™ micro-volume UV–Vis spectrophotometer (Thermo Scientific).
2. Linearize approximately 25–50 μg of plasmid pFKPuro with *Spe*I restriction endonuclease. Confirm complete digestion of pFKPuro by standard agarose gel electrophoresis using a small aliquot of the digestion (approximately 100–200 ng along with a size standard and an aliquot of uncut pFKPuro plasmid). Purify the remaining plasmid digestion solution by standard phenol:chloroform extraction followed by ethanol

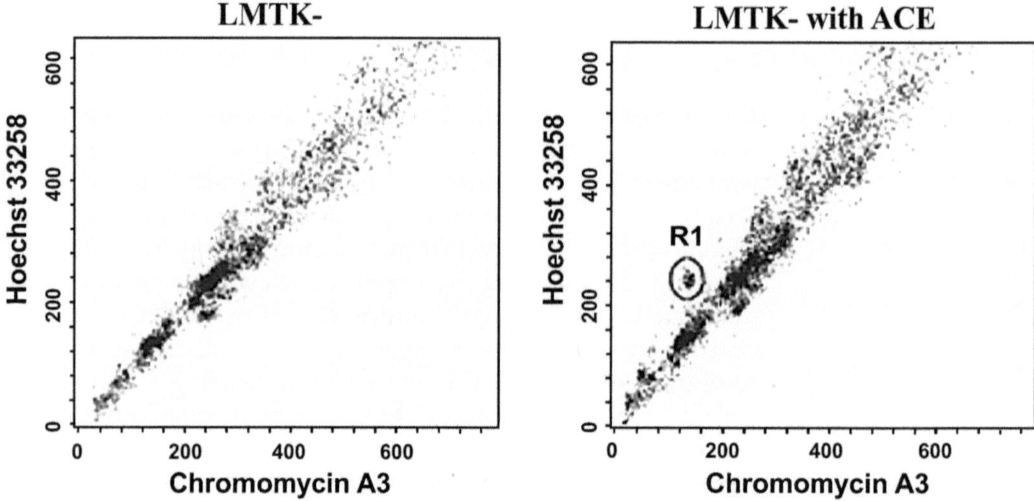

Fig. 2. Validation of ACE chromosome generation. The generated ACE chromosome is validated by a combination of FISH analysis and flow cytometry. For validation by flow cytometry, a bivariant flow histogram of the generated ACE chromosome is generated. Dual Hoechst 33258/chromomycin A3 staining of metaphase chromosomes from LMTK– cells exhibits a near diagonal line indicative of roughly equimolar binding of the two dyes, with larger chromosomes binding proportionally more of the two dyes. The flow karyogram of metaphase chromosomes, from ACE-containing LMTK cells, reveals the presences of the ACE chromosome (circled region R1). The relative AT-richness of the generated ACE chromosome binds more Hoechst 33258 dye than chromomycin A3 dye, thus permitting the identification of the ACE chromosome. This resolution of the ACE chromosome from the endogenous chromosomes provides the basis of ACE chromosome purification by flow cytometry/chromosome sorting. [Parts of this figure reprinted with permission from (13)].

precipitation of the DNA. Resuspend the digested plasmid to a concentration of 20–25 μg plasmid DNA per 300 μl dH_2O.

3. One day prior to transfection, plate approximately $1–1.5 \times 10^6$ LMTK– cells in a 10-cm cell culture dish. To maximize transfection efficiency, it is important to use a sub-confluent culture of LMTK– for the plating prior to transfection.
4. The following day, transfect cells with 20–25 μg of digested plasmid DNA using the Calcium Phosphate Transfection Kit according to the manufacturer's protocol (Invitrogen). Include a mock transfection control (*e.g.* using 20–25 μg of pUC19 or similar plasmid DNA *in lieu* of pFKPuro).
5. Cells are maintained for 2–3 days post-transfection at which point they are trypsinized and replated in media supplemented with 10 μg/ml puromycin.
6. Puromycin selection is maintained for 10–14 days with changes in media every 2–3 days until discreet colonies can be visualized. Discreet colonies should not be observable in the mock transfected cells.

7. After selection, colonies are picked using cloning rings and transferred into new dishes and expanded. At this point, selection should be maintained using media containing 10 μg/ml puromycin.
8. One aliquot of cells is frozen down and a second aliquot is expanded for subsequent FISH analysis to detect ACE chromosome generation.

3.2. Validation of ACE Chromosome Construction by FISH

1. Puromycin resistant colonies are expanded and subsequently metaphase arrested by treating cultures with 1 μg/ml colchicine for 16 h.
2. Colonies are analyzed cytogenetically *via* standard multi-color FISH analysis on the metaphase chromosomes (Note 2).
3. FISH analysis is performed to confirm colocalization of pFK-Puro with sequences associated with murine centromeric heterochromatin (mouse major and mouse minor satellite repeats) against a background of DAPI stained chromosomes. Vectashield Mounting Medium with DAPI (Vector Laboratories) is routinely used in our laboratory for chromosome/DNA counterstaining.
4. For production of the FISH probes, pFKPuro is labeled with biotin-dUTP using the Biotin-Nick Translation Kit (Roche).
5. DIG-labeled mouse major satellite probes are generated by PCR using primers 5′-mmajor and 3′-mmajor and DIG-dUTP. Biotin-labeled mouse minor satellite probes are generated by PCR using primers 5′-mminor, 3′-mminor and biotin-dUTP. PCR-based labeling reactions are performed using the Taq PCR Master Mix Kit according to the manufacturer's protocol.
6. Hybridization of probes is visualized by fluorescence microscopy (Note 3).
7. Colocalization of pFKPuro with sequences associated with murine centromeric heterochromatin (mouse major and mouse minor satellite repeats) is a key validation point, which confirms a large scale amplification proximal to the pericentric heterochromatin on an existing acrocentric chromosome, due to integration of the transfected transgene and followed by subsequent ACE chromosome formation.

3.3. Validation of ACE Chromosome Construction by Flow Cytometry

1. Exponentially growing cultures containing ACE chromosomes are grown on 150-mm tissue culture dishes and arrested at metaphase as described above.
2. Mitotic arrested cells are harvested by trypsinization and the cells are swelled in hypotonic swelling solution (75 mM KCl) for 10 min at room temperature.

3. Cells are centrifuged at 100×*g* for 5 min and the resulting pellet is resuspended in 10 ml Polyamine Buffer.
4. The cell membranes were sheared by force using a 10-ml syringe with attached 22-gauge needle through which the cells are drawn up and down.
5. An equal volume (10 ml) of 2% hexylene glycol in 200 mM glycine buffer is added to the chromosome solution.
6. At this point, the release of the chromosomes is monitored by placing a drop of the chromosome suspension in 0.5 μl of 10 μg/ml propidium iodide on a slide and visualized by fluorescence microscopy. Released chromosomes are evaluated for their degree of condensation (tightly condensed to swollen-looking chromosomes).
7. The chromosome solution is centrifuged for 1 min at 100×*g* to remove the cellular debris.
8. The chromosome containing supernatant was transferred to a new tube and the chromosomes are stained with Hoechst 33258 (2.5 μg/ml) and chromomycin A3 (50 μg/ml) with $MgCl_2$ added to a final concentration of 2.5 mM.
9. Stained chromosomes are stored at 4°C for a minimum of 2 h.
10. At this point, stained chromosomes are ready for flow cytometric analysis. Fifteen minutes prior to flow cytometry, the chromosome solution is adjusted to 10 mM sodium citrate, 25 mM sodium sulfite in order to enhance chromosome resolution during flow cytometric analysis as previously described (26).
11. For flow cytometric analysis, a bivariate flow karyotype of metaphase chromosomes is generated using dual laser excitation. Hoechst 33258 is excited by a UV-laser beam and detected using a 420 nm band-pass filter and chromomycin A3 is excited by a second laser set at 458 nm and fluorescence is detected using 475 nm long-pass filter (Note 4).

3.4. Retrofitting of ACE Chromosomes

1. For targeting onto the ACE platform chromosome, a DsRed red fluorescent protein variant expression cassette (DsRed1-N1) flanked by *lox*P sites was inserted into *Not*I rDNA fragment derived from plasmid pFKPuro resulting in plasmid pBSFKLoxDsRedLox (Note 5). This plasmid is digested with restriction enzymes *Pml*I and *Eco*RV to release an approximately 11 kb fragment containing the DsRed expression cassette with flanking ends of homology to the amplified rDNA regions on the ACE chromosome (Fig. 3).
2. The digested plasmid is subjected to agarose gel electrophoresis and the approximately 11 kb fragment is gel purified using

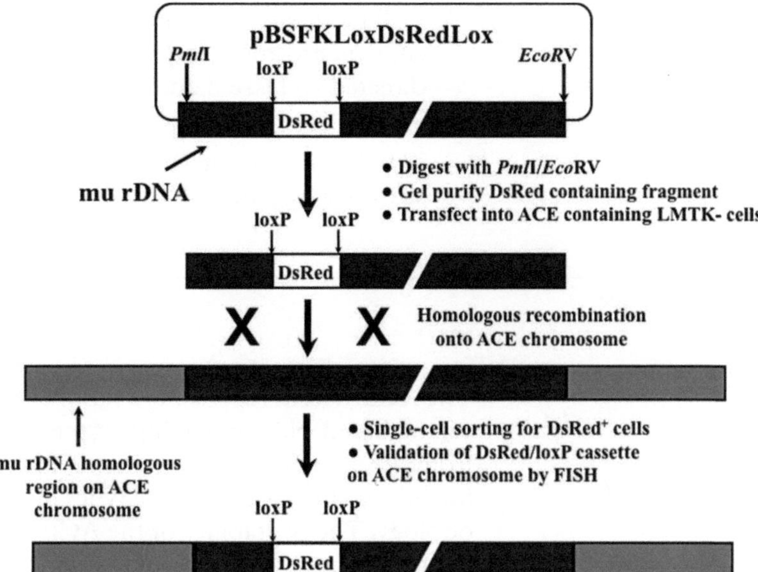

Fig. 3. Retrofitting of the ACE platform chromosome to contain *loxP* sites. Overview of retrofitting *loxP* sites on the ACE chromosome. Plasmid pBSFKLoxDsRedLox contains a fragment corresponding to a portion of the rDNA insert from the parent plasmid pFKPuro [further details of this plasmid can be found in (13)]. A *loxP*–DsRed–*loxP* expression cassette was inserted into the subcloned rDNA fragment. The resulting rDNA fragment containing the *loxP*–DsRed–*loxP* expression cassette was digested with restriction enzymes *PmlI* & *Eco*RV and purified from the vector backbone prior to its transfection into LMTK– cells containing the validated ACE chromosome. After expansion of the transfected cells, repeated rounds of cell sorting is performed to enrich the number of DsRed fluorescent cells. Subsequent single cell sorting yields isolated, DsRed$^+$ clones. [Parts of this figure reprinted with permission from (13)].

the Qiaex II Gel Extraction Kit according to the manufacturer's instructions (Qiagen, Inc.).

3. The digested fragment (approximately 2 μg of purified fragment) is transfected into the LMTK– cell line containing the generated ACE platform chromosome using Lipofectamine Plus transfection reagent according to the manufacturer's instructions (Invitrogen).
4. After transfection (48–72 h post-transfection), transfected cells are expanded and subjected to sequential rounds of flow sorting to enrich for DsRed stably transfected expressing cells (enrichment for greater than 50% of the cell population expressing the DsRed transgene). For flow cytometry, a cell sorter equipped with a 488-nm laser for excitation and a 585/42 band-pass filter will permit optimal detection of the DsRed fluorescence (Note 6).
5. At this point, DsRed expressing transfectants from the enriched population can be single cell sorted and expanded for analysis (Note 7).

6. FISH validates integration of the DsRed transgene on the ACE platform chromosome, with colocalization of probes directed to DsRed and the ACE chromosome (Fig. 3).
7. The ACE platform chromosome containing the *loxP* recombination site is now amenable to further user-specific, downstream applications (Note 8).

4. Conclusion

The ACE system represents a rational and tractable chromosome-vector based bioengineering system that can be readily purified from host cells by flow cytometry/chromosome sorting and introduced into recipient cell types by standard transfection methodology. The portability of the ACE chromosome vector is a particularly cogent feature placing it on a par with other vector-based gene expression systems. Additional salient features of the ACE chromosome technology include:

- ACE platform chromosomes can be generated and stably maintained without drug selection in a wide variety of mammalian cells including primary cells and stem cells.
- ACE platform chromosomes can be tractably engineered *via* user designed, site-specific recombination systems.
- ACE platform chromosomes circumvent gene delivery, packaging size limitations such that single genes, multiple genetic factors, or large genetic payloads (BAC/PAC genomic DNA vectors) can be expressed.
- ACE chromosomes are stable in numerous cell types including stem cells.
- ACE chromosomes are an effective bioengineering vector system for transgenic animal production.
- Importantly, engineering of ACE chromosomes circumvents insertional mutagenesis of the host chromosomes.

Taken together, these attributes suggest that the ACE chromosome technology provides a singular cytoreagent, bioengineering tool applicable to a variety of broad cell-based applications. Furthermore, bioengineered ACE platform chromosomes containing user-defined recombination sites represent an ideal circuit board to which an array of genetic factors can be plugged-in and expressed for various research and therapeutic applications. Toward the goal of using the ACE chromosome in a therapeutic venue, the ability to bioengineer large segments of genomic DNA such as fragments containing long-range genetic elements required for appropriate regulation of gene expression, developmentally regulated multi-gene loci, or multiple copies of two or more genes in fixed stoichiometry may potentially circumvent many of the current limitations associated with plasmid and viral-based gene expression systems.

5. Notes

1. ACE platform chromosome construction is not limited to the use of LMTK– cells. Other mammalian cells lines including human/hamster hybrid cell lines, other mouse cell lines, and human cell line have been utilized for ACE chromosome construction (17, 21 and Perkins et al., unpublished).
2. The use of FISH for the characterization of ACE chromosome structure is the basic cytogenetic technique that is critically important for validating and characterizing ACE platform chromosomes. Furthermore, during routine passaging of cell lines containing ACE platform chromosomes, cell aliquots are routinely tested for mitotic stability and presence of exogenous of ACE chromosome in the host cells. Readers are directed to (17, 27–29) for a description of this technique.
3. An example of using FISH to validate the generation of an ACE chromosome can be found in (13). In this example the generated ACE chromosome exhibits colocalization of digoxigenin (DIG)-labeled mouse major satellite repeat probe and biotinylated pFKPuro probe. The mouse major satellite probe hybridizes with the pericentric heterochromatin of all other chromosomes. The biotinylated pFKPuro probe strongly hybridizes to two regions on this ACE chromosome. For analysis, all chromosomes are counterstained with DAPI. Further FISH characterization of a flow-sorted ACE chromosome is described in (13).
4. Dual Hoechst (AT selectivity) and Chromomycin A3 (CG selectivity) staining of ACE chromosomes allows for identification and subsequent sorting of the ACE chromosomes. Host chromosomes that are relatively AT/GC balanced will show a 45° plane of size distributed, normal chromosomes in a bivariate flow histogram (Fig. 2). During ACE chromosome generation, the ACE chromosome will contain a high proportion of AT-rich heterochromatin and because of this high ratio of AT/GC DNA, the newly generated ACE chromosome will sit above the 45° plane of normal mammalian chromosomes in the bivariate flow histogram. This unique feature of ACE chromosomes permits a high production rate and purity sorted ACE chromosomes. Additional details surrounding

the flow cytometric analysis and sorting of ACE chromosomes is described in (23).

5. As described in Stewart et al. (13), the red fluorescent protein variant DsRed1-N1 was utilized for retrofitting the ACE chromosome. Since this description, new fluorescent protein variants of different colors have been produced which exhibit faster fluorescent protein maturation and lower cell toxicity and are ideal for ACE chromosome engineering (*e.g.* see (30) for a description of new fluorescent protein variants).
6. If a cell sorter is not available, ACE chromosomes can be retrofitted in a similar fashion utilizing drug selectable markers.
7. Single cell DsRed + transfectants can also be isolated *via* cloning by limiting dilution.
8. In addition to Cre/lox site-specific recombination systems, a number of site-specific recombinase systems are available include FLP and PhiC31 systems (reviewed in (31)). We have also described an ACE platform chromosome system incorporating an engineered variant of the bacterial lambda integrase amenable to multiple loading onto the platform chromosome (5).

Acknowledgments

This effort was supported by a grant from the Georgia Cancer Coalition (E.P.).

References

1. Basu, J. and Willard, H.F. (2006). Human artificial chromosomes: potential applications and clinical considerations. *Pediatr. Clin. North Am.* **53**, 843–853.
2. Duncan, A. and Hadlaczky, G. (2007). Chromosomal engineering. *Curr. Opin. Biotechnol.* **18**, 420–424.
3. Grimes, B.R. and Monaco, Z.L. (2005). Artificial and engineered chromosomes: developments and prospects for gene therapy. *Chromosoma* **114**, 230–241.
4. Irvine, D. V., Shaw, M. L., Choo, K. H. A., and Saffery, R. (2005). Engineering chromosomes for delivery of therapeutic genes *Trends Biotechnol* **23**, 575–583.
5. Lindenbaum, M., Perkins, E., Csonka, E., Fleming, E., Garcia, L., Greene, A., Gung, L., Hadlaczky, G., Lee, E., Leung, J., MacDonald, N., Maxwell, A., Mills, K., Monteith, D., Perez, C.F., Shellard, J., Stewart, S., Stodola, T., Vandenborre, D., Vanderbyl, S. and Ledebur, H.C. (2004). A mammalian artificial chromosome engineering system (ACE System) applicable to biopharmaceutical protein production, transgenesis and gene-based cell therapy. *Nucleic Acids Res.* **32**, e172–e172.
6. Co, D. O., Borowski, A. H., Leung, J. D., van der Kaa, J., Hengst, S., Platenburg, G. J., et al. (2000). Generation of transgenic mice and germline transmission of a mammalian artificial chromosome introduced into embryos by pronuclear microinjection. *Chrom. Res.* **8**, 183–191.

7. Katona, R. L., Cserpan, I., Fatyol, K., Csonka, E., and Hadlaczky, G. (2005). Transgenic mice, carrying an expressed anti-HIV ribozyme in their genome, show no sign of phenotypic alterations *Acta Biol Hung* **56**, 67–74.
8. Katona, R.L., Sinkó, I., Holló, G., Szucs, K.S., Praznovszky, T., Kereső, J., Csonka, E., Fodor, K., Cserpán, I., Szakál, B., Blazsó, P., Udvardy, A. and Hadlaczky, G. (2008). A combined artificial chromosome-stem cell therapy method in a model experiment aimed at the treatment of Krabbe's disease in the Twitcher mouse. *Cell. Mol. Life Sci.* **65**, 3830–3838.
9. Perkins, E., Perez, C., Lindenbaum, M., and Greene, A. (2006). Chromosome-based platforms. February 2, 2006. US Patent Application No. 200330119104. Chromos Molecular Systems, Vancouver, BC Canada.
10. Perkins, E., Perez, C., Lindenbaum, M., Greene, A., Leung, J., Fleming, E., and Stewart, S. (2003). Chromosome-based platforms. June 26, 2003. US Patent Application No. 20050181506. Chromos Molecular Systems, Vancouver B.C., Canada.
11. Perkins, E., Perez, C., Lindenbaum, M., Greene, A., Leung, J., Fleming, E., Stewart, S., and Shellard, J. (2005). Chromosome-based platform. August 18, 2005. US Patent Application No. 20060024820. Chromos Molecular Systems, Vancouver, BC, Canada.
12. de Jong, G., Telenius, A., Vanderbyl, S., Meitz, A., and Drayer, J. (2001). Efficient *in-vitro* transfer of a 60-Mb mammalian artificial chromosome into murine and hamster cells using cationic lipids and dendrimers. *Chromosome Res.* **9**, 475–485.
13. Stewart, S., MacDonald, N., Perkins, E., DeJong, G., Perez, C., and Lindenbaum, M. (2002). Retrofitting of a satellite repeat DNA-based murine artificial chromosome (ACes) to contain loxP recombination sites. *Gene Ther* **9**, 719–23.
14. Vanderbyl, S., MacDonald, G. N., Sidhu, S., Gung, L., Telenius, A., Perez, C., and Perkins, E. (2004). Transfer and stable transgene expression of a mammalian artificial chromosome into bone marrow-derived human mesenchymal stem cells. *Stem Cells.* **22**, 324–333.
15. Vanderbyl, S., MacDonald, N. and de Jong,G. (2001). A flow cytometry technique for measuring chromosome-mediated gene transfer. *Cytometry* **44**, 100–105.
16. Vanderbyl, S. L., Sullenbarger, B., White, N., Perez, C. F., MacDonald, G. N., Stodola, T., Bunnell, B. A., Ledebur, H. C., and Lasky, L. C. (2005). Transgene expression after stable transfer of a mammalian artificial chromosome into human hematopoietic cells. *Exp Hematol.* **33**, 1470–1476.
17. Csonka, E., Cserpán, I., Fodor, K., Holló, Gy., Katona, R., Kereső, J., Praznovszky, T., Szakál, B., Telenius, A., deJong, G., Udvardy, A., Hadlaczky, Gy. (2000). Novel Generation of Human Satellite DNA-based Artificial Chromosomes in Mammalian Cells. *J Cell Sci.* **113**, 3207–3216.
18. Hadlaczky, Gy. (2001). Satellite DNA-based artificial chromosomes for use in gene therapy. *Current Opinion in Molecular Therapeutics* **3**(2), 125–132.
19. Hadlaczky, Gy., Praznovszky, T., Cserpán, I., Kereső, J., Péterfy, M., Kelemen, I., Atalay, E., Szeles, A., Szelei, J., Tubak, V. (1991). Centromere formation in mouse cells cotransformed with human DNA and a dominant marker gene. *Proc. Natl. Acad. Sci. USA.* **88**, 8106–8110.
20. Holló, Gy., Kereső, J., Praznovszky, T., Cserpán, I., Fodor, K., Katona, R., Csonka, E., Fátyol, K., Szeles, A., Szalay, A.A., Hadlaczky, Gy. (1996). Evidence for a megareplicon covering megabases of centromeric chromosome segments. *Chromosome Res.* **4**, 240–247.
21. Kereső, J., Praznovszky, T., Cserpán, I., Fodor, K., Katona, R., Csonka, E., Fátyol, K., Holló, Gy., Szeles, A., Ross, A.R., Sumner, A.T., Szalay, A.A., Hadlaczky, Gy. (1996). *De novo* chromosome formations by large-scale amplification of the centromeric region of mouse chromosomes. *Chromosome Res.* **4**, 226–239.
22. Praznovszky, T. Kereső, J. Tubak, V., Cserpán, I., Fátyol, K., Hadlaczky, Gy. (1991). *De novo* chromosome formation in rodent cells. *Proc. Natl. Acad. Sci. USA.* **88**, 11042–11046.
23. dejong, G., Telenius, A. H., Telenius, H., Perez, C. F., Drayer, J. I., and Hadlaczky, G. (1999). Mammalian artificial chromosome pilot production facility: large-scale isolation of functional satellite DNA-based artificial chromosomes *Cytometry* **35**, 129–33.
24. Christmann, L., Eberhardt, D. M., Leavitt, M. C., and Harvey, A. J. (2006). AviGenics, Inc., USA. Artificial chromosomes and transchromosomic avians. USA. US Patent Application No. 20060174364.
25. Monteith, D.P., Leung, J.D., Borowski, A.H., Co, D.O., Praznovski, T., Jiric, F.R., Hadlaczky, Gy. and Perez, C.F. (2003). Pronuclear microinjection of purified artificial chromosomes for generation of transgenic mice: Pick-and-Inject Technique. In Mammalian Artificial Chromosomes: *Methods and Protocols.* Eds.: Sgaramella, V. and Eridani, S. *Methods in Mol. Biol.* **240**, 227–242.

26. van den Engh, G., Trask, B., Lansdorp, P., and Gray, J. (1988). *Cytometry* **9**, 266–70.
27. Pinkel, D., Gray, J. W., Trask, B., van den Engh, G., Fuscoe, J., and van Dekken, H. (1986). Flow karyotyping and sorting of human chromosomes. *Cold Spring Harb Symp Quant Biol* **51** Pt 1, 151–7.
28. Pinkel, D., Straume, T., and Gray, J. W. (1986). Cytogenetic analysis using quantitative, high-sensitivity, fluorescence hybridization. *Proc Natl Acad Sci USA* **83**, 2934–8.
29. Telenius, H., Szeles, A., Kereső, J., Csonka, E., Praznovszky, T., Imreh, S., Maxwell, A., Perez, C.F., Drayer, J.I., Hadlaczky, Gy. (1999). Stability of a functional murine satellite DNA-based artificial chromosome across mammalian species. *Chromosome Res.* **7**, 3–7.
30. Shaner, N. C., Campbell, R. E., Steinbach, P. A., Giepmans, B. N., Palmer, A. E., and Tsien, R. Y. (2004). Improved monomeric red, orange and yellow fluorescent proteins derived from *Discosoma* sp. red fluorescent protein. *Nat Biotechnol* **22**, 1567–72.
31. Raymond, C., and Soriano, P. (2007). High-efficiency FLP and PhiC31 site-specific recombination in mammalian cells. *PloS One* **2**, e162.

Chapter 10

Chromosome Engineering with Lambda-Integrase Mediated Recombination System: The ACE System

Tünde Praznovszky

Abstract

Mammalian satellite DNA-based artificial chromosomes (SATACs) are unique among the mammalian artificial chromosomes. These reproducibly generated de novo chromosomes are stably maintained in different species, readily purified from the host cell's chromosomes and can be introduced into a variety of recipient cells. An artificial chromosome expression system (ACE system) has been developed on these SATACs to extend them for chromosome engineering. This system includes a Platform ACE containing multiple acceptor sites, specially designed targeting vector (ATV), and an ACE-integrase expression vector (pCXLamIntROK). Gene of interest are cloned into targeting vector (ATV), and site-specific loading of genes onto Platform ACE is facilitated by ACE-integrase mediated recombination. ACE system is suitable for multiple or subsequent loading of useful genes onto the same chromosome vector. This chapter describes the detailed procedure of chromosome engineering using the ACE system.

Key words: Platform ACE, Site-specific integration, Mutant lambda-integrase, Targeting vector ATV

1. Introduction

Although genetic manipulation of mammalian cells and animals have been achieved successfully by plasmid and viral gene delivery, mammalian artificial chromosomes (MACs), as potential vectors offer significant improvement in gene transfer. MACs provide stable, nonintegrating introduction of large payloads of genetic information, and for this reason several groups generated artificial chromosomes using different approaches (1–12).

We developed a satellite DNA-based artificial chromosome (SATAC) (13–16). This chromosome had some basic advantages compared to previously published MACs. The generation of SATACs was reproducible in a variety of host cells, they were

Gyula Hadlaczky (ed.), *Mammalian Chromosome Engineering: Methods and Protocols*, Methods in Molecular Biology, vol. 738,
DOI 10.1007/978-1-61779-099-7_10, © Springer Science+Business Media, LLC 2011

composed of known DNA sequences, and could be purified close to 100% purity by high-speed flow cytometry. Several cell lines were established and also transgenic animals were produced using isolated and purified SATACs (17).

Disadvantage of SATAC-based gene delivery system was that de novo generation of SATACs was necessary for each individual application. To overcome this problem, a mammalian artificial chromosome engineering system (ACE system) was developed (18), based on lambda integrase catalyzed site/specific recombination.

De novo SATAC was generated with multiple lambda integrase specific recognition/acceptor sites (Platform ACE). A plasmid construct pCXLamIntROK (pACE Integrase) provided integrase expression for site-specific loading of exogenous DNA sequences into Platform ACE. For targeting, a vector (ATV) was constructed, containing a lambda integrase recombination site upstream of a promoterless selectable marker gene. In the course of cotransfection of Platform ACE-carrying cell lines with ATV and pACE Integrase, site-specific integration of the ATV molecules took place (see Note 8). The selectable marker gene of ATV acquired a promoter located on the Platform ACE, and cells carrying correctly targeted ACE became resistant to the selective drug (Fig. 1).

Selected resistant clones were analysed by PCR using a primer pair specific to sequences of Platform ACE and to the selectable marker gene (Fig. 2).

Conventional, two-color fluorescent in situ hybridization (19) analyses were carried out exclusively on PCR screened, resistant clones. The integrity of ACE and site-specific integration on Platform ACE was also demonstrated (Fig. 3). FISH on metaphase spreads reveals not only the targeted integrations, but allows the detection of the nonspecific integration of the ATV in the genome. In mouse cells, the efficiency of targeting was more than 90%, but in hamster cells it was usually below 50%.

By the protocol provided here site-specific loading of useful gene(s) onto Platform ACEs at 20–90% integration efficiency was achieved within a reasonably short period of time, i.e. in about 2 months. In addition, the copy number of the introduced gene of interest can be variable on the loaded Platform ACEs, allowing selection of cell lines with the desirable level of transgene expression. Considering that Platform ACE contains multiple acceptor sites, second round targeting of the already engineered ACE can be achieved using the same protocol, with an ATV, carrying a different promoterless selectable marker gene.

This lambda integrase based chromosome engineering system has already been used successfully to generate stable, high MAb expressing CHO cell lines (20, 21) and in a combined artificial chromosome-stem cell gene therapy model experiment (22).

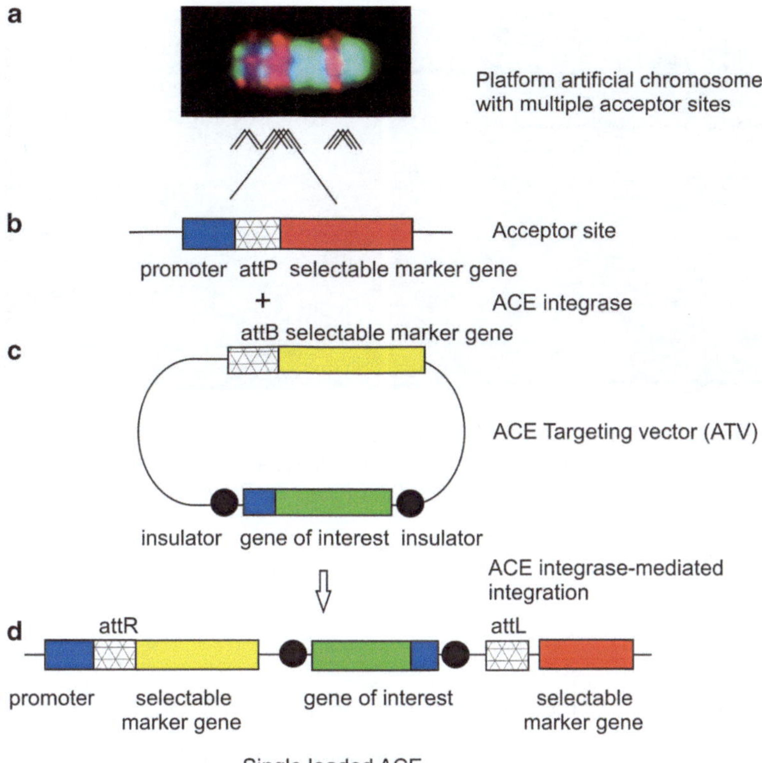

Fig. 1. *Site-specific loading of a transgene onto the Platform ACE.* (**a**) The satellite DNA-based artificial chromosome generated by large-scale amplification contains multiple integration sites. (**b**) The recombination acceptor sites for the ACE-integrase, attP is located between a selectable marker gene and its promoter. (**c**) The ACE targeting vector (ATV) carries a gene of interest bordered with insulator sequences, and the attB integrase specific site upstream a promoterless selectable marker gene. (**d**) Expression of ACE-integrase catalyzes the recombination between attP and attB sites resulting in the site-specific integration of ATV into Platform ACE. Thus, the promoterless selectable marker gene of ATV will be driven by the promoter on Platform ACE and drug resistance will be provided by the integrated marker gene of ATV.

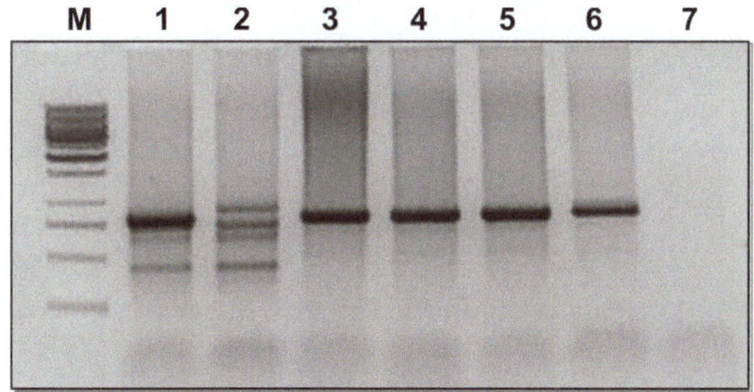

Fig. 2. Detection site-specific integration in targeted cell lines by PCR. PCR was carried out using 193AF as the forward primer specific to Platform ACE and a reverse primer specific to the selectable marker gene of ATV; the templates were genomic DNAs of different transformed clones. Site-specific targeting to Platform ACE was detected in the genome of clones represented in lanes 1, 3, 4, and 5. In lane 2, the PCR showed no site-specific integration of ATV. Lane 6 is the positive, and lane 7 is the negative control of PCR, respectively. The Marker is the 1 kB DNA ladder (Fermentas).

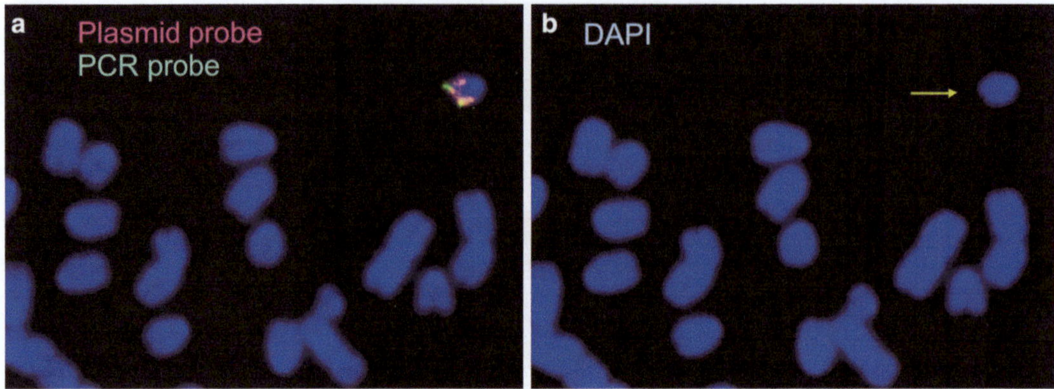

Fig. 3. Analysis of targeted cell line by two-color FISH. (**a**) DIG labeled (*red*) plasmid DNA probe was used to detect ACE, and biotin labeled transgene-specific PCR sequences (*green*) was the probe to show the integration of therapeutic gene into ACE. The chromosomes were counterstained with DAPI (*blue*). (**b**) The corresponding DAPI stained metaphase spread.

2. Materials

2.1. Culture of Cell Lines

1. CO_2 tissue culture incubator (37°C, 5% CO_2) (ShelLab).
2. Inverted light microscope Olympus CK30.
3. Bürker counting chamber (Roth).
4. MEM alpha medium (Gibco).
5. DMEM medium (Gibco).
6. FBS (Lonza).
7. D2 Trypsine (Serra).
8. Puromycin (Sigma), the stock solution 11 mg/ml is prepared in triple distilled water, aliquots are kept –20°C.
9. 6-Well TC Test plates (Orange).
10. Cell culture dishes (Greiner).

2.2. Purification and Measuring of Plasmid DNAs

1. EndoFree Plasmid Maxi Kit (Qiagen).
2. Nano Drop ND-1000 Spectrophotometer.
3. Sterile TE buffer (10 mM Tris–HCl and 1 mM EDTA at pH 7.5).
4. pCXLamIntROK (pACE Integrase), lambda integrase expression plasmid.
5. pATV targeting vector.
6. 1.5-ml test tubes (Eppendorf).

2.3. Cotransfection of Platform ACE Containing Cells

1. PBS (Oxoid).
2. SuperFect Transfection Reagent (Qiagen).
3. Biofuge Pico (Heraeus).

2.4. Selection of Transformants

1. 8-channel micropipette (Eppendorf).
2. 96-Well Cell Culture Cluster (Costar).
3. 24-Well Cell Culture Cluster (Costar).
4. 6-Well TC Test plates (Orange).
5. 50-ml PS Test tube sterile (Greiner).
6. CO_2 Incubator.

2.5. Analyses of Transformants

1. Wizard Genomic DNA purification Kit (Promega).
2. PTC-150 MiniCycler (MJ Research).
3. GoTaq Flexi DNA Polymerase (Promega).
4. 193AF primer (5′-ACCCCCTTGCGCTAATGCTCTGTTA).
5. 1 kB DNA Ladder (Fermentas).
6. Biotin-Nick Translation Mix (Roche).
7. DIG-Nick Translation Mix (Roche).
8. Fluorescence microscope, Olympus Vanox-S or similar.
9. Image analysis system, Quips XL Genetics Workstation system or similar.
10. High sensitivity CCD camera, Photometrics KAF 1400-G2 CCD or similar.

3. Methods

3.1. Culturing of Cell Lines

1. Culture LMTK cell line containing Platform ACE (B19-38) in monolayer culture in plastic culture dishes with DMEM, 10% FBS, streptomycin–penicillin, and supplemented with 5 μg/ml of puromycin. Feed the mouse cells every 3 days.
2. Culture CHO-DG44 derived cells containing the Platform ACE (Y19-13DSFS) in monolayer in MEM alpha medium with 5% FBS, streptomycin–penicillin, and 10 μg/ml puromycin. Feed the hamster cells every 3 days.

Seed the cells at a density of 3×10^5 cells per well of a 6-well culture dish, from both cell lines 1 day before transfection (see Note 1). Determine the cell concentration using Bürker counting chamber.

3.2. Purification and Measuring of Plasmid DNAs

1. Purify targeting vector (ATV) and pACE Integrase expression plasmid with EndoFree Plasmid Maxi Kit (see Note 2).
2. Measure the DNA concentration with spectrophotometer. On the day of transfection, dilute the plasmid DNAs in TE buffer (pH 7.5) to 0.2 mg/ml concentration.

3.3. Cotransfection of Platform ACE Containing Cells

1. Mix 1 μg of targeting vector (ATV) and 1 μg of pACE Integrase plasmid DNA in a sterile Eppendorf tube in cell growth medium without serum and antibiotics to a total volume of 100 μl (see Note 3). Spin down the tube for a few seconds.
2. Add 10 μl of SuperFect Transfection Reagent to the DNA solution and mix gently by pipetting up and down five times. Incubation of the mixture at room temperature for 10 min allows transfection–complex formation.
3. During this time remove the growth medium from the 60–80% confluent cells and wash the cells once with 2 ml of sterile, prewarmed PBS.
4. Add 600 μl of cell growth medium (containing serum) to the transfection complexes in the Eppendorf tube and mix by pipetting up and down twice. The mix should immediately be added to the cells.
5. Incubate the cells with the transfection complex for 2 h under normal growth conditions, 37°C and 5% CO_2. During this time, transient expression of lambda integrase is expected to result in targeting the ATV onto the Platform ACE.
6. After 2 h remove the medium containing the remaining complexes from the cells, and wash the cells three times with 2 ml of PBS (see Note 4).
7. Add fresh cell growth medium containing serum to the cells and incubate under normal growth conditions.

3.4. Selection of Transformants

1. After 24 h, trypsinize the transfected cells, resuspend them in 5 ml of growth medium in a 50-ml sterile tube, and count the cell number using the Bürker chamber.
2. Dilute the cells in growth medium to a density of 50 cells/μl and distribute them into 96-well dishes with the 8-channel pipette; 50 μl of cell suspension is added to each well (see Note 5).
3. 48 h after transfection add 150 μl of growth medium into each well supplemented with antibiotic to reach the final selection level of the drug.
4. When the resistant colonies reach a cell number of about 50–60, transfer them into individual wells of a 24-well tissue culture dish. When the cells nearly become confluent, harvest them using trypsin treatment and distribute each clonal suspension into two wells of a 6-well dish.

3.5. Analyses of Transformants

1. From one of the wells of the 6-well dish, purify genomic DNA of the resistant colonies using Wizard Genomic DNA Purification Kit (Promega).
2. In PCR analysis of targeting, use 193AF primer, specific to the Platform ACE and a targeting vector specific primer (Fig. 2) (see Note 6).

3. Freeze one plate of clones giving the site-specific PCR product and culture a twin plate for further analyses.
4. Perform conventional single and two-color FISH (16) on metaphase spreads of selected cell lines. Use plasmid DNA labeled with Biotin-Nick Translation Mix to analyse the presence and integrity of ACEs. Resistant clones that contain other integration sites and/or cytological abnormalities (e.g. double minutes) should be excluded (see Note 7). In two-color FISH experiments, use DIG-labeled plasmid DNA and biotinylated PCR DNA of the integrated gene, to prove the site-specific integration on ACE (Fig. 3).

4. Notes

1. In transfection experiments, use rapidly growing cells. Before transfection experiments, feed cells every day. The seeding number on the day before transfection depends on the cell type. The cells should be at 60–80% confluency on the day of transfection.
2. Plasmid DNAs purified by alkaline lysis method are also suitable for transfection of mouse and Chinese hamster cell lines.
3. Use freshly prepared plasmid DNA in transfection experiments. Using old plasmid DNAs the number of colonies may decrease dramatically.
4. Other transfection reagents like Lipofectamine 2000 (Invitrogen) and ExGen 500 (Fermentas) also proved to be effective in targeting experiments, for both mouse and hamster cell lines. In our hands, SuperFect transfection reagent gave the best results.
5. Do not plate more than 2,500 cells/well of a 96-well plate. Higher number of cells may result in multiple colonies in a single well.
6. In PCR experiments with GoTaq polymerase the $MgCl_2$ concentration should be optimized when a different reverse primer is used.
7. The transgene expression should be achieved from the engineered ACE. When other integration is detected by FISH, the source of transgene expression is uncertain, and therefore, these cell lines must be excluded from further experiments.
8. Successful targeting of more than 110 kB exogenous DNA sequences was carried out onto Platform ACE using the above lambda integrase specific recombination system (Praznovszky et al. unpublished).

References

1. Carine, K., Solus, J., Waltzer, E., Manch-Citron, J., Hamkalo, B., and Scheffler, I. (1986) Chinese hamster cells with a minichromosome containing the centromere region of human chromosome 1. *Somat. Cell Mol. Genet.* **12**, 479–491.
2. Harrington, J. J., Van Bokkelen, G., Mays, R. W., Gustashaw, K., and Willard, H. F. (1997) Formation of de novo centromeres and construction of first-generation human artificial microchromosomes. *Nat. Genet.* **15**, 345–355.
3. Warburton, P. E., and Cooke, H. J. (1997) Hamster chromosomes containing amplified human alpha-satellite DNA show delayed sister chromatid separation in the absence of de novo kinetochore formation. *Chromosoma* **106**, 145–159.
4. Ikeno, M., Grimes, B., Okazaki, T., Nakano, M., Saitoh, K., Hoshino, H., et al. (1998) Construction of YAC-based mammalian artificial chromosomes. *Nat. Biotech.* **16**, 431–439.
5. Henning, K. A., Novotny, E. A., Compton, S. T., Guan, X. Y., Liu, P. P., and Ashlock, M. A. (1999) Human artificial chromosome generated by modification of a yeast artificial chromosome containing both human alpha satellite and single-copy sequences. *Proc. Natl. Acad. Sci. USA* **96**, 592–597.
6. Ebersole, T. A., Ross, A., Clark, E., McGill, N., Schindelhauer, D., Cooke, H., and Grimes, B. (2000) Mammalian artificial chromosome formation from circular alphoid input DNA does not require telomere repeats. *Hum. Mol. Genet.* **9**, 1623–1631.
7. Farr, C., Fantes, J., Goodfellow, P., and Cooke, H. (1991) Functional reintroduction of human telomeres into mammalian cells. *Pro. Natl. Acad. Sci.* USA **88**, 7006–7010.
8. Farr, C. J., Stevanic, M., Thomson, E. J., Goodfellow, P. N., and Cooke, H. J. (1992) Telomere-associated chromosome fragmentation: applications in genome manipulation and analysis. *Nat. Genet.* **2**, 275–282.
9. Farr, C. J., Bayne, R. A., Kipling, D., Mills, WW., Critcher, R., and Cooke, H. J. (1995) Generation of a human X-derived minichromosomeusing telomere-associated chromosome fragmentation. *EMBO J.* **14**, 5444–5454.
10. Heller, R., Brown, K. E., Burgtorf, C., and Brown, W. R. (1996) Minichromosomes derived from the human Y chromosome by telomere directed chromosome breakage. *Proc. Natl. Acad. Sci. USA* **93**, 7125–7130.
11. Au, H. C., Mascarello, J. T., and Scheffler, I. E. (1999) Targeted integration of a dominant neoR marker into a 2- to 30-Mb human minichromosome and transfer between cells. *Cytogenet. Cell Genet.* **6**, 1375–1382.
12. Sun, T. Q., Fernstermacher, D. A., and Vos, J. M. (1994) Human artificial episomal chromosomes for cloning large DNA fragments in human cells. *Nat. Genet,* **8**, 33–41.
13. Kereső, J., Praznovszky, T., Cserpán, I., Fodor, K., Katona, R., Csonka, E., Fátyol, K., Holló, G., Szeles, A., and et al. (1996) *De novo* chromosome formations by large-scale amplification of the centromeric region of mouse chromosomes. *Chromosome Res.* **4**, 226–239.
14. Praznovszky, T., Kereső, J., Tubak, V., Cserpán, I., Fátyol, K., and Hadlaczky, G. (1991) *De novo* chromosome formation in rodent cells. *Proc. Natl. Acad. Sci. USA.* **88**, 11042–11046.
15. Csonka, E., Cserpán, I., Fodor, K., Hollo, G., Katona, R., Kereső, J., Praznovszky, T., Szakál, B., Telenius, A., de Jong, G., and et al. (2000) Novel generation of human satellite DNA-based artificial chromosomes in mammalian cells. *J. Cell. Sci.* **113**, 3207–3216.
16. Hadlaczky, G. (2001) Satellite DNA-based artificial chromosomes for use in gene therapy. *Curr. Opin. Mol. Ther.* **3**, 125–132.
17. Co, D. O., Borowski, A. H., Leung, J. D., van der Kaa, J., Hengst, S., Platenburg, G. J., et al. (2000) Generation of transgenic mice and germline transmission of a mammalian artificial chromosome introduced into embryos by pronuclear microinjection. *Chrom. Res.* **8**, 183–191.
18. Lindenbaum, M., Perkins, E., Csonka, E., Fleming, E., Garcia, L., Greene, A., Gung, L., Hadlaczky, G., Lee, E., Leung, J., MacDonald, N.,Maxwell, A., Mills, K., Monteith, D., Perez, C. F., Shellard, J., Stewart, S., Stodola, T., Vadenborre, D., Vanderbyl, S., and Ledebur,Jr. H. C. (2004) A mammalian artificial chromosome engineering system (ACE System) applicable to biopharmaceutical protein production, transgenesis and gene-based cell therapy. *Nucleic Acids Res.* **32**, e172.
19. Pinkel, D., Straume, T., and Gray, J. W. (1986) Cytogenetic analysis using quantitative, high-sensitivity, fluorescence hybridization. *Proc. Natl. Acad. Sci. USA* **83**, 2937–2938.
20. Kennard, M. L., Goosney, D. L., Monteith, D., Roe, S., Fischer, D., Mott, J. (2009) Auditioning of CHO host cell lines using the artificial chromosome expression (ACE) technology. *Biotechnol. Bioeng.* **104**, 526–539.

21. Kennard, M. L., Goosney, D. L., Monteith, D., Zhang, L., Moffat, M., Fischer, D., Mott, J. (2009) The generation of stable, high MAb expressioning CHO cell lines based on the artificial chromosome expression (ACE) technology. *Biotechnol. Bioeng.* **104**, 540–553.

22. Katona, R. L., Sinkó, I., Holló, G., Székely Szűcs, K., Praznovszky, T., Kereső, J., Csonka, E., Fodor, K., Cserpán, I., Szakál, B., Blazsó, P.,Udvardy, A., and Hadlaczky, G. (2008) A combined artificial chromosome-stem cell therapy method in a model experiment aimed at the treatment of Krabbe's disease in the Twitcher mouse. *Cell. Mol. Life Sci.* **65**, 3830–3838.

23. Vanderbyl, S., MacDonald, G. N., Sidhu, S., Gung, I., Telenius, A., Perez, C., and Perkins, E. (2004) Transfer and stable transgene expression of a mammalian artificial chromosome into bone marrow-derived human mesenchymal stem cells. *Stem Cells.* **22**, 324–333.

Chapter 11

Dendrimer Mediated Transfer of Engineered Chromosomes

Robert L. Katona

Abstract

Gene therapy encounters important problems such as insertional mutagenesis caused by the integration of viral vectors. These problems could be circumvented by the use of mammalian artificial chromosomes (MACs) that are unique and high capacity gene delivery tools. MACs were delivered into various target cell lines including stem cells by microcell-mediated chromosome transfer (MMCT), microinjection, and cationic lipid and dendrimer mediated transfers. MACs were also cleansed to more than 95% purity before transfer with an expensive technology. We present here a method by which MACs can be delivered into murine embryonic stem (ES) cells with a nonexpensive, less tedious, but still efficient way.

Key words: Mammalian artificial chromosomes, Dendrimer, Murine embryonic stem cell, Gene therapy

1. Introduction

MACs (mammalian artificial chromosomes) have an almost unlimited therapeutic transgene-carrying capacity and offer a stable, nonintegrating vector system without the drawbacks of other gene delivery systems, such as viruses, plasmid vectors, and bacterial and yeast artificial chromosomes (1). The Platform ACE System (2) was previously developed for the reliable and systematic engineering of MACs with large cDNAs or genomic sequences.

Embryonic and adult stem cells present opportunities for disease modeling, pharmacological screening, and cell-based therapies. Embryonic stem cells can generate all somatic cell types and adult stem cells can be differentiated into many somatic cell types, therefore they represent a potential and viable target for MAC-based gene therapy.

Gyula Hadlaczky (ed.), *Mammalian Chromosome Engineering: Methods and Protocols*, Methods in Molecular Biology, vol. 738, DOI 10.1007/978-1-61779-099-7_11,

The delivery of engineered MACs into mammalian cells is a fundamental challenge for gene therapy applications. Microcell-mediated chromosome transfer (MMCT) has been the most widely used technique for transferring MACs to various cell types (3–6). This method, however, is tedious and generally inefficient (10^{-7} to 10^{-5}). Gene-loaded 21ΔqHACs also have been transferred into human primary fibroblasts (7) and into human hematopoietic stem cells (HSCs) (8) at clinically relevant frequencies of 1.26×10^{-4} and 4.0×10^{-4}, respectively. Nevertheless, despite these increased transfer efficiencies, during the process of microcell formation the host cell generates a heterogeneous population of microcells encapsulating endogenous chromosomes as well as MACs. Consequently, this transfer technique increases the probability that host chromosomes and chromosomal fragments will be cotransferred with MACs to the clinical target cells, which may result in unknown and potentially deleterious effects to the patient.

While isolated MACs have been microinjected into the pronuclei of murine embryos to generate transgenic mice (9), very little success has been made in microinjecting mammalian cells in vitro due to the large outer diameter sizes (2.3–3.2 μm) of the microinjection needles (10). Alternatively, purified MACs have been transferred to various cells, including human mesenchymal stem cells (MSCs) (11) and HSCs (12) using commercially available cationic reagents and dendrimers (13) with high transfer efficiencies (10^{-2} to 10^{-4}). Recently, MACs loaded with a therapeutic transgene have been transferred to murine ESCs by lipid-mediated chromosome transfer, establishing stem cell clones carrying intact MACs, which was verified by fluorescent in situ hybridization. Subsequently, these stem cell clones were used to produce chimeric mice and demonstrated therapeutic effect of the transgene that was expressed by the MAC. This was the first demonstration of a new technology called combined artificial chromosome stem cell therapy (14).

2. Materials (see Note 1)

2.1. Culture of Artificial Chromosome Production Cell Line

1. Platform ACE-carrying Y2913D-SFS Chinese hamster ovary (CHO DG44) cell line was obtained from Chromos Molecular Systems Inc. It was cultured in MEM alpha (Gibco, 22571), 5% FCS, streptomycin-penicillin (Gibco, 15070-063), and 10 μg/ml of Puromycin (Sigma, P-7255).
2. Puromycin (Sigma, P-7255) was dissolved in sterile water in 10 mg/ml concentration and stored in aliquots at −20°C.
3. Geneticin (Gibco, 10131-027) was dissolved in sterile water in 50 mg/ml concentration, and stored at 4°C. For selection of stably transfected colonies, a 400 μg/ml final concentration was applied.

2.2. Mouse Embryonic Stem Cell Culture

1. Geneticin resistant mouse embryonic fibroblast cells were purchased from Millipore (PMEF-N).
2. KO-DMEM culture media (10829-018), Penicillin–streptomycin solution (15140148), and nonessential amino acids (11140050) were obtained from Gibco.
3. Embryomax serum (ES-009-B) and beta-mercaptoethanol (ES-007-E) were purchased from Chemicon.
4. Glutamax-I (35050061) was from Invitrogen.
5. ESGRO (Chemicon, ESG1107).
6. R1 ES cell line was kindly given to us by Andras Nagy (15).
7. Tryple-Express (GIBCO, 12605).

2.3. Chromosome Isolation

1. Recipe of GH buffer: 100 mM Glycine and 1% Hexylene–glycol; adjust pH to 8.4–8.6 with [$Ca(OH)_2$].
2. Colchicine (BDH).

2.4. Chromosome Delivery into ES Cells

1. Superfect reagent (301305) was purchased from Qiagen.

2.5. Selecting, Picking, and Establishment of Drug Resistant ES Cell Lines

1. 96-well Round bottom plate (Costar, 3799).

2.6. Karyotype Analysis of ES Cell Lines

1. Karyomax (Invitrogen).
2. Wright's stain (powder) (Sigma, W3000).
3. Trypsin (powder) (Difco Laboratories, 215240).
4. Diaton-CT (Diagon, 15101).
5. pHydrion buffer capsule (Micro Essential Laboratory, 270-7.00).
6. Stock Wright's stain: 0.3 g of powdered Wright's stain is dissolved in 100 ml of methanol, stir with a stirring rod on stir plate for 15 min, allow the solution to remain for 15 min and pour through a filter paper into a brown bottle (light sensitive); store at 4°C.
7. Stock buffer: one pHydrion capsule is dissolved in 100 ml of water. Store at 4°C.
8. Working buffer for Wright's stain: 5 ml of stock buffer in 95 ml of water.

2.7. Examination of Pluripotency in Established ES Cell Lines Carrying Artificial Chromosomes

1. StemTAG™ Alkaline Phosphatase Staining Kit (Cell Biolabs Inc., CBA-300).
2. Human embryonic stem cell marker antibody panel plus (R&D Systems, SC009).
3. Colonies were photographed under a Zeiss Axiovision Z1 fluorescent microscope at 4× and 10× objective magnifications.

3. Methods

3.1. Culture of Artificial Chromosome Carrying Cell Line

1. The Platform ACE-carrying Y2913D-SFS cells were thawed from a stock that was previously frozen down from a confluent 10-cm tissue culture dish and seeded into a new 10-cm tissue culture dish.
2. They were cultured until they reached confluency and then the culture medium was removed by aspiration, and the cells were washed with 5 ml of PBS.
3. PBS was aspirated and 2 ml of Tryple-Express was applied onto the cells. The dish was transfered into a CO_2 cell culture incubator for 5 min.
4. Cells were detached from the bottom of the dish by pipetting up and down with a 5-ml glass pipet. When cells were appropriately detached, they were transfered into a 50-ml centrifugation tube containing 5 ml of MEM alpha with supplements.
5. Cells were collected by centrifugation at $165 \times g$ for 5 min.
6. Supernatant was discarded and cells were resuspended in 40 ml of MEM alpha with supplements and they were dispensed in 10-ml aliquots into four 10-cm tissue culture dishes.

3.2. Culture of ES Cells

1. Thaw inactivated feeder cells, and plate on gelatin treated dishes or flasks at a density of 1.25×10^6 cells/25 cm^2 (see Notes 2 and 3).
2. Thaw ES cells at 37°C, wash once by pelleting, resuspend in complete ES medium, and plate on the feeder cells at a density of $1–1.5 \times 10^6$ ES cells/25 cm^2 (day 0).
3. Change the growth medium (complete ES medium supplemented with 1,000 U/ml LIF) every day. Split the cells on day 3 using the following harvesting procedure: remove the growth medium; rinse the cells with PBS without Ca^{2+} or Mg^{2+} (5–6 ml/25 cm^2); add Tryple-Express (0.5 ml/25 cm^2), and incubate the cells at 37°C for 4–5 min; add 5 ml of complete ES medium (25 cm^2), and pipet the cells up and down with a transfer pipet until they are a single-cell suspension without any significant clumps of cells (>8 cells).
4. Count the cells and dilute in complete ES medium for passage at a density of $1–1.5 \times 10^6$/ml and plate on feeder cells.

3.3. Isolation of Mitotic Chromosomes

1. Block cells in mitosis with 5 μg/ml of Colchicine for 6 h. Collect mitotic cells and count them.
2. Spin down cells at $950 \times g$, for 10 min at 4°C.

3. Wash cells in 20 ml of GH buffer. Fix a small amount of (methanol:acetic acid, 3:1) cells for the determination of mitotic index.
4. Spin down cells at 950 × *g*, for 10 min at 4°C.
5. Resuspend cells in 10 ml of GH buffer.
6. Incubate cells at 37°C for 10 min in water bath.
7. Incubate cells on ice for 5 min.
8. Add 100 μl of 10% Triton X-100 to cells (Final concentration is 0.1% Tx).
9. Incubate them on ice for 5 min.
10. Press cells through a 23-G needle three times (Check lysis of cells and release of chromosomes by microscopy).
11. Dilute cells to 20–50 ml volume with GHTx (GH + 0.1% Tx).
12. Spin down debris at 165 × *g*, for 10 min at 4°C.
13. Collect supernatant into a new tube and spin down chromosomes at 650 × *g*, for 20 min at 4°C.
14. At this point you can wash chromosomes with GH buffer or with the buffer you need for your further experiments (like culture media or PBS) at 650 × *g*, for 20 min at 4°C.

3.4. Chromosome Delivery into ES Cells

1. Collect about 30 million Rl ES cells in 7 ml of ES medium.
2. Collect about 40 million chromosomes from the cell line containing the mouse artificial chromosome and resuspend in 600 μl of KO-DMEM only.
3. Add 220 μl of Superfect reagent to the chromosomes and mix. Allow the chromosomes to remain with the Superfect reagent for 10 min at room temperature.
4. Add this mixture to the Rl ES cells and plate them into a 10-cm bacterial dish onto which Rl ES cells cannot attach.
5. Transfection is done for 4 h (see Note 4).
6. Collect cells by centrifugation at 165 × *g* for 5 min, discard supernatant, and resuspend cells in 40 ml of ES medium and then plate into four 10-cm dishes with feeder cells (see Note 5).
7. Selection with Geneticin (400 μg/ml) is started 24 h later.

3.5. Selecting, Picking, and Establishment of Drug Resistant ES Cell Lines

1. Selection with Geneticin (400 μg/ml) is started 24 h after transfection of mouse embryonic stem (mES) cells take place.
2. Refresh the medium containing Geneticin every day until the majority of cells are dead, then refresh every other day.

3. Five to seven days after the selection has started the ES colonies should become visible and they should be picked between day 8 and day 12 in most cases. The size and quality of the colonies are the critical things here, not the day. Pick them when the size is large enough and the colonies look optimal.
4. Fill the wells of a 96-well round bottom plate with 25 μl of Tryple-Express per well and have a flat bottom 96-well plate seeded with feeder cells and 200 μl of ES cell medium plus selective drugs per well ready.
5. Pick ES cell colonies by using a P20 pipette set at 3 μl. Picking is done under the microscope using the 4× objective. Put the pipette tip close to the ES cell colony and scrape the colony off the bottom of the dish. Suck the scraped colony up into the pipet tip and transfer it to a well containing Tryple-Express.
6. When you pick your first colony set a timer for 20 min and check it every time you get to the end of a row – if there is not enough time to do another full row, proceed to the next step (see Note 6).
7. Incubate cells for 4 min at 37°C, then pipet the cells up and down the wells with a multichannel pipette to make a single cell suspension (foam is acceptable.)
8. Transfer the cells to the 96-well plate containing the feeders. Take up some medium from the 96-well plate containing feeders and use it to wash the wells of the U-bottom plate and transfer the cells to the 96-well plate containing feeders. Keep the selective medium on the clones until the next passage.
9. When a clone reaches 70–80% confluency and the medium starts to turn orange-yellow, it is ready to be split. Most clones need to be split in 2–5 days.
10. To split, wash the well once with PBS, add 25 μl of Tryple-Express and allow the plate to remain for 3–5 min in the incubator.
11. Pipet up and down ES colonies repeatedly to break up clumps, then split the cells evenly over two wells (wells on adjacent rows – not the same row) in a 96-well plate containing feeders and 200 μl of ES cell medium (no selection) per well. At this step it is best to trypsinize the wells with similar densities together so that at the next step an entire row will reach confluence at once rather than scattered wells. This is also the step at which you should begin numbering your clones for future reference.

3.6. Karyotype Analysis of ES Cell Lines

1. The following protocol was developed for one well of a 24-well plate.
2. Grow mES cells on feeders to 60–70% confluency and cells require fresh media change the night prior to harvest.

3. Block cells in mitosis by adding 20 μl/ml of colcemid. Cells are incubated with colcemid for 4–5 h.
4. After colcemid exposure time, supernatant is removed and reserved by using a pipet.
5. mES culture is rinsed with 500 μl of PBS. After rinsing, the supernatant is removed and added to the previously reserved cells.
6. To lift cells from the dish, 500 μl of Tryple-Express is added. The cells are gently transferred to reserved supernatant from prior rinses.
7. Rinse the dish with 500 μl of PBS to retrieve the remaining cells. The rinse is added to reserved supernatant and cells.
8. Collect cells by centrifugation at 165 × *g* for 5 min at room temperature.
9. Carefully remove the supernatant. "Flick" the tube to disturb the cell pellet prior to adding 1 ml of the hypotonic solution (0.075 M KCl). Allow the cells to remain in this solution for 8–9 min at room temperature. This solution will lyse the cells and release nuclei.
10. Add one to two drops of fixative solution (3:1, methanol:acetic acid) to the hypotonic solution and cells prior to centrifugation (165 × *g* for 6 min) (see Note 7).
11. Remove supernatant, disturb the cell pellet, and add 1–5 ml fixative (depending on total cell number). This first fix should remain for 15 min.
12. Centrifuge again at 165 × *g* for 6 min. Remove supernatant, disturb the pellet and resuspend with another round of fixative solution. The second fix should remain for 10 min (see Note 8).
13. Final suspension for making slides should be cloudy, but not milky (see Note 9).
14. Drop cells on to slides while standing at a sink of hot running water (see Note 10).
15. Place the freshly prepared slide on to a slightly warm hot plate to help evaporate the fixative and draw the chromosomes apart.
16. G-Banding set-up – use for Coplin jars as follows:
    (a) 1st jar – 0.12 mg of Trypsin dissolved in 50 ml of Diaton-CT.
    (b) 2nd jar – 10 ml of stock Wright's stain in 40 ml of working buffer.
    (c) 3rd jar – distilled water for rinsing.
    (d) 4th jar – distilled water for rinsing.

17. Banding technique: allow slides to air dry at least 24 h prior to staining for best results (see Note 11). To stain the slide, place the slide in the 1st Coplin jar of Trypsin solution for 10–12 s. Immediately move slide from the Trypsin solution to the 2nd Coplin jar of Wright's stain solution for 2 min. After the staining time lapses, move the slide through two rinses in Coplin jars three and four. Allow the slide to stand on end or place in a slide rack to dry.
18. Examine unmounted slides with a 100× oil immersion lens. Take photographs and analyze metaphase spreads with a suitable karyotyping software (Fig. 1).

3.7. Examination of Pluripotency in Established ES Cell Lines Carrying Artificial Chromosomes

1. ES colonies were stained for the presence of alkaline phosphatase activity by the StemTAG™ Alkaline Phosphatase Staining Kit as suggested by the manufacturers. After staining, colonies were photographed under bright field microscope at 4× and 10× objective magnifications.
2. ES colonies were stained with the Human embryonic stem cell marker antibody panel plus as suggested by the manufacturers. After staining, colonies were photographed under a Zeiss Axiovision Z1 fluorescent microscope at 4× and 10× objective magnifications.

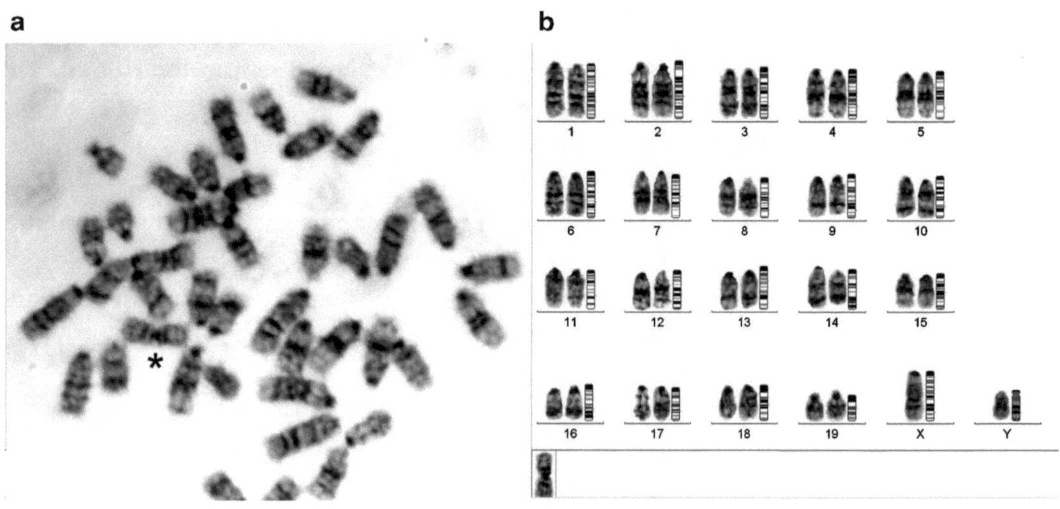

Fig. 1. Dendrimer-mediated transfer of an MAC into mouse ES cell. Cells of mouse ES cell line carrying MAC (14) were blocked in the metaphase stage of cell cycle and fixed for karyotype analysis as described. A G-banded metaphase spread of a cell from this ES cell line (**a**) was analyzed by Metasystems Ikaros software on a Zeiss Axiovision Z1 microscope and the full mouse karyotype was assembled in order (**b**). This ES cell contained an intact diploid set of mouse chromosomes and the MAC labeled with (*) in the Figure.

4. Notes

1. Unless stated otherwise, all solutions should be prepared in sterile water.
2. The dishes or flasks of feeder cells may be used as early as 6 h later, but are generally used the next day.
3. Feeder cells can also be used for up to 14 days later, depending on how intact the feeder cell layer appears.
4. When you transfer MACs into ES cells with Superfect reagent, do not perform transfection less than 4 h and more than 12 h. Less than 4 h would result in no transfection at all and more than 12 h will damage the cells and then again, you will not get transfected ES cells.
5. After the transfection experiment with MACs is finished, do not agitate the ES cells too much. Handle them very gentle, because you can detach transfection reagent complexed with MACs from cell membranes and that blocks the further progression of the transfection process.
6. The maximum time the ES cells should remain in Tryple-Express before incubation is 20 min. Do not keep them for longer periods of time in Tryple Express, because viability and stability of the cells will suffer.
7. Addition of one to two drops of fixative solution (3:1, methanol:acetic acid) to hypotonic solution and cells prior to centrifugation will prevent clumping of the cells.
8. When fixing cells for a karyotype analysis, a third round of fixative will usually provide better morphology, but is not necessary.
9. Final suspension for making slides should be cloudy, but not milky. This will just take practice and judgment. If too thick, metaphases will not spread well. If too thin, you will spend all day looking for metaphases on your slide!
10. You really have to drop cells on to slides while standing at a sink of hot running water. This can help tremendously, because humidity is a major factor to get well spread out chromosomes.
11. Before G-banding, slides should be "aged" for 3–21 days by leaving them in a closed box at room temperature. Fresh slides give poor G-band resolution. Maximum G-band resolution is achieved at ~10 days after slide preparation.

References

1. Hadlaczky G. Satellite DNA-based artificial chromosomes for use in gene therapy. Curr Opin Mol Ther. 2001;3:125–32.
2. Lindenbaum M, Perkins E, Csonka E, Fleming E, Garcia L, Greene A, et al. A mammalian artificial chromosome engineering system (ACE System) applicable to biopharmaceutical protein production, transgenesis and gene-based cell therapy. Nucleic Acids Res. 2004;32:e172.
3. Schor SL, Johnson RT, Mullinger AM. Perturbation of mammalian cell division. II. Studies on the isolation and characterization of human mini segregant cells. J Cell Sci. 1975;19:281–303.
4. Fournier RE, Ruddle FH. Microcell-mediated transfer of murine chromosomes into mouse, Chinese hamster, and human somatic cells. Proc Natl Acad Sci USA. 1977;74:319–23.
5. Meaburn KJ, Parris CN, Bridger JM. The manipulation of chromosomes by mankind: the uses of microcell-mediated chromosome transfer. Chromosoma. 2005;114:263–74.
6. Doherty AMO, Fisher EMC. Microcell-mediated chromosome transfer (MMCT): small cells with huge potential. Mamm Genome. 2003;14:583–92.
7. Kakeda M, Hiratsuka M, Nagata K, Kuroiwa Y, Kakitani M, Katoh M, et al. Human artificial chromosome (HAC) vector provides long-term therapeutic transgene expression in normal human primary fibroblasts. Gene Ther. 2005;12:852–6.
8. Yamada H, Kunisato A, Kawahara M, Tahimic CG, Ren X, Ueda H, et al. Exogenous gene expression and growth regulation of hematopoietic cells via a novel human artificial chromosome. J Hum Genet. 2006;51:147–50.
9. Co DO, Borowski AH, Leung JD, van der Kaa J, Hengst S, Platenburg GJ, et al. Generation of transgenic mice and germline transmission of a mammalian artificial chromosome introduced into embryos by pronuclear microinjection. Chromosome Res. 2000;8:183–91.
10. Monteith DP, Leung JD, Borowski AH, Co DO, Praznovszky T, Jirik FR, et al. Pronuclear microinjection of purified artificial chromosomes for generation of transgenic mice: pick-and-inject technique. Methods Mol Biol. 2004;240:227–42.
11. Vanderbyl S, MacDonald GN, Sidhu S, Gung L, Telenius A, Perez C, et al. Transfer and stable transgene expression of a mammalian artificial chromosome into bone marrow-derived human mesenchymal stem cells. Stem Cells. 2004;22:324–33.
12. Vanderbyl SL, Sullenbarger B, White N, Perez CF, MacDonald GN, Stodola T, et al. Transgene expression after stable transfer of a mammalian artificial chromosome into human hematopoietic cells. Exp Hematol. 2005;33:1470–6.
13. de Jong G, Telenius A, Vanderbyl S, Meitz A, Drayer J. Efficient in-vitro transfer of a 60-Mb mammalian artificial chromosome into murine and hamster cells using cationic lipids and dendrimers. Chromosome Res. 2001;9:475–85.
14. Katona RL, Sinkó I, Holló G, Szucs KS, Praznovszky T, Kereső J, et al. A combined artificial chromosome-stem cell therapy method in a model experiment aimed at the treatment of Krabbe's disease in the Twitcher mouse. Cell Mol Life Sci. 2008;65:3830–8.
15. Nagy A, Rossant J, Nagy R, Abramow-Newerly W, Roder JC. Derivation of completely cell culture-derived mice from early-passage embryonic stem cells. Proc Natl Acad Sci USA. 1993;90:8424–8.

Chapter 12

Engineered Chromosomes in Transgenics

Peter Blazso, Ildiko Sinko, and Robert L. Katona

Abstract

Horizontal gene transfer or simply transgenic technology has evolved much since 1980. Gene delivery strategies, systems, and equipments have become more and more precise and efficient. It has also been shown that even chromosomes can be used besides traditional plasmid and viral vectors for zygote or embryonic stem cell transformation. Artificial chromosomes and their loadable variants have brought their advantages over traditional genetic information carriers into the field of transgenesis. Engineered chromosomes are appealing vectors for gene transfer since they have large transgene carrying capacity, they are non-integrating, and stably expressing in eukaryotic cells. Embryonic stem cell lines can be established that carry engineered chromosomes and ultimately used in transgenic mouse chimera creation. The demonstrated protocol describes all the steps necessary for the successful production of transgenic mouse chimeras with engineered chromosome bearer embryonic stem cells.

Key words: Transgenic mouse, Transgenesis, Blastocyst injection, Embryonic stem cell, Chimera, ACE system, Chromosome engineering, Artificial chromosome

1. Introduction

Transgenic technology and molecular cloning have gone through almost 40 years of development since Cohen and his colleagues managed to transfer a *Salmonella typhimurium* gene into an *Escherichia coli* in 1973 (1). Horizontal gene transfer has become possible and widely applied in multicellular eukaryotes as well to help us understand developmental and other general physiological processes better. The first successful mouse transgenesis by Gordon et al. in 1980 opened the way to a deeper knowledge of human diseases and laid the basis for gene and cellular therapies (2). Vectors and gene delivery methods have also evolved and nowadays even chromosomes can be used besides DNA fragments and plasmids as genetic information carriers.

Gyula Hadlaczky (ed.), *Mammalian Chromosome Engineering: Methods and Protocols*, Methods in Molecular Biology, vol. 738,
DOI 10.1007/978-1-61779-099-7_12, © Springer Science+Business Media, LLC 2011

Many advantages make chromosomes the optimal vectors in transgenesis. There are "built-in" biochemical mechanisms in eukaryotic cells that precisely maintain the optimal structure and function of chromosomes – such as DNA replication and repair, regulated expression of genes, etc. – throughout the lifespan of these cells. Moreover, simply the presence of a supernumerary chromosome in the nucleus does not necessarily affect the life of the cell or the whole organism significantly or at all (3, 4).

During the past 20 years many strategies have been worked out to create or engineer chromosomes artificially that can be used for gene transfer (5, 6). To date, loadable variants of artificial chromosomes have also been generated (7). Two strategies have been developed further into ready-to-use, loadable expression systems such as the ΔHAC (Human Artificial Chromosome) and ACE (Artificial Chromosome Engineering) systems (8, 9).

ACE system is the successor of the satellite DNA-based artificial chromosome (SATAC) technology. SATAC and ACE technologies have already been tested and proven to perform well in various applications. The most notable advantages of these technologies over classical (viral or plasmid) methods have been demonstrated in the following areas: industrial scale protein synthesis, transgenesis of animals, and gene-based cellular therapy (9–13). Furthermore, it is also shown that ACE system has large carrying capacity. It is a non-integrating, stable and safe vector next to traditional gene carriers.

In order to generate transgenic mice, chromosome vectors need to get into zygotes or cells of early embryos. Several techniques exist to achieve this goal. In one of them, artificial chromosome containing microcells fused with embryonic stem (ES) cells yield the transformed pluripotent cells necessary for transgenesis (14, 15). In other approaches, the adaptation of traditional pronuclear microinjection ("pick-and-inject" technique) (16) or lipid mediated transfection of ES cells are found to be successful (13). Once the transformation gives rise to genetically modified zygotes or ES cells, they can be used to produce transgenic founders. Transformed zygotes are directly injected into the oviducts of surrogate mothers to generate transgenic pups. In case of ES cells, all developing bodies will be chimeric because their cells originate from more than one embryo: the recipient embryo and the transformed ES cells. Injection of embryonic stem cells into the cavity of blastocyst is the most commonly applied method for chimera generation (17, 18). Since ES cells are able to integrate into eight-cell embryos, another procedure has also been developed based on this phenomenon. In this approach, the eight-cell embryo or morula and transformed ES cells are aggregated after the enzymatic breakdown of zona pellucida (17, 19, 20). Though this technique does not need expensive tools or special skills such as capillary fabrication and handling, it still has drawbacks. Aggregation chimeras are very sticky, extremely sensitive to environmental changes and aggregated ES cells can not be selected previously by morphological criteria (17). The combination of these methods together with a laser-assisted zona pellucida puncture resulted in a very efficient third procedure. In this case when selected ES cells are injected under the zona pellucida of a laser-pierced eight-cell-stage embryo, the developing mouse can completely be derived from the injected ES cells without chimerism (21, 22). ACE carrier ES cells are injected with a specialized capillary either into eight-cell embryos or into blastocysts in the traditional way to create chimeras. Finally, these transgenic chimeras are implanted into the uterine horns of pseudopregnant foster mothers for further development.

In this chapter, transgenic chimera production by conventional blastocyst injection with ACE transformed ES cells is discussed in detail. This protocol presents how to

1. Mate embryo donor and recipient females
2. Fabricate and position proper capillaries that are used either for holding or injecting blastocysts
3. Flush, collect, and select healthy blastocysts
4. Prepare the injection-ready suspension of ACE bearer ES cells
5. Inject blastocysts and
6. Transfer chimeric embryos into recipient mothers

2. Materials

2.1. Mating Donor and Recipient Mice

1. Eight- to ten-week-old C57BL/6 blastocyst donor females (see Note 1).
2. Two- to ten-month-old C57BL/6 males.
3. Six- to eight-week-old CD1 foster mothers (see Note 2).
4. Two- to twelve-month-old, vasectomized CD1 or other outbred male (see Note 3).
5. Conventional animal house with stabilized 12 h–12 h dark–light cycle, relatively constant air temperature (between 20 and 25°C), and relative humidity.

2.2. Capillary Making

1. Modeling clay or plasticine.
2. Clean, glass Petri dishes.
3. Borosilicate glass capillary with no internal filament (e.g., GC120T-15 from Warner Instruments [formerly Clark Electromedical Instruments] or TW100-6 from World Precision Instruments, Inc.). It is possible to buy holding and

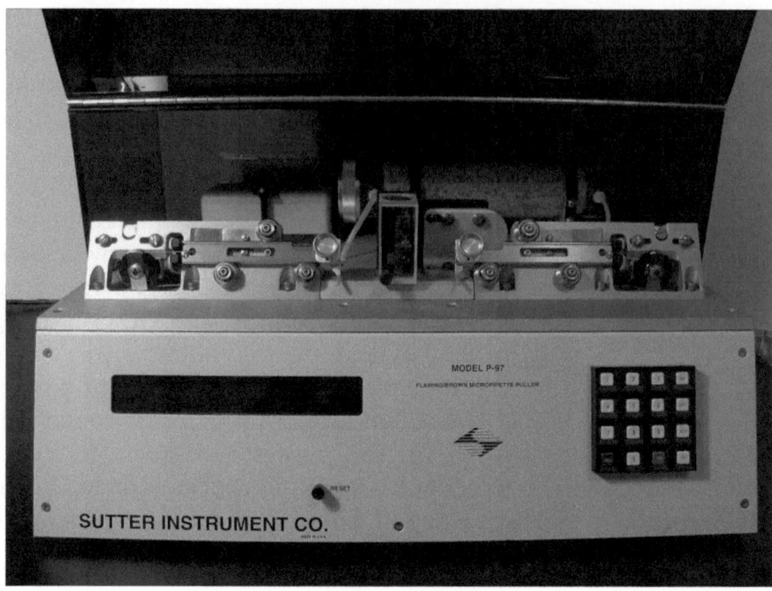

Fig. 1. Sutter P-97 Flaming/Brown Micropipette Puller.

injection capillaries from vendors (e.g., Holding and Stem Cell Micropipets from Humagen Fertility Diagnostics, TransferTipsES, Eppendorf VacuTip).

4. Sutter P-97 Flaming/Brown Micropipette Puller (see Fig. 1) with suitable 3 mm through filament (e.g., Science Bioproducts FT330B).
5. Scalpel with disposable blade.
6. A 3–5 cm long piece of silicon or rubber tube, ~2 cm Ø cut in half through its longitude axis like a half-pipe or tunnel.
7. Stereo dissecting microscope with adjustable zoom (10× to 45×), upper and scattered lower white light (e.g., Leica ZOOM 2000, see Fig. 2).
8. Bunsen burner.
9. Capillary microforge (e.g., World Precision Instruments MF200 complete microforge system including H602 Stereo Microscope).
10. Diamond pen.

2.3. Blastocyst Collection

1. KSOM medium (Millipore). Store in 3–4 ml sterile aliquots at 2–8°C.
2. M2 medium (Sigma-Aldrich). Store in 3–4 ml sterile aliquots at 2–8°C.
3. Embryo tested, sterile filtered mineral oil (Sigma-Aldrich).
4. 70% ethanol.

Fig. 2. Leica ZOOM 2000 stereo microscope.

5. Paper towels.
6. 60 mm and 35 mm Ø sterile, untreated, plastic cell-culture dishes.
7. CO_2 incubator with 37°C temperature and 5% CO_2 content.
8. Stereo dissecting microscope (see above).
9. Two small tweezers and two small scissors.
10. Fine, high precision, biological forceps (e.g., Dumont Dumostar 5, cat. #10788) or watchmaker's forceps.
11. 1 ml, Luer-slip, concentric, all plastic syringe (e.g., National Scientific, cat. #S7510-1).
12. Sterile G28 hypodermic needles blunted with sand paper.
13. Mouth pipette device (e.g., Sigma-Aldrich A5177) with transfer capillaries (see Subheading 3.2.1).

2.4. ES Cell Preparation

1. ES cells: ACE transformed R1 cells (23) grown on 60-mm gelatin-coated tissue culture dish without feeder cells.
2. ES cell medium (the same used for ES cell culturing), store at 4°C.

3. M2 (Sigma-Aldrich) or ES cell injection medium: 200 μl of 1 M HEPES in 10 ml of ES cell culture medium, store it for a maximum of 1 week at 4°C.
4. A box of fresh ice.
5. Phosphate-buffered saline (PBS) without Ca^{2+} or Mg^{2+}.
6. Trypsin, 0.25% in PBS with 0.2% EDTA or TrypLE™ Express (Invitrogen).
7. CO_2 incubator with 37°C temperature and 5% CO_2 content.
8. 37°C water bath.
9. Laminar flow hood for cell-culture laboratory.
10. Pipettes.
11. 12-ml sterile centrifuge tubes.
12. Centrifuge for 12-ml tubes.

2.5. Blastocyst Injection

1. M2 medium (see also Subheading 2.3).
2. ES cell injection medium (see also Subheading 2.3).
3. ES cell suspension in M2 or ES cell injection medium (see also Subheading 3.4).
4. Isolated blastocysts (see also Subheading 3.3).
5. Mouth pipette device (e.g., Sigma-Aldrich A5177) with transfer capillaries (see Subheading 3.2.1).
6. 6-cm Petri (cell culture) dishes or glass depression slides. Clean the slides with ethanol and let them dry before use.
7. Holding and injection pipettes (see Subheading 3.2.3).
8. Stereo dissecting microscope (see above).
9. Inverted light microscope with phase contrast optics (see also Note 4), 10×, 20× objectives, and a 10× eyepiece (e.g., Olympus IMT-2, see Fig. 3).
10. Two mechanical micromanipulators (e.g., Narishige or Leitz) fixed stably on each side of the microscope.
11. Two attachable, precision air pressure controllers (or injectors) (e.g., Narishige IM-9A) with 2 mm Ø plastic connector tubes and metallic capillary holders (see Fig. 4).

2.6. Embryo Transfer

1. M2 medium (Sigma-Aldrich). Store in 3–4 ml sterile aliquots at 2–8°C.
2. 70% ethanol.
3. Nembutal® (pentobarbital). Store it tightly sealed in dark at 2–8°C. Avoid freezing. Handle with care because it is sedative and toxic. You may need permission to use this drug. Check the related regulations in your country.

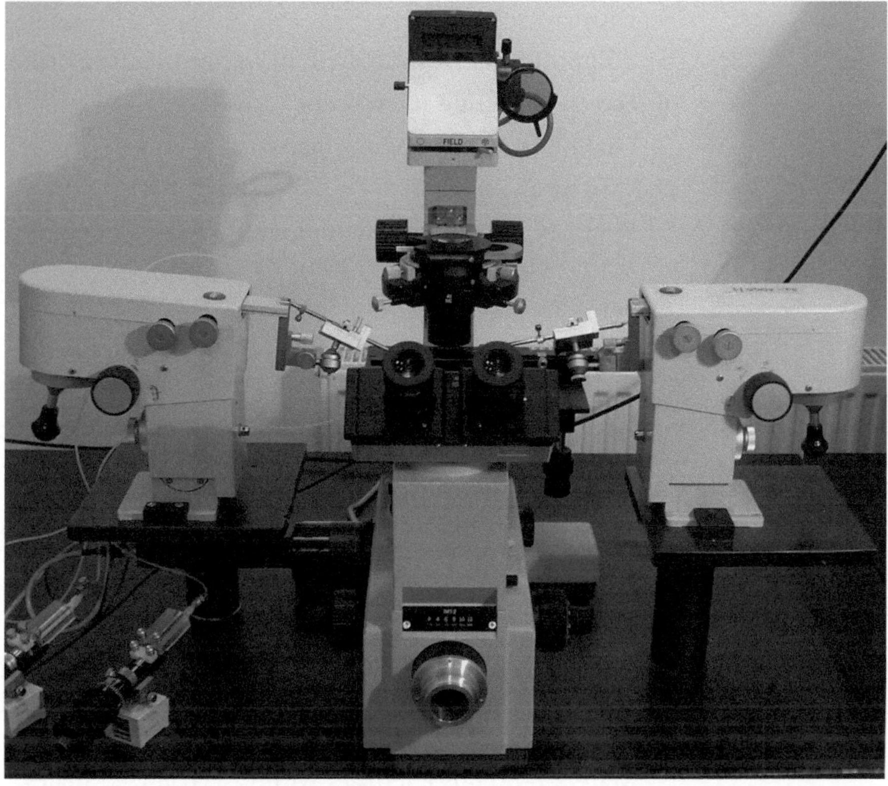

Fig. 3. Olympus IMT-2 inverted light microscope with phase contrast optics and two Leitz micromanipulators on both sides.

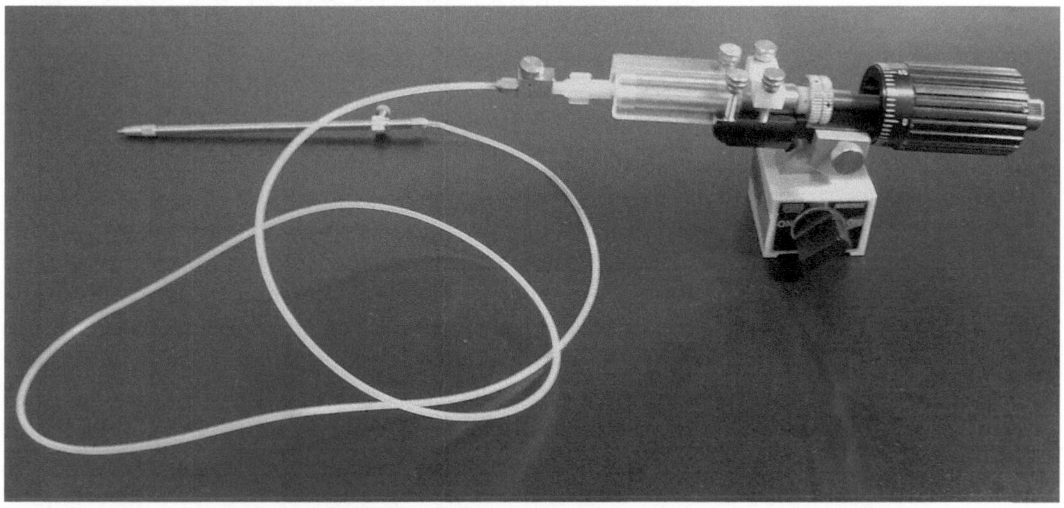

Fig. 4. Narishige IM-9A air pressure controller with tubing and capillary holder.

4. Paper towels.
5. 35 mm Ø sterile, untreated, plastic cell-culture dishes.
6. Stereo dissecting microscope (see above).
7. One hooked and one normal small tweezer.
8. Two small scissors.
9. Fine, high precision, biological forceps (e.g., Dumont Dumostar 5, cat. #10788) or watchmaker's forceps.
10. Serrefine clamp, 28 mm or smaller.
11. 7-mm wound clip and suitable wound clip applier.
12. Sterile, G28 hypodermic needles.
13. Mouth pipette device (e.g., Sigma-Aldrich A5177) with transfer capillaries (see Subheading 3.2.1).

3. Methods

3.1. Mating Donor and Recipient Mice

1. Mate C57BL/6 donor female mice with C57BL/6 males later in the afternoon or in the evening. Put only one male and one female together in one cage.
2. Mate 25–30 CD1 females with vasectomized males. Put two to three females and one vasectomized male in one cage.
3. Check for post coital plugs (firm, white coagulates) in the vagina in all females next morning. If a plug is found then separate plugged female and note the date and male partner (see also Note 3). These females are considered to be 0.5 day after copulation or *post coitum* (d.p.c.).
4. If no more plugged females are needed, separate the remaining ones from male partners. If they do not become pregnant they can be reused after 3 weeks.

3.2. Capillary Making

Micropipettes or microcapillaries are the most important equipments used in blastocyst injection because the physical manipulations like fixing, grabbing, transferring or injecting cells or embryos are done with these tools. They are made of glass and hence they are very fragile. Since their diameter is in the microscopic range they tend to clog. Because of these facts the experimenter should keep these fine materials in a safe, clean, and dust-free place. A clean Petri dish with cover and a streak of modeling clay attached to the bottom of the dish is suitable. For storage the thick, stem part of the micropipette is pushed into the attached clay or plasticine streak so that it does not plug the thick end of the capillary. A small, plastic tube filled with 70% ethanol and closed with a rubber plug is also a good alternative. The plug itself holds the capillary in the alcohol bath (e.g., Humagen's capillary containers).

3.2.1. Pulling a Transfer Capillary

1. Light the Bunsen burner and grab one glass capillary at its ends with both hands. Put each end in between your pointing fingers and thumbs.
2. Place approximately 1 cm section of the middle of the capillary into the flame for a few seconds and rotate it until it starts to glow.
3. Remove the pipette from the flame suddenly and pull apart its ends at the same time quickly. Your pull should be even and completely straight (see also Note 5).
4. Bend the capillary until it breaks at its middle. It gives two micropipettes. Put down one of them and carry on with the other.
5. Very carefully scratch the thin part with the diamond pen 2–3 cm away from the thicker part of your pipette.
6. Bend the thin end until it breaks down. The thin tip should break at the scratched point.
7. Check your pipette tip under the stereo microscope. The rim or lip of the thin end should be even and near circular without any chips or major cracks.
8. Hold the other half of your broken capillary and repeat from step 5.

3.2.2. Preparing Capillaries for Blastocyst Holding

1. Take a transfer capillary and put it in the microforge.
2. Position the thin end opposite to the heating filament viewing from every angle.
3. Switch on heating and move the tip closer to filament.
4. Follow the melting of the end of your pipette through the eyepiece and stop heating as soon as the desired end diameter is reached, which should be 10–20 μm or about one fifth of a blastocyst (see Note 9).
5. Turn the capillary perpendicular to its current position in the microforge and position the filament close to its side and 2–3 mm away from its end.
6. Switch on heating and move the tip closer to filament.
7. Watch the tip bending and stop heating when a 20–30° deviation from the original axis is reached. The angle depends on the settings of your micromanipulator and capillary holder. For microinjection, the tip of the holding pipette should be parallel to the bottom of the injection chamber (see Fig. 5a).

3.2.3. Preparing Needles for ES Cell Injection

1. Place one glass capillary in the puller and fasten it on both sides of the filament.
2. Perform a RAMP test if it has not already been done. Note RAMP value (see also Note 6).

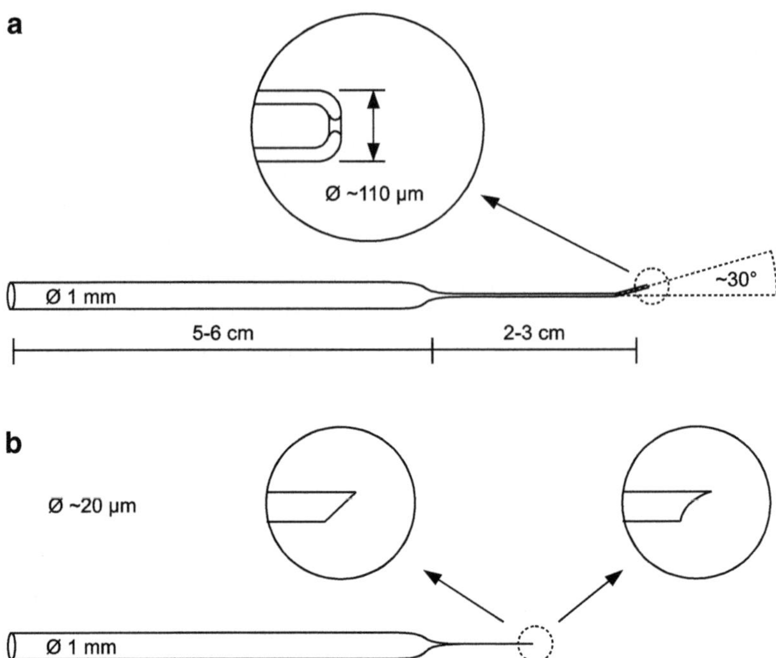

Fig. 5. Pipette geometry. (**a**) Scheme demonstrates the bent holding pipette, the shape of its tip and the 20–30° of deviation. (**b**) ES cell injection needle is a fine, straight, sharp needle. Its tip resembles the tip of a hypodermic needle or it can be slightly concave.

3. Replace capillary in the puller and put in a new one.
4. Set the following parameters: HEAT = RAMP + 10, PULL = 50, VELOCITY = 75, TIME = 25, PRESSURE = 300. Press “PULL” (see also Note 7).
5. Loosen the screws, carefully take out the capillary and check its tip under the dissection microscope. It should be long enough to conveniently hold 10–15 ES cells in one row. Its diameter must be around 20–25 μm, just like the size of an ES cell. See Fig. 5b for more information on the shape and size of the tip.
6. If needed, modify one parameter at a time. Increase or decrease the value by 10 and repeat pulling from step 4.
7. If the needle’s geometry looks right, put it in your left hand and with its tip touch the surface of the silicone or rubber halfpipe placed on the table with its convex side up.
8. Hold the scalpel in your right hand and slightly bend the tip on the halfpipe.
9. Touch the tip smoothly with the blade of the scalpel and break it. During this process the edge of the blade and the tip together must make a sharp angle (see Note 8).
10. Check the needle with real ES cells before blastocyst injection (see also Subheading 3.5).

3.3. Blastocyst Collection

1. Put an aliquot of M2 into the 37°C CO_2 incubator. Put 4–5, equally 30–40 µl drops of KSOM onto the bottom of a 60-mm cell culture dish. KSOM drops should be well separated from each other. Put the dish into the CO_2 incubator for 1 h and let the medium be warmed up and equilibrated. Completely cover the drops with embryo tested, sterile mineral oil.
2. Cover a small area on the table with two to three sheets of paper towel. Wipe the scissors and tweezers with 70% ethanol.
3. Right before killing the donor female suck up 1 ml of warm M2 medium into a syringe, attach the blunted needle and inject 100–150 µl of M2 medium onto the bottom of a new, sterile 60-mm dish. This will keep the uterus wet.
4. Euthanize the plugged donor mother by cervical dislocation. It should be on the third day after plug detection (3.5 d.p.c.).
5. Put the mouse onto the paper towel and lay it on its back. Wet and wipe its abdomen with 70% ethanol.
6. Pinch the abdominal skin with a forceps approximately 1 cm above the anus and while holding the skin make a cut with scissors through the pinched skin right next to the forceps.
7. Replace the scissors to another forceps, pinch the skin on other lip of the cut and peel the skin by pulling the tweezers apart firmly but carefully.
8. Wipe the hair off the scissors and tweezers. Cut the abdominal membrane and expose the abdominal organs. Look for the V-shape uterus behind the intestines.
9. Grab the cervix with a tweezer. Lift it up while cutting off the vaginal side and the highly vascular uterus serosa membrane with scissors. Cut the uterine horns at their connections to the oviducts.
10. Put the extracted uterus into the 60-mm dish with M2 medium and put it under the dissection microscope.
11. Chip off the uterine horns very close to the cervix (which you should leave there for orientation).
12. Hold the ovarian end of a uterus horn with a watchmaker's forceps in your left hand. Stick the blunted needle of the syringe containing M2 medium into the ovarian end of the uterine horn and flush the embryos into dish with about 300–400 µl of medium. It is worth to flush it from the cervical end as well. Repeat it with the other uterine horn.
13. Check the flushed blastocysts for intact zona pellucida and well detectable blastocoel cavity. Collect them with a mouth pipette into the attached transfer capillary (see also Note 10).

14. Transfer collected blastocysts into one of the prewarmed and equilibrated KSOM drops and put them back in the incubator.

3.4. ES Cell Preparation

Anything that can have direct contact with the media or ES cell culture should be handled under an UV-sterilized laminar flow hood in order to minimize the risk of ES cell culture infection.

1. Prewarm trypsin or TrypLE™ Express and ES cell medium in the 37°C water bath. Put ES cell injection medium on ice.
2. Fill the sterile 12-ml tube with 5 ml of ES cell medium.
3. Remove medium from the ES cell culture and rinse the cells once with PBS.
4. Pipette 1 ml of trypsin or TrypLE™ Express onto the cells and incubate at 37°C for 3 min.
5. Resuspend the ES cells by pipetting them eight to ten times with a 1 ml pipette, because a one-cell suspension is required.
6. Transfer the suspension into the 12-ml centrifuge tube containing ES cell medium.
7. Centrifuge it for 5 min at 165 × *g* and discard supernatant (ES cell medium).
8. Resuspend the cells completely in 500 μl of ice-cold ES cell injection medium.
9. Put the tube on ice for 30 min.
10. Discard about 3/4 of the medium without stirring up the settled cells from the bottom (see Note 11).
11. Add 500 μl of ice-cold ES injection medium and resuspend the cells again. These cells should be kept on ice and must be used for injection within 3 h (see Note 12).

3.5. Blastocyst Injection

1. Orient the holding capillary on the left side of the microscope by positioning the bent end of the holder horizontally in the injection chamber so that the bent part of the holding pipette is sharply in focus throughout its total length.
2. Orient the injection capillary on the right side of the microscope by positioning it opposite to the holding capillary. The opening of the tip of the pipette should look down and the injection pipette should not tilt more than 5°.
3. Put 50 μl of an ES cell injection medium drop into a sterile Petri dish. Dip the capillaries into this drop and let the small amount of medium flow up into the lumen of the capillaries.
4. Microinjection of blastocysts is performed on a 6-cm Petri (cell culture) dish lid or in a depression slide. Put 150 μl of blastocyst injection or M2 media onto the dish lid or into the

pit of the depression slide and add a 10–20 μl drop of ES cell suspension to the medium. Check the concentration of the ES cells and then cover the surface with embryo-tested mineral oil. This setup is also called the "injection chamber."

5. Add the blastocysts using a mouth pipette (see Note 13). Keep the blastocysts grouped and centered on the plate.
6. Put the injection chamber onto the microscope stage. First, focus on the blastocysts then adjust the height of the manipulators to bring the instruments in center and focus.
7. Select a blastocyst for injection. Move it in between the two capillaries by positioning the microscope stage.
8. Immobilize and fix the blastocyst on the tip of the holding pipette with a gentle suction. Select the correct state for injection by releasing, turning, then fixing the blastocyst again. You can help this maneuver by carefully touching the zona pellucida of the embryo on the other side with the tip of the injection needle. The inner cell mass should be opposite down and close to the opening of the holding pipette, but a slightly tilted position is desirable. The best place for penetration into the blastocoel cavity is between two trophectoderm cells.
9. Load individual ES cells into the injection pipette. The cells selected should have good refraction (which is a sign of viabil ity) (see Note 14 and Fig. 6a). Keep the volume of injected media as low as possible. Maximally 12–15 healthy-looking cells should be picked up for one blastocyst injection (see Note 15).
10. Suck up the ES cells into the tip of the injection capillary. Keep them together and very close to the opening of the capillary.
11. Move the injection needle with the joystick close to the junction between the trophoblast cells (see Fig. 6b). Bring the embryo into sharp focus by adjusting height on the right micromanipulator.
12. With one smooth push bring the tip of the injection pipette into the embryo between the trophoblast cells without collapsing or puncturing the opposite wall. The injection needle should not touch the inner cell mass (see Fig. 6c).
13. When the tip of the injection pipette is clearly visible in the middle of the blastocoel, carefully expel the ES cells (12–15 cells per blastocyst) with a slight, positive pressure (see Note 16).
14. When every cell has left the capillary, immediately take the pressure off the injection needle. Carefully and slowly withdraw the needle from the embryo (see Note 17 and Fig. 6d).
15. Separate injected and non injected blastocyst inside the drop. The injected blastocysts usually collapse after the process, but few hours later they start to expand again.

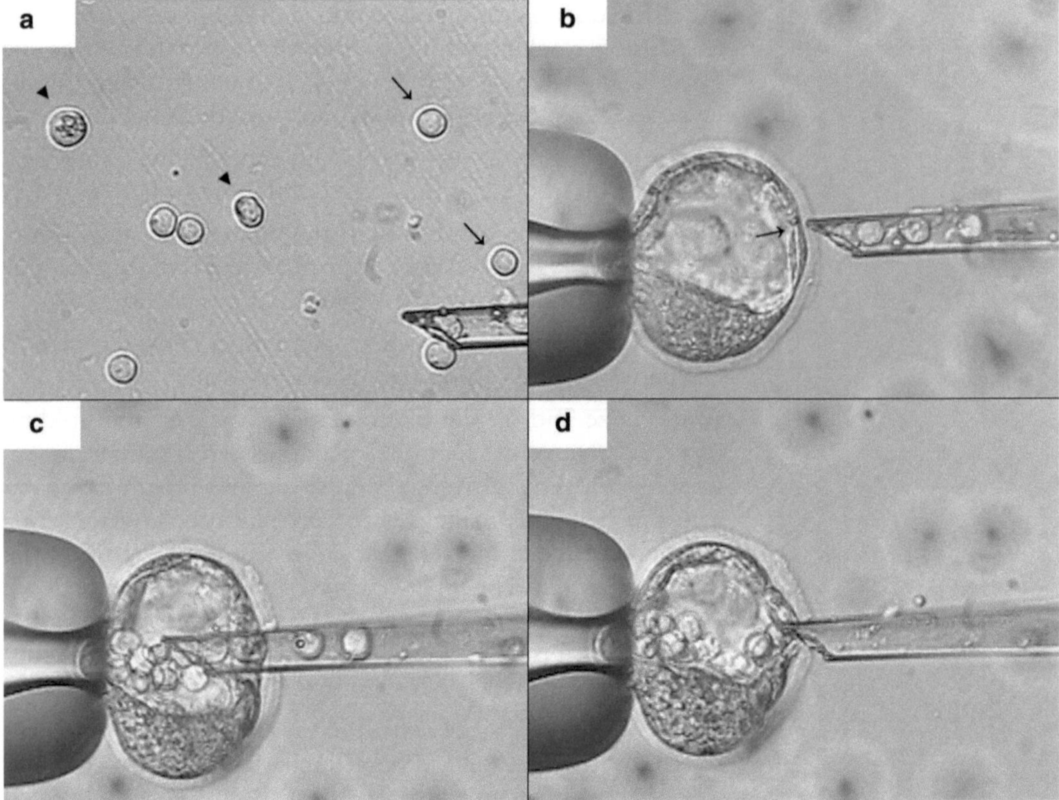

Fig. 6. Blastocyst injection. (**a**) ES cell selection. *Dark cells* (*arrowhead*) are not suitable for injection. *Round cells* with good refraction (*arrow*) should be selected. (**b**) During the injection, the inner cell mass should "look down" in a slightly tilted position. Penetrate the blastocoel cavity between two trophectoderm cells (*arrow*). (**c**) Smoothly but confidently push the needle into the blastocyst between the trophoblast cells without touching the opposite wall of the blastocoel cavity. (**d**) Wait a few seconds at the trophoblast lining to balance the positive pressure inside the embryo.

16. Repeat the injection process with a new blastocyst.
17. When all the blastocysts have been injected in the injection chamber put them back into an empty KSOM drop under mineral oil and culture them in a CO_2 incubator until the embryo transfer.
18. Repeat all the steps with another group of blastocysts each time in a new chamber.
19. Perform the embryo transfer on the day of injection or culture the embryos overnight if no (more) foster mothers are at hand.

3.6. Embryo Transfer

1. Wait at least half hour after the injection procedure. If there is not enough recipient female, you can culture the injected embryos overnight and transfer them on the following day.
2. Anesthetize recipient females by intraperitoneal injection of about 0.3–0.4 ml of 10× diluted commercially available

Nembutal solution. Usually the final dose should be around 40–50 mg/kg (see also Note 18).

3. Place the narcotized mouse on a small, firm platform (e.g., the lid of a 10 cm Petri dish) covered with a piece of paper towel for easier handling. The animal should be lying on its stomach.
4. Wipe the hair in the middle of its back with 70% ethanol. Cut the skin here with a pair of scissors. The wound should be about 1 cm long along the longitudinal axis.
5. Locate the ovarian fat pad through the wound in one side of the abdominal cavity beneath the body wall (see Note 19 and Fig. 7a).
6. Pinch the body wall with the hooked forceps right above the ovary and cut an approximately 5 mm long incision. Try to prevent cutting any of the blood vessels.

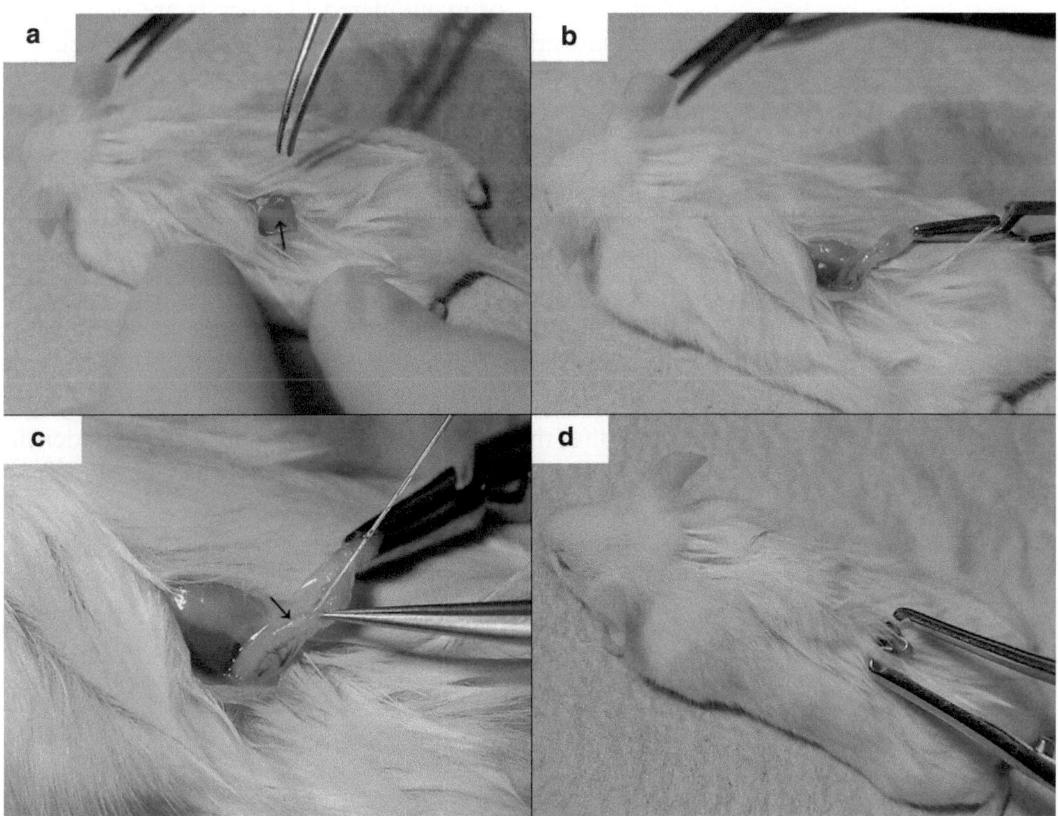

Fig. 7. Embryo transfer. (**a**) Look for the fat pad (*arrow*) through the skin wound. (**b**) Grab the fat pad with the blunted tweezers, pull it out, replace the tweezers with the serrefine and lay it onto the opposite side of the animal's back. (**c**) Stick the loaded transfer capillary into the uterine horn (*arrow*) under the stereomicroscope. (**d**) Close the wound with a wound clip.

7. Grab the ovarian fat pad with the blunt forceps. Pull out the ovary, oviduct, and upper part of the uterus. Fix the fat pad with the serrefine, pull it out and lay it onto the opposite side of the animal's back (see Fig. 7b). Gently pick up the platform with the mouse and place them onto the stage of the stereo microscope.
8. Transfer the embryos from the microdrop culture into a large drop (100 μl) of ES injection medium and wash the embryos through three large drops to get rid of the injection chamber's oil.
9. Take up a small amount of medium into the tip of transfer pipette, then a small air bubble, then medium, and then air bubble again. Draw up the embryos with the next portion of medium, then sip up a small air bubble again and finally "close" the load with a small amount of medium (see Fig. 8). Place the pipette in a safe place such that it does not touch anything until you are ready for embryo injection.
10. Transfer at most ten embryos into one uterus. Survival and implantation rate of manipulated blastocysts are better than in the conventional transgenic procedures where you can even transfer up to 15 embryos in one uterine horn.
11. Hold a fine forceps in your left hand, the loaded embryo transfer pipette and the hypodermic needle in your right hand. The transfer pipette should be between your thumb and index finger and the needle should be pressed against your index finger by your middle finger.
12. Grab the oviduct with the forceps close to the ovary and cut a tiny hole on it by inserting the hypodermic needle into the uterus lumen. Avoid piercing the opposite wall of the uterus otherwise the medium used for flushing will leak there.
13. Keep holding the oviduct. Pull out and drop the hypodermic needle. Shift the transfer pipette in between your right forefinger and thumb.
14. Insert the embryo transfer pipette into the preformed hole (see Fig. 7c). Blow the content of the pipette into the uterus. Air bubbles, which separated the liquid content in the pipette, will show up in the uterus indicating the success of the transfer (see Note 20).

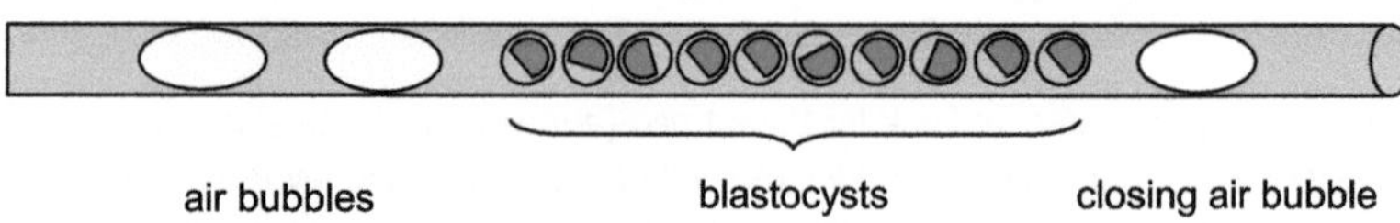

Fig. 8. Pipette loaded with blastocysts ready for transfer.

15. Release the fat pad, let the organs slide back and gently push the uterus back into the peritoneum.
16. Repeat the transfer on the other side if desired.
17. Close the skin of the animal with a wound clip (see Fig. 7d). Place it into a new, clean cage in a quiet warm place until the mouse recovers from the anesthesia. You may put two females into the same cage. They will help each other raising a joint litter (see also Note 21)
18. As soon as the newborn mice get hairy you can see the degree of chimerism based on the coat color patches. R1 ES cells are derived from agouti blastocysts (23) and practically host blastocysts should be derived from a white or non-agouti colored mice (e.g., Twitcher mice) (13).

4. Notes

1. Donor females should provide large number of blastocysts with permissive cellular environment or "nest" for manipulated ES cells. The coat-color genotype of the recipient embryo should differ from the genotype of donor ES cells and this way the chimerism can simply be detected. The most common strains used for blastocyst donors are BALB/c and C57BL/6.
2. Surrogate mothers must tolerate implanted embryos and be able to feed and nurture newborns until they are self supporters. Outbred strains (like CD1) are preferred because they are good mothers, generally more resistant to environmental stress, mate more frequently, and usually stronger than inbred ones.
3. Males can be vasectomized and checked for sterility in house (17) but they are also available from vendors of laboratory mice (e.g., The Jackson Laboratory). Copulation with a vasectomized male also produces post coital plug and results in pseudopregnancy where subsequent hormonal changes in the female are necessary for efficient implantation of manipulated embryos.
4. Phase contrast optics provides clearer and better vision of 3D structures and it is very important for ES cell selection based on morphology.
5. Your pull is correct if you can get an evenly thin, completely straight, 5–10 cm long middle part with a diameter about the size of a blastocyst (90–120 μm). This technique needs some time, patience, and practice as much as injection needle fabrication.
6. RAMP test is a heating test and is used to automatically adjust the optimal heating power of the filament. It depends strongly

on the type of applied filament and capillary. This also means if you want to use another type of filament or capillary, you *must* redo RAMP test otherwise you might ruin your filament or the puller machine. While you work with the *same filament* and the *same type of capillary* you can safely omit further RAMP tests.

7. Pulling parameters affect different physical aspects of the pulling process, thus they alter different characteristics of the forming capillary. Higher temperature or HEAT will produce a longer taper with smaller tip diameter. Greater PULL values correlate with longer capillary without changing its diameter significantly. Higher VELOCITY means finer tip diameter. Increased TIME provides shorter taper. PRESSURE elevation shortens taper and widens tip diameter. More information on the puller can be found on the manufacturer's website: http://www.sutter.com.
8. Fabricating the right injection needle is not an easy task. It definitely needs some experience to prepare good capillaries consistently, so it might need a lot of patience, trial-and-error steps and practice in order to do it well. Do not get frustrated if you fail to create perfect needles for the first time. It is completely normal. Practice a lot, because this is one of the most crucial steps in this protocol. Your success with injection depends largely on the shape of your needle's tip.
9. The shrunk tunnel at the end of your capillary should be the straight continuation of your pipette's lumen. The surface of the tip should be even, plain and perpendicular to the axis of the capillary. In case of any deviation from the characteristics described here, the microforge manipulations should be repeated after breaking off the end of pipette or with a new transfer pipette.
10. The best stage to inject is when the blastocyst is not fully expanded but a skilled person is capable of injecting the early and the fully expanded blastocyst, too. Inject the early stage blastocysts later because they might expand by the end of the microinjection procedure.
11. This trick can remove the majority of the floating dead cells and cellular debris from the suspension.
12. Cooling down ES cells and keeping them in HEPES containing medium before injection significantly slows down the pH change of the medium outside the incubator and prevents cells from getting aggregated.
13. The number of embryos that can be injected in one go strongly depends on the skill of the person. However you can follow this rule of thumb. Put as much embryo into the injection chamber as you can inject within 30 min.

14. Cells that appear darker than the others are dying therefore should not be used.
15. Cells in the injection pipette should be loaded very close to the tip and they should stay closely together side-by-side (like a straight queue of cells).
16. Just like in other microinjection procedures the success of blastocyst injection strongly depends on the quality of the injection capillary. The needle turns out to be usable during the initial phase of the injection procedure. It has to be wide enough to pick up the ES cells without any damage. However, it should not be too wide in order to prevent the medium overfill of the blastocoel cavity. The sharpness of the tip is essential to pass through between two adjacent trophoblast cells easily.
17. When a huge amount of media is injected into the blastocoel or the injected blastocyst is expanded, ES cells might start to flow out when you try to withdraw the capillary after the injection. You can eliminate this event and balance the pressure inside the embryo if you wait a few seconds at the border of trophoblast cells while pulling out the capillary.
18. The exact dosage of Nembutal should be tested preliminarily, but usually a 40–50 mg/kg dose is sufficient.
19. It can be seen as a tiny white patch lateral to the vertebral column and close to the lower rim of the rib cage.
20. The embryo transfer is also a complex procedure. Just one bad movement and you can lose your all day's work. It is advised to practice without embryos previously as much as needed. When you gain routine with the transfer you can start the injection. In case when no or not enough 2.5 d.p.c. recipient female is available you can use 0.5 d.p.c recipients for oviduct transfer. Even for a skilled person the uterus transfer is easier to perform. We have not recognized any difference between the oviduct or uterus transfer regarding the number of pups or the degree of chimerism however others did (24).
21. If one of the foster mothers does not get pregnant, separate it from the other a few days before the delivery.

References

1. Cohen, S. N., Chang, A. C. Y., Boyer, H. W., and Helling, R. B. (1973) Construction of Biologically Functional Bacterial Plasmids In Vitro *Proceedings of the National Academy of Sciences of the United States of America* **70**, 3240–3244.
2. Gordon, J. W., Scangos, G. A., Plotkin, D. J., Barbosa, J. A., and Ruddle, F. H. (1980) Genetic transformation of mouse embryos by microinjection of purified DNA *Proceedings of the National Academy of Sciences of the United States of America* 77, 7380–7384.

3. Sung, P., Chang, S., Wen, K., Chang, C., Yang, M., Chen, L., Chao, K., Huang, C. F., Li, Y., and Lin, C. (2009) Small supernumerary marker chromosome originating from chromosome 10 associated with an apparently normal phenotype *Am. J. Med. Genet. A* **149A**, 2768–2774.
4. Liehr, T., Ewers, E., Kosyakova, N., Klaschka, V., Rietz, F., Wagner, R., and Weise, A. (2009) Handling small supernumerary marker chromosomes in prenatal diagnostics *Expert Rev. Mol. Diagn* **9**, 317–324.
5. Irvine, D. V., Shaw, M. L., Choo, K. H. A., and Saffery, R. (2005) Engineering chromosomes for delivery of therapeutic genes *Trends Biotechnol* **23**, 575–583.
6. Grimes, B. R., and Monaco, Z. L. (2005) Artificial and engineered chromosomes: developments and prospects for gene therapy *Chromosoma* **114**, 230–241.
7. Duncan, A., and Hadlaczky, G. (2007) Chromosomal engineering *Curr. Opin. Biotechnol* **18**, 420–424.
8. Katoh, M., Ayabe, F., Norikane, S., Okada, T., Masumoto, H., Horike, S., Shirayoshi, Y., and Oshimura, M. (2004) Construction of a novel human artificial chromosome vector for gene delivery *Biochem. Biophys. Res. Commun* **321**, 280–290.
9. Lindenbaum, M., Perkins, E., Csonka, E., Fleming, E., Garcia, L., Greene, A., Gung, L., Hadlaczky, G., Lee, E., Leung, J., MacDonald, N., Maxwell, A., Mills, K., Monteith, D., Perez, C. F., Shellard, J., Stewart, S., Stodola, T., Vandenborre, D., Vanderbyl, S., and Ledebur, H. C. (2004) A mammalian artificial chromosome engineering system (ACE System) applicable to biopharmaceutical protein production, transgenesis and gene-based cell therapy *Nucleic Acids Res* **32**, e172.
10. Kennard, M. L., Goosney, D. L., Monteith, D., Zhang, L., Moffat, M., Fischer, D., and Mott, J. (2009) The generation of stable, high MAb expressing CHO cell lines based on the artificial chromosome expression (ACE) technology *Biotechnol. Bioeng* **104**, 540–553.
11. Kennard, M. L., Goosney, D. L., Monteith, D., Roe, S., Fischer, D., and Mott, J. (2009) Auditioning of CHO host cell lines using the artificial chromosome expression (ACE) technology *Biotechnol. Bioeng* **104**, 526–539.
12. Co, D. O., Borowski, A. H., Leung, J. D., van der Kaa, J., Hengst, S., Platenburg, G. J., Pieper, F. R., Perez, C. F., Jirik, F. R., and Drayer, J. I. (2000) Generation of transgenic mice and germline transmission of a mammalian artificial chromosome introduced into embryos by pronuclear microinjection *Chromosome Res* **8**, 183–191.
13. Katona, R. L., Sinkó, I., Holló, G., Szucs, K. S., Praznovszky, T., Kereso, J., Csonka, E., Fodor, K., Cserpán, I., Szakál, B., Blazsó, P., Udvardy, A., and Hadlaczky, G. (2008) A combined artificial chromosome-stem cell therapy method in a model experiment aimed at the treatment of Krabbe's disease in the Twitcher mouse *Cell. Mol. Life Sci* **65**, 3830–3838.
14. Suzuki, N., Nishii, K., Okazaki, T., and Ikeno, M. (2006) Human artificial chromosomes constructed using the bottom-up strategy are stably maintained in mitosis and efficiently transmissible to progeny mice *J. Biol. Chem* **281**, 26615–26623.
15. Oshimura, M., and Katoh, M. (2008) Transfer of human artificial chromosome vectors into stem cells *Reprod. Biomed. Online* **16**, 57–69.
16. Monteith, D. P., Leung, J. D., Borowski, A. H., Co, D. O., Praznovszky, T., Jirik, F. R., Hadlaczky, G., and Perez, C. F. (2004) Pronuclear microinjection of purified artificial chromosomes for generation of transgenic mice: pick-and-inject technique *Methods Mol. Biol* **240**, 227–242.
17. Nagy, A., Gertsenstein, M., Vintersten, K., and Behringer, R. (2002) Manipulating the Mouse Embryo: A Laboratory Manual, Third Edition. Cold Spring Harbor Laboratory Press.
18. Reid, S. W., and Tessarollo, L. (2009) Isolation, microinjection and transfer of mouse blastocysts *Methods Mol. Biol* **530**, 269–285.
19. Wood, S. A., Allen, N. D., Rossant, J., Auerbach, A., and Nagy, A. (1993) Non-injection methods for the production of embryonic stem cell-embryo chimaeras *Nature* **365**, 87–89.
20. Tanaka, M., Hadjantonakis, A., Vintersten, K., and Nagy, A. (2009) Aggregation chimeras: combining ES cells, diploid, and tetraploid embryos *Methods Mol. Biol* **530**, 287–309.
21. Poueymirou, W. T., Auerbach, W., Frendewey, D., Hickey, J. F., Escaravage, J. M., Esau, L., Doré, A. T., Stevens, S., Adams, N. C., Dominguez, M. G., Gale, N. W., Yancopoulos, G. D., DeChiara, T. M., and Valenzuela, D. M. (2007) F0 generation mice fully derived from gene-targeted embryonic stem cells allowing immediate phenotypic analyses *Nat. Biotechnol* **25**, 91–99.
22. Dechiara, T. M., Poueymirou, W. T., Auerbach, W., Frendewey, D., Yancopoulos, G. D., and Valenzuela, D. M. (2009) VelociMouse: fully ES cell-derived F0-generation mice obtained from the injection

of ES cells into eight-cell-stage embryos *Methods Mol. Biol* **530**, 311–324.

23. Nagy, A., Rossant, J., Nagy, R., Abramow-Newerly, W., and Roder, J. C. (1993) Derivation of completely cell culture-derived mice from early-passage embryonic stem cells *Proc. Natl. Acad. Sci. U.S.A* **90**, 8424–8428.

24. Ramírez, M. A., Fernández-González, R., Pérez-Crespo, M., Pericuesta, E., and Gutiérrez-Adán, A. (2009) Effect of stem cell activation, culture media of manipulated embryos, and site of embryo transfer in the production of F0 embryonic stem cell mice *Biol. Reprod* **80**, 1216–1222.

Chapter 13

Transfer of Stem Cells Carrying Engineered Chromosomes with XY Clone Laser System

Ildiko Sinko and Robert L. Katona

Abstract

Current transgenic technologies for gene transfer into the germline of mammals cause a random integration of exogenous naked DNA into the host genome that can generate undesirable position effects as well as insertional mutations. The vectors used to generate transgenic animals are limited by the amount of foreign DNA they can carry. Mammalian artificial chromosomes have large DNA-carrying capacity and ability to replicate in parallel with, but without integration into, the host genome. Hence they are attractive vectors for transgenesis, cellular protein production, and gene therapy applications as well. ES cells mediated chromosome transfer by conventional blastocyst injection has a limitation in unpredictable germline transmission. The demonstrated protocol of laser-assisted microinjection of artificial chromosome containing ES cells into eight-cell mouse embryos protocol described here can solve the problem for faster production of germline transchromosomic mice.

Key words: Eight-cell mouse embryo, Laser-assisted microinjection, ES cell, Artificial chromosomes, Germline transmission, Chimeric mice

1. Introduction

Transfer of foreign DNA into the host genome by genetically engineered chromosome vectors is a major advantage of their, almost unlimited, transgene-carrying capacity and the ability to replicate synchronously with the host genome. Introducing genes into artificial chromosomes eliminates the disadvantages of commercial transgenic technologies where the exogenous DNA is integrated into the host genome and can cause variegated gene expression and insertional mutagenesis. The previously described Platform ACE System developed for the reliable and systematic engineering of mammalian artificial chromosomes with large cDNAs or genomic sequences became vectors for transgenesis and gene therapy application.

Gyula Hadlaczky (ed.), *Mammalian Chromosome Engineering: Methods and Protocols*, Methods in Molecular Biology, vol. 738, DOI 10.1007/978-1-61779-099-7_13, © Springer Science+Business Media, LLC 2011

There are several methods to introduce artificial chromosomes into the host genome. One is microinjection of the chromosome into the pronuclei of a mouse zygote. It has been reported as a "pick-and-inject" technique (1). This procedure has 13% transgenesis frequency.

Chromosomal vectors can be transfected into embryonic stem cells. Microinjection of these mutant cells into the blastocyst stage of mouse embryos provides founder (F0) mice that are only partially derived from genetically modified ES cells. Because the coat color of the ES cells is different from that of the host mice coat color, the chimeric mice can be detected visually by the presence of the ES cell color in the background of host color. The percent of the ES cell color does not reflect the germline transmission availability of the ES cell derived cells. If mutant ES cells partially differentiate into germ cells, the founder chimeric mice can transmit the extra chromosomes to the next F1-generation of mice. To establish a stable transgenic line, another round of breeding between F1 heterozygous mice produces homozygous F2-generation mice. It normally takes a minimum of 9 months from the birth of F0 chimeras to obtain F2 mice.

The meiotic stability and germline transmission of foreign chromosomes in chimeric mice is variable. In most of the cases, male germline transmission was low or not detectable (2–6). Female chimeras carrying chromosomes containing ES cells show germline transmission more frequently (3–5). In some cases, both the genders showed efficient germline transmission (4, 5, 7), but others recognized no transmission of chromosomes of either sex of the chimeras (4). In our previous experiments, we also did not obtain germline transmission chimeras produced by microinjection of ES cells that express the GALC gene from the therapeutic artificial chromosomes.

The possible explanation for the sex difference observed for the germline transmission of extra (trans-) chromosomes in the literature resides in the sex differences in gametogenesis. The presence of one or more unpaired chromosome can disrupt the meiotic process in male mice during the first meiotic division of gametogenesis (8). In female meiosis, an unpaired X chromosome can segregate as an intact chromosome only in part of the first meiotic division and could affect the alignment and segregation of other chromosomes as well. Most of the oocytes from XO females were able to complete the first meiotic division (9).

The degree of chimerism could also depend on other factors. One is the activation of ES cells before injecting them into the embryo. The culture conditions and the site of the embryo transfer can also affect the successful germline chimerism (10).

The possible solution for the problem of germline transmission is the new method that introduces ES cells into the eight-cell mouse embryos (11). This method could shorten the breeding period to obtain homozygous transchromosomal mice as well. This laser assisted microinjection technique of either inbred or hybrid ES cells into eight-cell stage embryos efficiently yields F0 generation mice that are fully ES cell-derived and healthy, exhibit 100% germline transmission, and allow immediate phenotypic analysis, thus greatly accelerating the gene function assignment. This method works not only with standard male XY ES cells, but also with their XO congenic clones, thus rapidly and efficiently yielding not only male but also female F0 founder mice.

In this chapter, transgenic, chimeric mice production by eight-cell embryo injection with ACE transformed ES cells is discussed in detail.

2. Materials

2.1. Mating Donor and Recipient Mice

1. Eight- to ten-week-old donor females (see Note 1).
2. Two- to ten-month-old males (same strain as donor females).
3. Two- to four-month-old CD1 foster mothers (see Note 2).
4. Two- to twelve-month-old, vasectomized CD1 or other outbred male (see Note 3).
5. Conventional animal house with stabilized 10-h dark–14-h light cycle, relatively constant air temperature (between 20 and 25°C), and relative humidity.

2.2. Capillary Making

1. Modeling clay or plasticine.
2. Clean, glass Petri dishes.
3. Borosilicate glass capillary with no internal filament (e.g., GC120T-15 from Warner Instruments [formerly Clark Electromedical Instruments] or TW100-6 from World Precision Instruments, Inc.). It is possible to buy holding and injection capillaries from vendors (e.g., Holding and Stem Cell Micropipets from Humagen Fertility Diagnostics, TransferTipsES, Eppendorf VacuTip).
4. Sutter P-97 Flaming/Brown Micropipette Puller with suitable 3-mm trough filament (e.g., Science Bioproducts FT330B).
5. Scalpel with a disposable blade.
6. A 3–5 cm long piece of silicon or rubber tube, ~2 cm Ø cut in half trough its longitude axis like a halfpipe or tunnel.
7. Stereo dissecting microscope with adjustable zoom (10× to 45×), upper and scattered lower white light (e.g., Leica ZOOM 2000).

8. Bunsen burner.
9. Capillary microforge (e.g., World Precision Instruments MF200 Complete microforge system including H602 Stereo Microscope).
10. Diamond pen.

2.3. Eight-Cell Embryo Collection

1. KSOM medium (Millipore). Store in 3–4 ml sterile aliquots at 2–8°C.
2. ES cell injection medium: 200 μl of 1 M HEPES in 10 ml of ES cell culture medium, store maximum for 1 week at 4°C.
3. Embryo tested sterile filtered mineral oil (Sigma-Aldrich).
4. 70% ethanol.
5. Paper towels.
6. 60 and 35 mm Ø sterile, untreated, plastic cell-culture dishes.
7. CO_2 incubator with 37°C temperature and 5% CO_2 content.
8. Stereo dissecting microscope with adjustable zoom (10× to 45×), upper and scattered lower white light (e.g., Leica ZOOM 2000).
9. Two small tweezers and two small scissors.
10. Fine, high precision, biological forceps (e.g., Dumont Dumostar 5, cat. #10788), or watchmaker's forceps.
11. 1 ml, Luer-slip, concentric, all plastic syringe (e.g., National Scientific, cat. #S7510-1).
12. Sterile G30 hypodermic needles blunted with sand paper.
13. Mouth pipette device (e.g., Sigma-Aldrich A5177) with transfer capillaries (see Subheading 3.2.1).

2.4. ES Cell Preparation

1. ES cells: ACE transformed R1 cells grown on 60-mm gelatin-coated tissue culture dish without feeder cells.
2. ES cell medium (the same used for ES cell culturing), store at 4°C.
3. ES cell injection medium.
4. A box of fresh ice.
5. Phosphate-buffered saline (PBS) without Ca^{2+} or Mg^{2+}.
6. Trypsin, 0.25% in PBS with 0.2% EDTA or TrypLE™ Express (Invitrogen).
7. CO_2 incubator with 37°C temperature and 5% CO_2 content.
8. 37°C water bath.
9. Laminar flow hood for cell-culture laboratory.
10. Pipettes.
11. Centrifuge.
12. 12-ml sterile centrifuge tubes.

2.5. Eight-Cell Embryo Injection

1. High quality, inverted light microscope with phase contrast optics (see Note 4) and 10× and 20× objectives, and a 10× eyepiece (e.g., Olympus IMT-2).
2. Two mechanical micromanipulators (e.g., Narishige or Leitz) fixed stably on each side of the microscope.
3. Two attachable, precision air pressure controllers (or injectors) (e.g., Narishige IM-9A) with 2 mm Ø plastic, connector tubes, and metallic capillary holders.
4. Computer-controlled infrared laser fire (1,480 nm) device with 40× objective (Hamilton Thorne Biosciences, Beverly, MA). This system allows for the alignment, temperature control, and delivery of a pulse precisely to ablate a small portion of the zona pellucida without damaging the embryo.

2.6. Embryo Transfer

1. ES cell injection media.
2. 70% ethanol.
3. Paper towels.
4. 35 mm Ø sterile, untreated, plastic cell-culture dishes.
5. Stereo dissecting microscope with adjustable zoom (10× to 45×), upper and scattered lower white light (e.g., Leica ZOOM 2000).
6. Two small tweezers and two small scissors.
7. Fine, high precision, biological forceps (e.g., Dumont Dumostar 5, cat. #10788) or watchmaker's forceps.
8. Serrefine 1.5 in. or smaller.
9. Sterile G27 hypodermic needles.
10. Mouth pipette device (e.g., Sigma-Aldrich A5177) with transfer capillaries (see Subheading 3.2.1).
11. Nembutal (pentobarbital sodium, CEVA SANTE ANIMALE) for anesthesia.

3. Methods

3.1. Mating Donor and Recipient Mice

1. For donors mate female mice with males later in the afternoon or in the evening. Put only one male and one female together in one cage (see Note 5).
2. Mate two or three CD1 females with one sterile vasectomized CD1 male to get the recipient females.
3. Check for post-coital plugs (firm, white coagulates) in the vagina in all females next morning. If a plug is found, then separate plugged female and note the date and male partner (see also Note 3). These females are considered to be 0.5 day after copulation or *post coitum* (dpc).

4. If no more plugged females are needed, separate the remaining ones from male partners. If they do not become pregnant they can be reused after 3 weeks.

3.2. Capillary Making

Micropipets or microcapillaries are one of the most important equipments used in microinjection procedures. The size and the structure of holding and injection capillaries should be correct for successful injection.

3.2.1. Pulling a Transfer Capillary

1. Heat the middle of a capillary in a microflame or Bunsen burner and rotate it until it starts to glow.
2. Remove the pipette from the flame suddenly and pull the two ends at the same time sharply. Your pull should be even and completely straight. See also Note 6.
3. Bend the capillary until it breaks at its middle. It gives two micropipettes. Put down one of them and carry on with the other.
4. Very carefully scratch the thin part with the diamond pen 2–3 cm away from the thicker part of your pipette.
5. Bend the thin end until it breaks down. The thin tip should break at the scratched point.
6. Check your pipette tip under the stereo microscope. The rim or lip of the thin end should be even and near circular without any chips or major cracks. If not score and bend again a few millimeters below the tip.
7. Hold the other half of your broken capillary and repeat from step 5.
8. Put the capillary on the microforge so that the tip is directly above the heating filament. Melt slightly the cut rim to polish cutting edges, but avoiding construction of internal diameter at the tip.
9. Bend the pipet about 15° at 2–3 mm above the end.
10. Store the pipets undisturbed in a suitable container.

3.2.2. Preparing Capillaries for Eight-Cell Embryo Holding

1. Make a capillary in a same way as the transfer capillary. The thin part should be at least 1 cm long and about 100 μm wide.
2. Put the capillary in the microforge and position the thin end opposite to the heating filament viewing from every angle.
3. Switch on heating and move the tip closer to the filament.
4. Follow the melting of the end of your pipette through the eyepiece and allow the inside diameter of the tip to shrink about 5–10 μm (less than one tenth of the original outside diameter). When the proper diameter is reached, switch off

the heating immediately or move the pipet away from the filament. See also Note 7.

5. Position the pipet to the far end of the heating filament such that the filament is to the side, 2–3 mm away from the tip.
6. Heat the filament and move it close to the pipet until it slowly starts to bend to the direction of the filament and stop heating when a 15–30° deviation from the original axis is reached. The angle depends on the settings of your micromanipulator and capillary holder. For microinjection, the tip of the holding pipette should be parallel to the bottom of the injection chamber.
7. Store the pipets undisturbed in a suitable container.

3.2.3. Preparing Needles for ES Cell Injection

1. Place one glass capillary in the puller and fasten it on both sides of the filament.
2. Perform a RAMP test if it has not already been done. Note the RAMP value (see also Note 8).
3. Replace capillary in the puller and put in a new one.
4. Loosen the screws, carefully take out the capillary and check its tip under the dissection microscope. It should be long enough to conveniently hold seven to nine ES cells in one row. Its diameter must be around 20–25 μm, just like the size of an ES cell.
5. If needed, modify one parameter at a time. Increase or decrease the value by 10 and repeat pulling from step 4.
6. If the needle's geometry looks right, put it in your left hand and with its tip touch the surface of the silicone or rubber halfpipe placed on the table with its convex side up.
7. Hold the scalpel in your right hand and slightly bend the tip on the halfpipe.
8. Touch the tip smoothly with the blade of the scalpel and break it. During this process the edge of the blade and the tip together must make a sharp angle.
9. Check the needle with real ES cells before injection.

3.3. Eight-Cell Embryo Collection

1. Put an aliquot of ES injection medium into the 37°C incubator. Put 4–5, equally 30–40 μl drops of KSOM onto the bottom of a 60-mm cell culture dish and cover with sterile, filtered, embryo tested mineral oil. Drops should be well separated from each other. Put the dish into the CO_2 incubator to warm up and equilibrate the medium. These should be performed at least 1 h before oviduct removal.
2. Cover a small area on the table with two to three sheets of paper towel. Wipe the scissors and tweezers with 70% ethanol.

3. On the second day morning after plug detection (day 2.5 dpc) euthanize the plugged donor mother by carbon dioxide asphyxiation or cervical dislocation.
4. Put the mouse onto the paper towel and lay it on its back. Wet and wipe its abdomen with 70% ethanol.
5. Pinch the abdominal skin with a forceps approximately 1 cm above the anus and while holding the skin, make a cut with scissors through the pinched skin right next to the forceps.
6. Replace the scissors to another forceps, pinch the skin on other lip of the cut and peel the skin by pulling the tweezers apart firmly, but carefully.
7. Open the abdominal wall in the whole length of the midline using a fresh pair of scissors and a pair of hooked forceps. Pull the coils of gut out of the way (to the direction of the head) and expose the retroperitoneal organs and locate the V-shape uterus.
8. Carefully cut out the oviducts from both sides with the fine pair of scissors and forceps; cutting between the ovary and the oviduct and the uterus. Avoid staining the excised tissue with blood.
9. Collect oviducts from all donor mice in a drop of prewarmed ES injection medium onto the bottom of a new, sterile 60-mm dish. This will keep the oviducts wet.
10. Suck up 1 ml of warm ES injection medium into a syringe and attach the blunted needle. Test the syringe to be sure that it is free of air bubbles and the medium is flowing smoothly before inserting the needle.
11. Take one oviduct into a clear drop and find the infundibulum (the opening of the oviduct at an ovarian site) under the stereomicroscope with gentle rotating with a fine forceps. Hold the infundibulum within the two arms of the forceps like a tube.
12. Insert gently a 1-ml syringe attached with the blunt-ended needle into the infundibulum following the direction of the lumen of the oviduct and clamp to stabilize the needle inside the infundibulum with the forceps. Flush the oviduct with approximately 0.1 ml of medium. If it is successful, you can clearly see the oviduct dilate while the medium flows through the oviduct. See also Note 9.
13. Repeat the flushing with all the oviducts.
14. Collect the uncompacted eight-cell embryos under a dissecting microscope and place them into a KSOM drop covered with mineral oil, and keep them at 37°C in the CO_2 incubator until the injection. In order to ensure germline transmission it is advisable to inject ES cells before compaction.

3.4. ES Cell Preparation

1. It is advised to use a cell line that is at the lowest passage number available (preferably less than passage number 16). Minimize the time that the ES cell line is cultured.
2. Prewarm trypsin or TrypLE™ Express and ES cell medium in the 37°C water bath. Put ES cell injection medium on ice.
3. Prefill the sterile 12-ml tube with 5 ml of ES cell medium.
4. Remove medium from the ES cell culture and rinse the cells once with PBS.
5. Pipette 1 ml of trypsin or TrypLE™ Express onto the cells and incubate at 37°C for 3 min.
6. Resuspend the ES cells by pipetting them eight to ten times with a 1-ml pipette, because a one-cell suspension is required.
7. Transfer the suspension into the 12-ml centrifuge tube prefilled with ES cell medium.
8. Centrifuge it for 5 min at 1,500 × *g* and discard the supernatant (ES cell medium).
9. Resuspend the cells completely in 500 μl of ice-cold ES cell injection medium.
10. Place the tube on ice for 30 min.
11. Discard about 3/4 of the medium without stirring up the settled cells from the bottom (see Note 10).
12. Add 500 μl of ice-cold ES injection medium and resuspend the cells again. These cells should be kept on ice and must be used for injection within 3 h (see Note 11).

3.5. Eight-Cell Embryo Injection

1. Orientate the holding capillary at the left side of the micromanipulator by positioning the bent end of the holder horizontally in the injection chamber such that the bent part of the holding pipet is sharply in focus at its total length.
2. Orientate the injection capillary at the right side of the manipulator by positioning it opposite to the holding capillary. The opening of the tip of the pipette should look down and the injection pipet should not tilt more than 5°.
3. Dip the capillaries into the ES injection medium drop and allow the media to flow up into the lumen of the capillaries. Take up the capillaries from the immersing drop.
4. Microinjection of eight-cell embryos is performed on a 6-cm Petri dish lid or in a depression slide. Put about 150 μl of ES injection media on the plate and add a 10–20 μl drop of ES cells to the media. Check the concentration of the ES cells and then cover the surface with embryo-tested mineral oil.
5. Put the embryos into the injection drop by a mouth pipette (see Note 12). Keep the embryos grouped and centered on the plate.

6. Put the injection chamber into the microscope stage, first focusing on the embryos, and then adjusting the height of the manipulators to bring the microcapillaries centered and in focus.
7. Select an embryo for injection. Adjusting the microscope stage, move it into the center between the two capillaries.
8. Using the holding pipette gently immobilize the embryo by suction. Select the correct position for injection by releasing, turning, then fixing the embryo again to identify a region for the laser perforation of the zona pellucida (Fig. 1a). You can help this maneuver by gently touching the zona pellucida of the embryo on the other side with the tip of the injection needle.
9. Load individual ES cells into the injection pipette (Fig. 1b). The cells selected should have good refraction (for viability) (see Note 13). Keep the volume of injected media as small as possible. A total of seven to nine healthy looking cells should have been picked up for one injection (see Note 14).
10. Select the site for drilling to maximize the distance between the nearest blastomere and target site in zona pellucida. It should be at 1 o'clock position of the embryo (Fig. 1a). The innermost isotherm ring (red) should be centered over the zona pellucida.
11. Perforate the zona pellucida with a single, 800 μs tangential pulse at 100% power; proceed using either a single left mouse click on the "Fire" button on the user interface or a foot pedal (Fig. 1c).
12. Bring the injection needle close to the hole on the zona pellucida with the joy stick; move into sharp focus while adjusting the manipulator height with the right hand.
13. Move the ES cell inside the injection capillary just close to the opening of the capillary.
14. Deposit the ES cells into the embryo to the maximum possible distance from the perforation site by positive pressure (Fig. 1d). Do not touch the blastomeres with the injection needle.
15. When the last cell has been expelled, stop the pressure of the injection needle (Fig. 1e). Carefully withdraw the needle from the embryo (see Note 15).
16. Separate the injected and non injected embryos inside the drop.
17. Repeat the injection process with a new embryo.
18. When all the embryos have been injected in the injection chamber, remove them and put them back into an empty

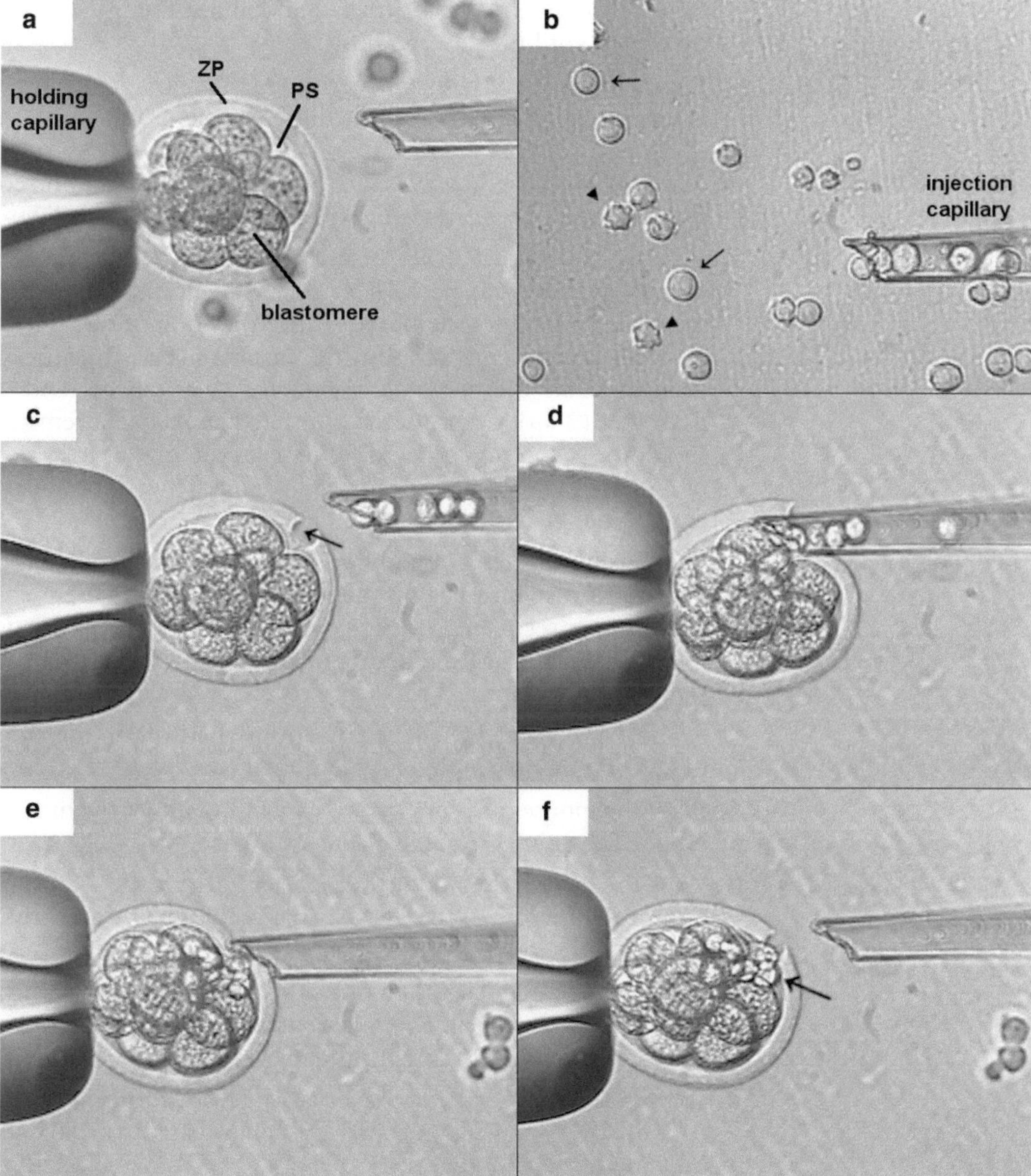

Fig. 1. Eight-cell embryo injection. (**a**) An uncompacted eight-cell embryo held on the holding pipette. Select the site for drilling to maximize the distance between the nearest blastomere and the target site on zona pellucida at 1 o'clock position. The zona pellucida (ZP) should be clearly in focus. Perivitelline space (PS). (**b**) Selection and collection of ES cells into the injection capillary. *Dark cells* (*arrowhead*) are not suitable for injection. *Round cells* with good refraction (*arrow*) should be selected. (**c**) Following one pulse of the laser, note a hole in the ZP (*arrow*) with the ×40 objective. (**d**) The injection pipette containing ES cells is gently pushed through the hole in the ZP and inject seven to nine ES cells into the PS. (**e**) To prevent the flow out of ES cells through the hole after the injection, apply a gentle downward pressure with the injection needle on the ZP just directly over the injected ES cells if necessary. (**f**) The injected eight-cell embryo. Note the ES cells located in the PS (*arrow*).

drop in the microculture media and culture them in a CO_2 incubator until the embryo transfer.

19. Repeat all the steps with another group of embryos each time in a new chamber with fresh medium and ES cells.

3.6. Embryo Transfer

1. Mate 30 6–8 weeks old CD1 females with vasectomized males; two to three female in one cage (see Notes 2 and 3).
2. Remove all mated females next day morning – they are the pseudopregnant females. These mice can be used as foster mothers for transgenic embryos, since they do not carry their own developing embryos, but the postcopulatory hormonal changes make the pregnancy possible. They can be used 2 days later (2.5 dpc) for uterus transfer. The recipient females should be 6–8 weeks old and 30 females can give at least five pseudopregnant mice. The remaining pseudopregnant females can be used for repeated mating after 2 weeks.
3. We usually transfer the embryos at least half hour after the injection procedure has finished (see Note 16). If there are not enough recipient females, you can culture the embryos overnight in KSOM medium and transfer them next day at the blastocyst stage.
4. Anesthetize recipient females by an intraperitoneal injection of 0.3–0.4 ml of 10× diluted Nembutal (see Note 17).
5. Place the mouse on a support for easy lifting under the microscope. The animal should lie on its stomach.
6. Wipe the hair on the middle of the back with alcohol. Cut the skin in the middle of the back with a regular pair of scissors. The wound should be about 2 cm long.
7. Locate the ovarian fat pad in one side of the abdominal cavity, beneath the body wall. It is apparent as white patch lateral to the vertebral column, close to the lower limit of the rib cage.
8. Pick up the body wall with the pair of hooked forceps, right over the ovary and cut an approx. 5 mm incision to avoid cutting any of the blood vessels.
9. Grab the ovarian fat pad with the blunt forceps, pull out the ovary, oviduct and the upper part of the uterus. Fix the fat pad with the serrefine and pull over the opposite side of the animal. Gently pick up the mouse and place it on the stage on the stereo microscope.
10. Transfer the embryos from the microdrop culture into a large drop of ES injection medium and wash the embryos in three drops to get rid of the oil in the transferring drop.
11. Take a small amount of medium into the tip of the transfer pipet, then a small volume of air, then medium, and then air again. Into the next volume of medium the embryos should

be drawn in; then take small volume of air and lastly a small volume of the medium. Place the pipet safely such that it does not touch anything until you are ready for injecting the embryos into the oviduct.

12. Transfer not more than eight to ten embryos into one uterus. The survival and implantation rate of micromanipulated embryos is better than the conventional transgenic procedures where you can transfer 15 embryos in one side as well.
13. Hold a fine forceps in your left hand, the loaded embryo transfer pipet and a 30G hypodermic needle in the right hand.
14. Grab the oviduct with a fine forceps close to the ovary and make a hole on it by inserting the hypodermic needle into the uterus lumen. Avoid puncturing the opposite wall of the uterus.
15. Let the hypodermic needle drop and move the transfer pipet in between the forefinger and thumb in your right hand, while not releasing the left hand.
16. Insert the embryo transfer pipet into the preformed hole. Blow the content of the transfer pipet into the uterus. Air bubbles, which separated the liquid content in the pipet will show up in the uterus, indicating the success of the transfer (see Note 18).
17. Release the fat pad, let the organs slide back and gently push the uterus back into the peritoneum.
18. Repeat the transfer on the other side if desired.
19. Close the skin of the animal with a wound clip and place it into a fresh cage on a quiet warm place until the mouse recovers from the anesthesia. You may put two females into the same cage. They will help each other in raising a joint litter. If one of the foster does not become pregnant remove from the cage a few days before the delivery.
20. When the delivered mice get hair you can see the degree of chimerism on the color of the mice. The injected R1 ES cells derived from agouti colored blastocysts and our host embryos derived from white or non-agouti colored Twitcher mice (12).

4. Notes

1. Donor females should provide a large number of embryos with permissive cellular environment or "nest" for manipulated ES cells. The coat-color genotype of the recipient embryo should differ from the genotype of donor ES cells and this way the chimerism can simply be detected. The most common strains used for embryo donors are C57BL/6,

C57BL/6Tyrc-2j (B6 albino), 129Sv/Ev, or SW. In our experiments, we used Twitcher heterozygote females (albino and non-agouti) mated with Twitcher heterozygous males.

2. Surrogate mothers must tolerate implanted embryos and be able to feed and nurture newborns until they are self supporters. Outbred strains (like CD1) are preferred because they are good mothers, generally more resistant to environmental stress, mate more frequently, and usually stronger than inbred ones.

3. Males can be vasectomized and checked for sterility in house (13), but they are also available from vendors of laboratory mice (e.g., The Jackson Laboratory). Copulation with a vasectomized male also produces post coital plug and results in pseudopregnancy where subsequent hormonal changes in the female are necessary for efficient implantation of manipulated embryos. 30–40 vasectomized males enough to produce recipient males. We used them simultaneously and set up mating with two to three females.

4. Phase contrast optics provides clearer and better vision of 3D structures and it is very important for ES cell selection based on morphology.

5. Some protocols recommend superovulation to increase the embryo recovery. We used naturally ovulated females because embryo yield by superovulation has the disadvantage of being less synchronous with regard to their developmental stage.

6. Your pull is correct if you can get an evenly thin, completely straight, 5–10 cm long middle part with a diameter about the size the embryo (90–120 μm). This technique needs some time, patience and practice as much as injection needle fabrication.

7. The shrunk tunnel at the end of your capillary should be the straight continuation of your pipette's lumen. The surface of the tip should be even, plain and perpendicular to the axis of the capillary. In case of any deviation from the characteristics described here, the microforge manipulations should be repeated after breaking off the end of pipette or with a new transfer pipette.

8. RAMP test is a heating test and is used to automatically adjust the optimal heating power of the filament. It depends strongly on the type of applied filament and capillary. This also means if you want to use another type of filament or capillary, you *must* redo the RAMP test, otherwise you might ruin your filament or the puller machine. While you work with the *same filament* and the *same type of capillary*, you can safely omit further RAMP tests. For our capillaries we set the following

parameters: HEAT = RAMP + 10, PULL = 50, VELOCITY = 75, TIME = 25, PRESSURE = 300. Press "PULL."

9. Washing the oviducts may need some practice in advance because of its small size. You have to practice to keep the infundibulum stably under the microscope in one hand and in the meantime, maneuver the syringe in the other hand. You need some time and a lot of patience to learn this technique. Practice as much as you need to get your hands stable under the microscope.
10. This trick can remove the majority of the floating dead cells and cellular debris from the suspension.
11. Cooling down ES cells and keeping them in HEPES containing medium before injection significantly slows down the pH change of the medium outside the incubator and prevents cells from getting aggregated.
12. The number of injected embryos at once is dependent on the skill of the person. We should say put as much as embryo at once into the injection chamber as you can inject through for not more than 30 min.
13. Cells that look dark are dying and should not be used. The healthy looking round cells are available for microinjection.
14. The cells in the injection pipette should be loaded up to the very tip so as to form a train of cells.
15. If a huge amount of media is injected into the perivitelline space of the embryo, it happens sometimes that, when we withdraw the capillary after the injection, the ES cells start to flow out through the hole. Applying a gentle downward pressure with the injection needle on the zona pellucida just directly over the injected ES cells could prevent this event (Fig. 1e).
16. Other investigators culture the injected eight-cell embryos overnight and put back only the healthy developing embryos. They found 90% of embryos suitable for transfer after the culture.
17. The dosage of the Nembutal should be tested previously.
18. The embryo transfer is not an easy procedure too. Just one bad movement and you can lose your all day's work. We advice to practice without embryos previously as much as needed. If the transfer procedure is in your hand, you can start the injection. In case when no or not enough 2.5 dpc recipient female is available, you can use 0.5 dpc recipients for oviduct transfer. Even for a good skilled person the uterus transfer is easier to perform. We have not recognized any difference between the oviduct or uterus transfer related to the number of pups or the degree of chimerism, but others find dissimilarities (10).

References

1. Monteith DP, Leung JD, Borowski AH, Co DO, Praznovszky T, Jirik FR, et al. (2004) Pronuclear microinjection of purified artificial chromosomes for generation of transgenic mice: pick-and-inject technique. *Methods Mol. Biol.* **240**: 227–242.
2. Hernandez, D., Mee, P.J., Martin, J.E., Tybulewicz, V.L.J., and Fisher, E.M.C. (1999) Transchromosomal mouse embryonic stem cell lines and chimeric mice that contain freely segregating segments of human chromosome 21. *Hum. Mol. Genet.* **8**: 923–933.
3. Shen, M.H., Mee, P.J., Nichols, J., Yang, J., Brook, F., Gardner, R.L., Smith, A.G., and Brown, W.R. (2000) A structurally defined mini-chromosome vector for the mouse germ line. *Curr. Biol.* **10**: 31–34.
4. Tomizuka, K., Yoshida, H., Uejima, H., Kugoh, H., Sato, K., Ohguma, A., Hayasaka, M., Hanaoka, K., Oshimura, M., and Ishida, I. (1997) Functional expression and germline transmission of a human chromosome fragment in chimaeric mice. *Nat. Genet.***16**: 133–143.
5. Tomizuka, K., Shinohara, T., Yoshida, H., Uejima, H., Ohguma, A., Tanaka, S., Sato, K., Oshimura, M., and Ishida, I. (2000) Double trans-chromosomic mice: Maintenance of two individual human chromosome fragments containing Ig heavy and kappa loci and expression of fully human antibodies. *Proc. Natl. Acad. Sci.* **97**: 722–727.
6. Co, D.O., Borowski, A.H., Leung, J.D., van der Kaa, K.J., Hengst, S., Platenburg, G.J., Pieper, F.R., Perez, C.F., Jirik, F.R., and Drayer, J.I. (2000) Generation of transgenic mice and germline transmission of a mammalian artificial chromosome introduced into embryos by pronuclear microinjection. *Chromosome. Res.* **8**: 183–191.
7. Voet T, Vermeesch J, Carens A, Dürr J, Labaere C, Duhamel H, David G, Marynen P. (2001) Efficient male and female germline transmission of a human chromosomal vector in mice. *Genome Res.* **11**(1): 124–36.
8. Burgoyne, P.S. and Mahadevaiah, S.K. (1993) Unpaired sex chromosomes and gametogenetic failure. In *Chromosomes Today, Volume 11* (eds. A.T. Summers and A.C. Chandley) pp. 243–263. Chapman and Hall, New York.
9. Hunt, P., LeMaire, R., Embury, P., Sheean, L., and Mroz, K. (1995) Analysis of chromosome behavior in intact mammalian oocytes: Monitoring the segregation of a univalent chromosome during female meiosis. *Hum. Mol. Genet.* **4**: 2007–2012.
10. Ramírez, M.A., Fernández-González, R., Pérez-Crespo, M., Pericuesta, E., Gutiérrez-Adán A. (2009) Effect of Stem Cell Activation, Culture Media of Manipulated Embryos, and Site of Embryo Transfer in the Production of F0 Embryonic Stem Cell Mice. *Biology of Reproduction* **80**: 1216–1222
11. Poueymirou W.T., Auerbach W., Frendewey D., et al. (2007) F0 generation mice fully derived from gene-targeted embryonic stem cells allowing immediate phenotypic analyses. *Nat Biotechnol.* **25**: 91–9.
12. Katona RL, Sinkó I, Holló G, Szucs KS, Praznovszky T, Kereso J, et al. (2008) A combined artificial chromosome-stem cell therapy method in a model experiment aimed at the treatment of Krabbe's disease in the Twitcher mouse. *Cell. Mol. Life Sci.* **65**(23): 3830–3838.
13. Nagy, A., Gertsenstein, M., Vintersten, K. & Behringer, R.. (2003) Manipulating the Mouse Embryo: a Laboratory Manual. 764 p. (Cold Spring Harbor Laboratory Press, Cold Spring Harbor, N.Y.).

Chapter 14

Mammalian Artificial Chromosomes and Clinical Applications for Genetic Modification of Stem Cells: An Overview

Robert L. Katona, Sandra L. Vanderbyl, and Carl F. Perez

Abstract

Modifying multipotent, self-renewing human stem cells with mammalian artificial chromosomes (MACs), present a promising clinical strategy for numerous diseases, especially ex vivo cell therapies that can benefit from constitutive or overexpression of therapeutic gene(s). MACs are nonintegrating, autonomously replicating, with the capacity to carry large cDNA or genomic sequences, which in turn enable potentially prolonged, safe, and regulated therapeutic transgene expression, and render MACs as attractive genetic vectors for "gene replacement" or for controlling differentiation pathways in progenitor cells. The *status quo* is that the most versatile target cell would be one that was pluripotent and self-renewing to address multiple disease target cell types, thus making multilineage stem cells, such as adult derived early progenitor cells and embryonic stem cells, as attractive universal host cells. We will describe the progress of MAC technologies, the subsequent modifications of stem cells, and discuss the establishment of MAC platform stem cell lines to facilitate proof-of-principle studies and preclinical development.

Key words: Mammalian artificial chromosomes, MACs, Embryonic stem cells, Adult stem cells, Gene therapy, Transgenic animals

1. Introduction

1.1. Background on Mammalian Artificial Chromosomes

Various groups worldwide have developed mammalian artificial chromosomes (MACs) and these technologies have been extensively reviewed (1–18). The procedures to generate MACs fall into three broad categories: (1) the de novo formation by centromere seeding ("bottom-up" approach) (19–30), (2) the truncation of normal chromosomes or the modification of naturally occurring minichromosomes ("top-down" approach) (31–42), and (3) the induction of the intrinsic large-scale amplification of

Gyula Hadlaczky (ed.), *Mammalian Chromosome Engineering: Methods and Protocols*, Methods in Molecular Biology, vol. 738, DOI 10.1007/978-1-61779-099-7_14, © Springer Science+Business Media, LLC 2011

pericentromeric heterochromatin to generate satellite DNA-based artificial chromosomes (SATACs and ACEs) (43–49). All three processes generate MACs with functioning centromeres to maintain nuclear location and to participate in mitotic/meiotic segregation; telomeres to preserve integrity and autonomy (e.g., nonintegrating and nontranslocating); and origins of DNA replication to duplicate genetic information required for transmission to daughter cells.

While various transgenes have been incorporated into centromere-seeded MACs (50, 51), they have restricted clinical applications due to the necessity of de novo chromosome formation. Furthermore, in all reported cases, the newly formed MACs are composed of significantly more DNA than the original transfected components, strongly suggesting that endogenous and unknown DNA had been incorporated. This issue raises safety questions which must be resolved before considering this technology for gene therapy. Nevertheless, this methodology has been valuable for the study of the structure and function of centromeres (52–54).

For MACs to function as genetic vectors, they should (1) possess large DNA payload capacity of 1 Mb or greater, (2) facilitate efficient gene(s) loading, (3) be capable of being transferred from one cell to another, and (4) enable stable, high-level transgene expression. The two MAC technologies that have made the most progress in addressing these stringent requirements are a class of truncated natural chromosomes (21ΔqHACs and 21ΔpqHACs) and satellite DNA-based artificial chromosomes (SATACs and ACEs).

21ΔqHACs and 21ΔpqHACs. By systematically deleting euchromatin from natural chromosomes, truncation-derived MACs preserve functioning centromeres, telomeres, and DNA origins of replication, while minimizing potential gene dosage effects in host cells. Of particular interests are the two MACs that were generated by telomere-directed truncation of the distal q ("long") arm of human acrocentric chromosome 21 generating 21ΔqHAC, and the subsequent truncation of the remaining distal p ("short") arm generating 21ΔpqHAC (42, 55). Of note, the remaining p arm of 21ΔqHAC is assumed to encode only an rDNA gene tandem array and pericentromeric heterochromatin, and hence should be genetically "neutral." A single loxP targeted integration site has been inserted into both ΔHACs, enabling the "loading" of various single transgene sequences by Cre-recombinase mediated DNA recombination. However, serial multiple loadings are hindered by the competing integration and excision processes that are inherent in the cre–loxP system, which increases the probability that the original integrated transgene would excise during the introduction of additional constructs. Therefore to attain greater transgene copy number, the single plasmid targeting vector must encode multiple transgenes, limiting the versatility of this recombination system. Transgenes that have been efficiently loaded and expressed from these vectors include hypoxanthine guanine phosphoribosyl transferase (HPRT), erythropoietin (EPO), enhanced green fluorescent protein (EGFP), antibody/cytokine receptor chimera, DNA-dependent protein kinase catalytic subunit (DNA-PKcs, critical for DNA repair), proinsulin, human dystrophin, CD40L, the human P53, and telomerase (42, 55–64).

ACEs. SATACs and ACEs primarily consist of heterochromatin and pericentromeric sequences, which are purported to be devoid of gene-coding sequences beyond rDNA (13, 43–48). There is no evidence in humans that amplification of these pericentromeric sequences is deleterious to an individual, as polymorphisms in the short arms of acrocentric chromosomes have been shown to consist of amplified pericentromeric heterochromatin and/or rDNA (65–73) and have been inherited with no adverse effects (74–78). Furthermore, supernumerary chromosomes derived from amplified pericentromeric heterochromatin and/or rDNA sequences of acrocentric chromosomes have been shown to be stably inherited and through 1–3 generations in humans without any deleterious effects of phenotypes (79–85).

Additionally, ACEs have been engineered with multiple (>50) site-specific recombination acceptor sites (*att*P), which enable the unidirectional loading of heterologous DNA (encoding an *att*B recombination donor site) via a mutant lambda integrase (48, 49). The DNA recombination is unidirectional in the context of the mammalian cells, given that the engineered integrase lacks the bacterial co-factors necessary for excision. The combination of the multiple *att*P sites on the ACE and the unidirectional mutant integrase enables multiple loadings (during a single transfection) or sequential loadings (via multiple transfections) of transgenes (48, 86). Transgenes that have been efficiently loaded and expressed from ACEs include EGFP, red fluorescent protein (RFP), EPO, monoclonal antibodies (Mabs), and therapeutic galactocerebrosidase transgenes to treat Krabbe disease in the Twitcher mice (48, 49, 86–89). In addition, specific MAb productivities of 55 pg/cell/day and 4 g/L yields in nonoptimized bioreactors have been attained after transferring MAb-ACEs to various industrial strains of CHO cells, demonstrating high level ACE-encoded transgene expression (87).

2. Methods

2.1. Transfer of MACs

The delivery of transgene-loaded MACs to primary mammalian cells is a fundamental challenge for gene therapy applications. While microcell-mediated chromosome transfer (MMCT) has been the most widely used technique for transferring MACs to various cell types, the method is tedious and generally inefficient (10^{-7} to 10^{-5}) (90–93). However, recently gene-loaded 21ΔqHACs have been transferred to human primary fibroblasts (58) and to human hematopoietic stem cells (HSCs) (94) at clinically relevant frequencies of 1.26×10^{-4} and 4.0×10^{-4}, respectively. Nevertheless, despite these increased transfer efficiencies, during the process of microcell formation, the host cell generates a heterogeneous population of microcells encapsulating endogenous chromosomes as well as MACs. Consequently, this transfer technique increases the probability that host chromosomes and chromosomal fragments will be cotransferred with MACs to the clinical target cells, which may result in unknown and potentially deleterious effects to the patient.

SATACs/ACEs are purified to near homogeneity by high-speed flow sorting due to their unique nucleotide composition (a predominance of AT base pairs to GC base pairs compared with endogenous host cell chromosomes) (95), and are the only technology that permits the production of purified gene-loaded MACs. While isolated SATACs have been microinjected into the pronuclei of murine embryos to generate transgenic mice (96), very little success has been made in microinjecting mammalian cells in vitro due to the large outer diameter sizes (2.3–3.2 μm) of the microinjection needles (97). Alternatively, purified ACEs have been transferred to various cells, including human mesenchymal stem cells (MSCs) (98) and HSCs (99) using commercially available cationic reagents and dendrimers (100) with high transfer efficiencies (10^{-2} to 10^{-4}). Recently, ACEs loaded with a therapeutic transgene have been transferred to murine ESCs by lipid-mediated chromosome transfer, establishing stem cell clones carrying intact ACEs, which was verified by fluorescent in situ hybridization (88).

2.2. Stability in Cells and Transgenic Animals

In general, MAC stability has been assessed in immortalized cell lines, such as human HT1080 fibrosarcoma cells and CHO rodent cells. The karyotypes of these cells vary greatly in culture, particularly, after rounds of MMCT and drug selection, which in turn may not replicate the human primary cell environment for the MAC centromere function. The transfer of 21ΔHACs and ACEs into primary human MSCs, HSCs, and fibroblasts demonstrated both autonomous chromosome maintenance and stable transgene expression (58, 94, 98, 99). Mouse ES cell lines carrying therapeutic ACEs maintained high quality karyotypes, preserved their pluripotent state (which was verified by the expression of four widely accepted pluripotency markers alkaline-phosphatase, Nanog, Sox2, and Oct3/4), and expressed the therapeutic transgene. Under nonselective culture conditions, 98.9% of the stem cells retained the therapeutic ACE through more than 100 cell generations.

When the colonies of these cells were differentiated to embryoid bodies (EBs), therapeutic ACE chromosomes were found in ~99% of the nuclei of the EBs (88).

Transgenic animals have been generated by pronuclear microinjection of purified ACEs (96), by blastocyst injection of murine ESCs modified with HACs (30) and ACEs (88), and by somatic cell nuclear transfer of HAC-modified bovine fibroblasts (101–103). Additionally, SATACs and HACs have been passed on to more than three generations through the mouse germline demonstrating meiotic stability (30, 96), providing further evidence that the MAC centromeres function properly in vivo. The SATAC transgenic founder mice and their progeny were healthy, robust, and free of neoplasms, which provided initial evidence that MACs are safe for in vivo applications (96).

2.3. Safety Issues

MACs are prospectively safer alternatives to viral gene therapy vectors, which have been shown to cause serious antigenic responses (104), induce systemic toxicity (105), demonstrate shedding in patient semen (106), be associated with chromosomal deletions/rearrangements (107, 108), and produce insertional oncogenesis (109, 110). These safety issues may be addressed with nonintegrating, autonomously replicating MACs, thereby providing mitotic stability and consistent vector/transgene copy number, as well as eliminating viral deleterious effects. As described earlier, 21ΔHACs and ACEs are ostensibly devoid of gene sequences that could lead to indirect or direct dosage effects in cells destined for patient implantation. Moreover, transferring flow sorted ACEs by cationic reagents assures that only the therapeutic ACEs is delivered to patients, reducing the risk of delivering host chromosomes and chromosomal fragments via MMCT.

2.4. Stem Cells as Therapeutic MAC Host Cells

Due to the fact that the applications for stem cell therapy are as diverse as the scope of target diseases, the host stem cell with the most potential would be one with multilineage capacity, such as adult derived early progenitor cells and embryonic stem cells. Initially human embryonic stem cells (hESCs) derived from embryos have encountered resistance due to ethical considerations. However, successful establishment of permanent ESC lines (111), isolation of pluripotent spermatogonial stem cells from adult testis tissue (112), and generation of fully pluripotent stem cells from cultured human primordial germ cells (113, 114), from single blastomere (115), and individualized ESC cell lines through somatic cell nuclear transplantation (116) seem to have addressed this issue. Another option for deriving multilineage cells was the formation of induced pluripotent stem (iPS) cells generated from mouse embryonic or adult fibroblasts by introducing four factors, (Oct3/4, Sox2, c-Myc, and Klf4), under ESC culture conditions. Unfortunately, the large-scale production of iPS cells, which are fully compatible with their natural counterparts, was hampered by the lack of concerted expression of these factors and that constitutive overexpression was detrimental for self-renewal (117).

2.5. ESCs, ESC-Derived Lineages

As discussed above, there are options for isolation of hESCs that address ethical issues related to isolation from embryos, including establishment of permanent ESC lines, or isolation from testis tissue, or primordial germ cells. However, direct transplantation of ESCs can result in the development of teratomas (118, 119) representing a serious safety issue. Similar concerns have also been raised regarding the direct therapeutic application of MSCs, namely, that transplanted MSCs continue to replicate in vivo and preferentially incorporate into the tumor stroma, constituting a significant fraction of the stromal tissue and possibly supporting tumor growth (120–126). The risk of tumorigenesis may be overcome with the use of stem cells partially differentiated ex vivo, prior to transplantation.

ESC differentiation models have established the potential to generate large numbers of lineage specific cells for cell replacement therapies. Mesoderm-derived lineages, including the hematopoietic lineages (127–131), vascular (132–134), and cardiac (135–141), are among the easiest to generate from ESCs and have been studied in considerable detail. Furthermore, methods have been established for selectively expanding multipotential hematopoietic cell populations (142), neutrophils (143), megakaryocytes (144), mast cells (145), eosinophils (146), dendritic cells (147), and erythroid cells (148, 149). Several studies have provided evidence for the generation of endoderm-derived cell types including populations that display characteristics of pancreatic islets (150–153), hepatocytes (154–156), thyrocytes (157), lung (158), and intestinal cells (159). Different protocols have also been established to promote neuroectoderm differentiation: oligodendrocytes capable of forming myelin sheaths around host neurons when transplanted into a myelin-deficient rat model of multiple sclerosis (160), glial precursors (161), midbrain dopaminergic neurons (162, 163), and primitive neural stem cells (164). In addition to the neural lineages, ESCs can also give rise to epithelial cells that will undergo differentiation to populations that express markers of keratinocytes (165). ES-cell-derived keratinocytes were able to form structures that resemble embryonic mouse skin, indicating that they possess some capacity to generate a functional tissue (166). Exhaustive studies will have to be conducted to determine whether induced ESC differentiation generates *bona fide* functional cells in vivo.

2.6. Adult Pluripotent Stem Cells and Induced Pluripotent Stem Cells

The adult derived blastomere-like and epiblast-like stem cell types (167) exemplify prime candidates with ES-like differentiation capacity. Multipotent stem cells have been derived from bone marrow (167, 168), dental pulp (169), adipose tissue (170), and amniotic fluid (171). In addition, MACs encoding synthetic or natural genetic networks may be used to program pluripotent stem cells (natural or MAC-generated induced pluripotent stem (iPS) cells) to follow specific lineage pathways to minimize the potential tumorigenicity issues.

2.7. Clinical Applications of MAC-Modified Stem Cells

Pluripotent or multipotent stem cells carrying MACs (MAC-SCs) encoding diverse combinations of transgenes may be considered for a variety of clinical applications. MAC-SCs expressing complex gene pathways could be converted into "bioreactors" generating blood cells (e.g., platelets, red blood cells), immune cells (e.g., artificial antigen presenting cells), and pancreatic islet cells. Xenogeneic or allogeneic transplantable organs may someday be produced in which MAC-encoded sequences could "down regulate" (by RNA interference technologies) the expression of host stem cell MHC class I/II loci and substitute MAC-encoded patient specific loci and appropriate transgenes to eliminate the lifetime need for immunosuppressive drugs post-transplantation. Recent discoveries that stem/progenitor cells preferentially home to tumor sites (122, 123, 125) offer the possibility for tumor therapy with artificial chromosomes armed with an arsenal of transgenes, including a drug–prodrug suicide system and cytokines (e.g., interferon-β).

2.8. Ex Vivo Therapy

At present, there is no ongoing clinical trial with artificial chromosomes. However, preliminary results of preclinical experiments (49, 57, 58, 89, 172) have made a case for the feasibility of MAC-based ex vivo cell therapies where constitutive or overexpression of therapeutic gene(s) is acceptable or required. The successful transfer of MACs to MSCs (98, 173, 174), HSCs (94, 99) and ESCs (88), and the subsequent MAC transgene expression in differentiated cells have opened a broad avenue of indications, including, lysosomal storage diseases, hematological diseases, immunodeficiency diseases, and cancer (49, 173).

The nonexhaustive list above indicates the broadness of diseases and disorders that could be targeted with specific strategies and protocols (e.g., central nervous system, see review (175)). In most of the cases, generation of stem cell derived therapeutic lineages require multiple transgenes, including a combined positive/negative selection system (176–178), in order to eliminate nondifferentiated stem cells. Platform ACEs by means of multiple acceptor sites have the capacity to accommodate such complex genetic accessories.

2.9. Surveillance and Safe-Guards

Additionally, MAC-SCs may also be engineered to express the recently reported artificial magnetic resonance gene encoding lysine rich-protein (LRP; (179)). The location of the MAC-SCs and any in vivo derived differentiated tissue may be detected by magnetic resonance imaging (MRI), providing a powerful noninvasive approach to monitor stem cell fate. Micro-PET analysis also can be used to follow stem cell fate in vivo (180). Finally, these therapies could be terminated by loading MACs with apoptosis-inducing genes that are expressed from inducible promoters (e.g., using tetracycline response elements). Therefore, the large payload capacity of MACs will ultimately enable reprogramming of pluripotent stem cells to specific lineages, therapeutic transgene expression, surveillance of tissue engraftment, and the precise termination of therapy.

3. Conclusions and Future Directions

While the imperfection of viral vectors still impede the long awaited breakthrough in gene therapy, mammalian artificial chromosomes are being developed as potentially safer genetic vectors for the therapeutic modification of human stem cells. ACEs and 21ΔHACs have made the greatest progress in demonstrating efficient gene loading and the subsequent transfer and stable transgene expression in embryonic and adult stem cells. Equally as important, these genetic vectors were maintained as autonomous chromosomes that facilitated transgene expression during stem cell differentiation and animal development.

These proof-of-concept studies are the first steps towards exploiting their potential for safer and coordinated control of stem cell differentiation and therapeutic transgene expression. To bring MAC-based stem cell therapies to the clinic, further advances in MAC technology should be made, including the demonstration of inducible promoter systems required for the fine-tuned control of therapeutic transgene expression and the potential activation of apoptotic transgenes for managed gene therapy termination; the loading and expression of genetic pathways that control the lineage specific differentiation of pluripotent stem cells; and the establishment of MAC platform stem cell lines demonstrating controlled differentiation and long-term transgene expression. While MAC-modified stem cells have great potential to address a wide spectrum of diseases, monogenic therapies without viable treatments will be the first indications addressed, where the constitutive overexpression of single transgenes, such as lysosomal storage or X-linked SCID diseases, can provide profound results. The success of these first trials will enable the engineering of genetic pathways and networks, which in turn will facilitate greater control of therapeutic transgene expression, tracking transplant fate, and transplant safety, thereby realizing the potential of modifying stem cells with mammalian artificial chromosomes for the treatment of human diseases.

References

1. Ascenzioni, F., Donini, P. and Lipps, H.J. (1997). Mammalian artificial chromosomes – vectors for somatic gene therapy. *Cancer Lett.* **118**, 135–142.
2. Basu, J. and Willard, H.F. (2006). Human artificial chromosomes: potential applications and clinical considerations. *Pediatr. Clin. North Am.* **53**, 843–853, viii-843–853, viii.
3. Bridger, J.M. (2004). Mammalian artificial chromosomes: modern day feats of engineering – Isambard Kingdom Brunel style. *Cytogenet. Genome Res.* **107**, 5–8.
4. Cooke, H. (2001). Mammalian artificial chromosomes as vectors: progress and prospects. *Cloning Stem Cells* **3**, 243–249.
5. Larin, Z. and Mejia, J.E. (2002). Advances in human artificial chromosome technology. *Trends Genet.* **18**, 313–319.
6. Lewis, M. (2001). Human artificial chromosomes: emerging from concept to reality in biomedicine. *Clin. Genet.* **59**, 15–16.
7. Lim, H.N. and Farr, C.J. (2004). Chromosome-based vectors for Mammalian cells: an overview. *Methods Mol. Biol.* **240**, 167–186.
8. Monaco, Z.L. and Moralli, D. (2006). Progress in artificial chromosome technology. *Biochem. Soc. Trans.* **34**, 324–327.
9. Ren, X., Tahimic, C.G.T., Katoh, M., Kurimasa, A., Inoue, T. and Oshimura, M. (2006). Human artificial chromosome vectors meet stem cells: new prospects for gene delivery. *Stem Cell Rev* **2**, 43–50.
10. Ehrhardt, A., Haase, R., Schepers, A., Deutsch, M.J., Lipps, H.J. and Baiker, A. (2008). Episomal vectors for gene therapy. *Curr Gene Ther* **8**, 147–161.
11. Macnab, S. and Whitehouse, A. (2009). Progress and prospects: human artificial chromosomes. *Gene Ther.* **16**, 1180–1188.
12. Duncan, A. and Hadlaczky, G. (2007). Chromosomal engineering. *Curr. Opin. Biotechnol.* **18**, 420–424.
13. Hadlaczky, G. (2001). Satellite DNA-based artificial chromosomes for use in gene therapy. *Curr. Opin. Mol. Ther.* **3**, 125–132.
14. Grimes, B.R. and Monaco, Z.L. (2005). Artificial and engineered chromosomes: developments and prospects for gene therapy. *Chromosoma* **114**, 230–241.
15. Basu, J. and Willard, H.F. (2005). Artificial and engineered chromosomes: nonintegrating vectors for gene therapy. *Trends Mol Med* **11**, 251–258.
16. Brown, W.R., Mee, P.J. and Hong Shen, M. (2000). Artificial chromosomes: ideal vectors? *Trends Biotechnol.* **18**, 218–223.
17. Lipps, H.J., Jenke, A.C., Nehlsen, K., Scinteie, M.F., Stehle, I.M. and Bode, J. (2003). Chromosome-based vectors for gene therapy. *Gene* **304**, 23–33.
18. Grimes, B.R., Warburton, P.E. and Farr, C.J. (2002). Chromosome engineering: prospects for gene therapy. *Gene Ther.* **9**, 713–718.
19. Harrington, J.J., Van Bokkelen, G., Mays, R.W., Gustashaw, K. and Willard, H.F. (1997). Formation of de novo centromeres and construction of first-generation human artificial microchromosomes. *Nat. Genet.* **15**, 345–355.
20. Ikeno, M., Grimes, B., Okazaki, T., Nakano, M., Saitoh, K., Hoshino, H., McGill, N.I., Cooke, H. and Masumoto, H. (1998). Construction of YAC-based mammalian artificial chromosomes. *Nat. Biotechnol.* **16**, 431–439.
21. Henning, K.A., Novotny, E.A., Compton, S.T., Guan, X.Y., Liu, P.P. and Ashlock, M.A. (1999). Human artificial chromosomes generated by modification of a yeast artificial chromosome containing both human alpha satellite and single-copy DNA sequences. *Proc Natl Acad Sci USA* **96**, 592–597.
22. de las Heras, J.I., D'Aiuto, L. and Cooke, H. (2004). Mammalian artificial chromosome formation in human cells after lipofection of a PAC precursor. *Methods Mol. Biol.* **240**, 187–206.
23. Basu, J., Willard, H.F. and Stromberg, G. (2007). Human artificial chromosome assembly by transposon-based retrofitting of genomic BACs with synthetic alpha-satellite arrays. *Curr Protoc Hum Genet* **Chapter 5**, Unit 5.18-Unit 5.18.
24. Nakashima, H., Nakano, M., Ohnishi, R., Hiraoka, Y., Kaneda, Y., Sugino, A. and Masumoto, H. (2005). Assembly of additional

heterochromatin distinct from centromerekinetochore chromatin is required for de novo formation of human artificial chromosome. *J. Cell. Sci.* **118**, 5885–5898.

25. Kaname, T., McGuigan, A., Georghiou, A., Yurov, Y., Osoegawa, K., De Jong, P.J., Ioannou, P. and Huxley, C. (2005). Alphoid DNA from different chromosomes forms de novo minichromosomes with high efficiency. *Chromosome Res.* **13**, 411–422.
26. Basu, J., Compitello, G., Stromberg, G., Willard, H.F. and Van Bokkelen, G. (2005). Efficient assembly of de novo human artificial chromosomes from large genomic loci. *BMC. Biotechnol* **5**, 21–21.
27. Grimes, B.R., Rhoades, A.A. and Willard, H.F. (2002). Alpha-satellite DNA and vector composition influence rates of human artificial chromosome formation. *Mol. Ther.* **5**, 798–805.
28. Ikeno, M., Suzuki, N., Hasegawa, Y. and Okazaki, T. (2009). Manipulating transgenes using a chromosome vector. *Nucleic Acids Res.* **37**, e44–e44.
29. Kouprina, N., Ebersole, T., Koriabine, M., Pak, E., Rogozin, I.B., Katoh, M., Oshimura, M., Ogi, K., Peredelchuk, M., Solomon, G., Brown, W., Barrett, J.C. and Larionov, V. (2003). Cloning of human centromeres by transformation-associated recombination in yeast and generation of functional human artificial chromosomes. *Nucleic Acids Res.* **31**, 922–934.
30. Suzuki, N., Nishii, K., Okazaki, T. and Ikeno, M. (2006). Human artificial chromosomes constructed using the bottom-up strategy are stably maintained in mitosis and efficiently transmissible to progeny mice. *J. Biol. Chem* **281**, 26615–26623.
31. Carine, K., Solus, J., Waltzer, E., Manch-Citron, J., Hamkalo, B.A. and Scheffler, I.E. (1986). Chinese hamster cells with a minichromosome containing the centromere region of human chromosome 1. *Somat. Cell Mol. Genet.* **12**, 479–491.
32. Farr, C.J., Bayne, R.A., Kipling, D., Mills, W., Critcher, R. and Cooke, H.J. (1995). Generation of a human X-derived minichromosome using telomere-associated chromosome fragmentation. *EMBO J.* **14**, 5444–5454.
33. Heller, R., Brown, K.E., Burgtorf, C. and Brown, W.R. (1996). Mini-chromosomes derived from the human Y chromosome by telomere directed chromosome breakage. *Proc Natl Acad Sci USA* **93**, 7125–7130.
34. Kuroiwa, Y., Tomizuka, K., Shinohara, T., Kazuki, Y., Yoshida, H., Ohguma, A., Yamamoto, T., Tanaka, S., Oshimura, M. and Ishida, I. (2000). Manipulation of human minichromosomes to carry greater than megabase-sized chromosome inserts. *Nat. Biotechnol* **18**, 1086–1090.
35. Au, H.C., Mascarello, J.T. and Scheffler, I.E. (1999). Targeted integration of a dominant neo(R) marker into a 2- to 3-Mb human minichromosome and transfer between cells. *Cytogenet. Cell Genet.* **86**, 194–203.
36. Mills, W., Critcher, R., Lee, C. and Farr, C.J. (1999). Generation of an approximately 2.4 Mb human X centromere-based minichromosome by targeted telomere-associated chromosome fragmentation in DT40. *Hum. Mol. Genet.* **8**, 751–761.
37. Shen, M.H., Mee, P.J., Nichols, J., Yang, J., Brook, F., Gardner, R.L., Smith, A.G. and Brown, W.R. (2000). A structurally defined mini-chromosome vector for the mouse germ line. *Curr. Biol.* **10**, 31–34.
38. Choo, K.H., Saffery, R., Wong, L.H., Irvine, D.V., Bateman, M.A., Griffiths, B., Cutts, S.M., Cancilla, M.R., Cendron, A.C. and Stafford, A.J. (2001). Construction of neo-centromere-based human minichromosomes by telomere-associated chromosomal truncation. *Proc Natl Acad Sci USA* **98**, 5705–5710.
39. Voet, T., Vermeesch, J., Carens, A., Durr, J., Labaere, C., Duhamel, H., David, G. and Marynen, P. (2001). Efficient male and female germline transmission of a human chromosomal vector in mice. *Genome Res.* **11**, 124–136.
40. Wong, L.H., Saffery, R. and Choo, K.H.A. (2002). Construction of neocentromere-based human minichromosomes for gene delivery and centromere studies. *Gene Ther.* **9**, 724–726.
41. Auriche, C., Donini, P. and Ascenzioni, F. (2001). Molecular and cytological analysis of a 5.5 Mb minichromosome. *EMBO Rep.* **2**, 102–107.
42. Katoh, M., Ayabe, F., Norikane, S., Okada, T., Masumoto, H., Horike, S., Shirayoshi, Y. and Oshimura, M. (2004). Construction of a novel human artificial chromosome vector for gene delivery. *Biochem. Biophys. Res. Commun.* **321**, 280–290.
43. Hadlaczky, G., Praznovszky, T., Cserpan, I., Kereso, J., Peterfy, M., Kelemen, I., Atalay, E., Szeles, A., Szelei, J. and Tubak, V. (1991). Centromere formation in mouse cells cotransformed with human DNA and a dominant marker gene. *Proc. Natl. Acad. Sci. U.S.A* **88**, 8106–8110.
44. Praznovszky, T., Kereső, J., Tubak, V., Cserpán, I., Fátyol, K. and Hadlaczky, G.

(1991). De novo chromosome formation in rodent cells. *Proc Natl Acad Sci USA* **88**, 11042–11046.

45. Kereso, J., Praznovszky, T., Cserpan, I., Fodor, K., Katona, R., Csonka, E., Fatyol, K., Hollo, G., Szeles, A., Ross, A.R., Sumner, A.T., Szalay, A.A. and Hadlaczky, G. (1996). De novo chromosome formations by large-scale amplification of the centromeric region of mouse chromosomes. *Chromosome Res.* **4**, 226–239.
46. Csonka, E., Cserpan, I., Fodor, K., Hollo, G., Katona, R., Kereso, J., Praznovszky, T., Szakal, B., Telenius, A., deJong, G., Udvardy, A. and Hadlaczky, G. (2000). Novel generation of human satellite DNA-based artificial chromosomes in mammalian cells. *J. Cell Sci* **113 (Pt 18)**, 3207–3216.
47. Hollo, G., Kereso, J., Praznovszky, T., Cserpan, I., Fodor, K., Katona, R., Csonka, E., Fatyol, K., Szeles, A., Szalay, A.A. and Hadlaczky, G. (1996). Evidence for a megareplicon covering megabases of centromeric chromosome segments. *Chromosome Res.* **4**, 240–247.
48. Lindenbaum, M., Perkins, E., Csonka, E., Fleming, E., Garcia, L., Greene, A., Gung, L., Hadlaczky, G., Lee, E., Leung, J., MacDonald, N., Maxwell, A., Mills, K., Monteith, D., Perez, C.F., Shellard, J., Stewart, S., Stodola, T., Vandenborre, D., Vanderbyl, S. and Ledebur, H.C. (2004). A mammalian artificial chromosome engineering system (ACE System) applicable to biopharmaceutical protein production, transgenesis and gene-based cell therapy. *Nucleic Acids Res.* **32**, e172–e172.
49. Perez, C.F., Vanderbyl, S.L., Mills, K. and Ledebur Jr, H.C. (2004). The ACE System: A versatile chromosome engineering technology with applications for genebased cell therapy. *Bioprocessing* **3**, 61–68.
50. Grimes, B.R., Schindelhauer, D., McGill, N.I., Ross, A., Ebersole, T.A. and Cooke, H.J. (2001). Stable gene expression from a mammalian artificial chromosome. *EMBO Rep.* **2**, 910–914.
51. Breman, A.M., Steiner, C.M., Slee, R.B. and Grimes, B.R. (2008). Input DNA ratio determines copy number of the 33 kb Factor IX gene on de novo human artificial chromosomes. *Mol. Ther.* **16**, 315–323.
52. Grimes, B.R., Babcock, J., Rudd, M.K., Chadwick, B. and Willard, H.F. (2004). Assembly and characterization of heterochromatin and euchromatin on human artificial chromosomes. *Genome Biol.* **5**, R89–R89.
53. Rudd, M.K., Mays, R.W., Schwartz, S. and Willard, H.F. (2003). Human artificial chromosomes with alpha satellite-based de novo centromeres show increased frequency of nondisjunction and anaphase lag. *Mol. Cell. Biol.* **23**, 7689–7697.
54. Higgins, A.W., Gustashaw, K.M. and Willard, H.F. (2005). Engineered human dicentric chromosomes show centromere plasticity. *Chromosome Res.* **13**, 745–762.
55. Oshimura, M. and Katoh, M. (2008). Transfer of human artificial chromosome vectors into stem cells. *Reprod. Biomed. Online* **16**, 57–69.
56. Ayabe, F., Katoh, M., Inoue, T., Kouprina, N., Larionov, V. and Oshimura, M. (2005). A novel expression system for genomic DNA loci using a human artificial chromosome vector with transformation-associated recombination cloning. *J. Hum. Genet* **50**, 592–599.
57. Kawahara, M., Inoue, T., Ren, X., Sogo, T., Yamada, H., Katoh, M., Ueda, H., Oshimura, M. and Nagamune, T. (2007). Antigen-mediated growth control of hybridoma cells via a human artificial chromosome. *Biochim. Biophys. Acta* **1770**, 206–212.
58. Kakeda, M., Hiratsuka, M., Nagata, K., Kuroiwa, Y., Kakitani, M., Katoh, M., Oshimura, M. and Tomizuka, K. (2005). Human artificial chromosome (HAC) vector provides long-term therapeutic transgene expression in normal human primary fibroblasts. *Gene Ther.* **12**, 852–856.
59. Otsuki, A., Tahimic, C.G., Tomimatsu, N., Katoh, M., Chen, D.J., Kurimasa, A. and Oshimura, M. (2005). Construction of a novel expression system on a human artificial chromosome. *Biochem. Biophys. Res. Commun* **329**, 1018–1025.
60. Suda, T., Katoh, M., Hiratsuka, M., Takiguchi, M., Kazuki, Y., Inoue, T. and Oshimura, M. (2006). Heat-regulated production and secretion of insulin from a human artificial chromosome vector. *Biochem. Biophys. Res. Commun.* **340**, 1053–1061.
61. Messina, G., Cossu, G., Oshimura, M., Hoshiya, H., Kazuki, Y., Abe, S., Takiguchi, M., Kajitani, N., Watanabe, Y., Yoshino, T., Shirayoshi, Y. and Higaki, K. (2009). A highly stable and nonintegrated human artificial chromosome (HAC) containing the 2.4 Mb entire human dystrophin gene. *Mol. Ther.* **17**, 309–317.
62. Yamada, H., Li, Y.C., Nishikawa, M., Oshimura, M. and Inoue, T. (2008). Introduction of a CD40L genomic fragment via a human artificial chromosome vector

permits cell-type-specific gene expression and induces immunoglobulin secretion. *J. Hum. Genet.* **53**, 447–453.

63. Inoue, K., Kajitani, N., Yoshino, T., Shirayoshi, Y., Ogura, A., Shinohara, T., Barrett, J.C., Oshimura, M., Kazuki, Y., Hoshiya, H., Kai, Y., Abe, S., Takiguchi, M., Osaki, M., Kawazoe, S., Katoh, M. and Kanatsu-Shinohara, M. (2008). Correction of a genetic defect in multipotent germline stem cells using a human artificial chromosome. *Gene Ther.* **15**, 617–624.
64. Oshimura, M., Tomizuka, K., Shitara, S., Kakeda, M., Nagata, K., Hiratsuka, M., Sano, A., Osawa, K., Okazaki, A., Katoh, M. and Kazuki, Y. (2008). Telomerasemediated lifespan extension of human primary fibroblasts by human artificial chromosome (HAC) vector. *Biochem. Biophys. Res. Commun.* **369**, 807–811.
65. Sands, V.E. (1969). Short arm enlargement in acrocentric chromosomes. *Am. J. Hum. Genet.* **21**, 293–304.
66. Warburton, D., Atwood, K.C. and Henderson, A.S. (1976). Variation in the number of genes for rRNA among human acrocentric chromosomes: correlation with frequency of satellite association. *Cytogenet. Cell Genet.* **17**, 221–230.
67. Verma, R.S., Dosik, H. and Lubs, H.A. (1977). Size variation polymorphisms of the short arm of human acrocentric chromosomes determined by R-banding by fluorescence using acridine orange (RFA). *Hum. Genet.* **38**, 231–234.
68. Sofuni, T., Tanabe, K. and Awa, A.A. (1980). Chromosome heteromorphisms in the Japanese. II. Nucleolus organizer regions of variant chromosomes in D and G groups. *Hum. Genet.* **55**, 265–270.
69. Trask, B., van den Engh, G., Mayall, B. and Gray, J.W. (1989). Chromosome heteromorphism quantified by high-resolution bivariate flow karyotyping. *Am. J. Hum. Genet.* **45**, 739–752.
70. Trask, B., van den Engh, G. and Gray, J.W. (1989). Inheritance of chromosome heteromorphisms analyzed by high-resolution bivariate flow karyotyping. *Am. J. Hum. Genet.* **45**, 753–760.
71. Stergianou, K., Gould, C.P., Waters, J.J. and Hultén, M.A. (1993). A DA/DAPI positive human 14p heteromorphism defined by fluorescence in-situ hybridisation using chromosome 15-specific probes D15Z1 (satellite III) and p-TRA-25 (alphoid). *Hereditas* **119**, 105–110.
72. Friedrich, U., Caprani, M., Niebuhr, E., Therkelsen, A.J. and JÃ¸rgensen, A.L. (1996). Extreme variant of the short arm of chromosome 15. *Hum. Genet.* **97**, 710–713.
73. Conte, R.A., Kleyman, S.M., Laundon, C. and Verma, R.S. (1997). Characterization of two extreme variants involving the short arm of chromosome 22: are they identical? *Ann. Genet.* **40**, 145–149.
74. Cooper, H.L. and Hirschhorn, K. (1962). Enlarged satellites as a familial chromosome marker. *Am. J. Hum. Genet.* **14**, 107–124.
75. Therkelsen, A.J. (1964). Enlarged short arm of a small acrocentric chromosome in grandfather, mother and child, the latter with Down's syndrome. *Cytogenetics* **10**, 441–451.
76. Yoder, F.E., Bias, W.B., Borgaonkar, D.S., Bahr, G.F., Yoder, I.I., Yoder, O.C. and Golomb, H.M. (1974). Cytogenetic and linkage studies of a familial 15pplus variant. *Am. J. Hum. Genet.* **26**, 535–548.
77. Bernstein, R., Dawson, B. and Griffiths, J. (1981). Human inherited marker chromosome 22 short-arm enlargement: investigation of rDNA gene multiplicity, Agband size, and acrocentric association. *Hum. Genet.* **58**, 135–139.
78. Schmid, M., Nanda, I., Steinlein, C. and Epplen, J.T. (1994). Amplification of (GACA)n simple repeats in an exceptional 14p+ marker chromosome. *Hum. Genet.* **93**, 375–382.
79. Mukerjee, D. and Burdette, W.J. (1966). A familial minute isochromosome. *Am. J. Hum. Genet.* **18**, 62–69.
80. Howard-Peebles, P.N. (1979). A familial bisatellited extra metacentric microchromosome in man. *J. Hered.* **70**, 347–348.
81. Blennow, E., Bui, T.H., Kristoffersson, U., Vujic, M., Annerén, G., Holmberg, E. and Nordenskjöld, M. (1994). Swedish survey on extra structurally abnormal chromosomes in 39 105 consecutive prenatal diagnoses: prevalence and characterization by fluorescence in situ hybridization. *Prenat. Diagn.* **14**, 1019–1028.
82. Brondum-Nielsen, K. and Mikkelsen, M. (1995). A 10-year survey, 1980-1990, of prenatally diagnosed small supernumerary marker chromosomes, identified by FISH analysis. Outcome and follow-up of 14 cases diagnosed in a series of 12,699 prenatal samples. *Prenat. Diagn.* **15**, 615–619.
83. Crolla, J.A., Howard, P., Mitchell, C., Long, F.L. and Dennis, N.R. (1997). A molecular and FISH approach to determining

karyotype and phenotype correlations in six patients with supernumerary marker(22) chromosomes. *Am. J. Med. Genet.* **72**, 440–447.

84. Adhvaryu, S.G., Peters-Brown, T., Livingston, E. and Qumsiyeh, M.B. (1998). Familial supernumerary marker chromosome evolution through three generations. *Prenat. Diagn.* **18**, 178–181.

85. Hadlaczky, G., Bujdosó, G., Koltai, E., Schwanitz, G. and Csonka, E. (2002). A natural equivalent to human satellite DNA-based artificial chromosome persists over 140 years, in a three-generation family. *European Journal of Human Genetics* **10 (Supplement)**, 143–143.

86. Kennard, M.L., Goosney, D.L., Monteith, D., Zhang, L., Moffat, M., Fischer, D. and Mott, J. (2009). The generation of stable, high MAb expressing CHO cell lines based on the artificial chromosome expression (ACE) technology. *Biotechnol. Bioeng.* **104**, 540–553.

87. Kennard, M.L., Goosney, D.L., Monteith, D., Roe, S., Fischer, D. and Mott, J. (2009). Auditioning of CHO host cell lines using the artificial chromosome expression (ACE) technology. *Biotechnol. Bioeng.* **104**, 526–539.

88. Katona, R.L., Sinkó, I., Holló, G., Szucs, K.S., Praznovszky, T., Kereső, J., Csonka, E., Fodor, K., Cserpán, I., Szakál, B., Blazsó, P., Udvardy, A. and Hadlaczky, G. (2008). A combined artificial chromosome-stem cell therapy method in a model experiment aimed at the treatment of Krabbe's disease in the Twitcher mouse. *Cell. Mol. Life Sci.* **65**, 3830–3838.

89. Adriaansen, J., Vervoordeldonk, M.J., Vanderbyl, S., de Jong, G. and Tak, P.P. (2006). A novel approach for gene therapy: engraftment of fibroblasts containing the artificial chromosome expression system at the site of inflammation. *J Gene Med* **8**, 63–71.

90. Schor, S.L., Johnson, R.T. and Mullinger, A.M. (1975). Perturbation of mammalian cell division. II. Studies on the isolation and characterization of human mini segregant cells. *J. Cell. Sci.* **19**, 281–303.

91. Fournier, R.E. and Ruddle, F.H. (1977). Microcell-mediated transfer of murine chromosomes into mouse, Chinese hamster, and human somatic cells. *Proc Natl Acad Sci USA* **74**, 319–323.

92. Meaburn, K.J., Parris, C.N. and Bridger, J.M. (2005). The manipulation of chromosomes by mankind: the uses of microcell-mediated chromosome transfer. *Chromosoma* **114**, 263–274.

93. Doherty, A.M.O. and Fisher, E.M.C. (2003). Microcell-mediated chromosome transfer (MMCT): small cells with huge potential. *Mamm. Genome* **14**, 583–592.

94. Yamada, H., Kunisato, A., Kawahara, M., Tahimic, C.G., Ren, X., Ueda, H., Nagamune, T., Katoh, M., Inoue, T., Nishikawa, M. and Oshimura, M. (2006). Exogenous gene expression and growth regulation of hematopoietic cells via a novel human artificial chromosome. *J. Hum .Genet* **51**, 147–150.

95. de Jong, G., Telenius, A.H., Telenius, H., Perez, C.F., Drayer, J.I. and Hadlaczky, G. (1999). Mammalian artificial chromosome pilot production facility: large-scale isolation of functional satellite DNA-based artificial chromosomes. *Cytometry* **35**, 129–133.

96. Co, D.O., Borowski, A.H., Leung, J.D., van der Kaa, J., Hengst, S., Platenburg, G.J., Pieper, F.R., Perez, C.F., Jirik, F.R. and Drayer, J.I. (2000). Generation of transgenic mice and germline transmission of a mammalian artificial chromosome introduced into embryos by pronuclear microinjection. *Chromosome Res.* **8**, 183–191.

97. Monteith, D.P., Leung, J.D., Borowski, A.H., Co, D.O., Praznovszky, T., Jirik, F.R., Hadlaczky, G. and Perez, C.F. (2004). Pronuclear microinjection of purified artificial chromosomes for generation of transgenic mice: pick-and-inject technique. *Methods Mol. Biol.* **240**, 227–242.

98. Vanderbyl, S., MacDonald, G.N., Sidhu, S., Gung, L., Telenius, A., Perez, C. and Perkins, E. (2004). Transfer and stable transgene expression of a mammalian artificial chromosome into bone marrow-derived human mesenchymal stem cells. *Stem Cells* **22**, 324–333.

99. Vanderbyl, S.L., Sullenbarger, B., White, N., Perez, C.F., MacDonald, G.N., Stodola, T., Bunnell, B.A., Ledebur, H.C. and Lasky, L.C. (2005). Transgene expression after stable transfer of a mammalian artificial chromosome into human hematopoietic cells. *Exp. Hematol.* **33**, 1470–1476.

100. de Jong, G., Telenius, A., Vanderbyl, S., Meitz, A. and Drayer, J. (2001). Efficient in-vitro transfer of a 60-Mb mammalian artificial chromosome into murine and hamster cells using cationic lipids and dendrimers. *Chromosome Res.* **9**, 475–485.

101. Robl, J.M., Kasinathan, P., Sullivan, E., Kuroiwa, Y., Tomizuka, K. and Ishida, I. (2003). Artificial chromosome vectors and

expression of complex proteins in transgenic animals. *Theriogenology* **59**, 107–113.

102. Kuroiwa, Y., Kasinathan, P., Choi, Y.J., Naeem, R., Tomizuka, K., Sullivan, E.J., Knott, J.G., Duteau, A., Goldsby, R.A., Osborne, B.A., Ishida, I. and Robl, J.M. (2002). Cloned transchromosomic calves producing human immunoglobulin. *Nat. Biotechnol* **20**, 889–894.
103. Kazuki, Y., Shinohara, T., Tomizuka, K., Katoh, M., Ohguma, A., Ishida, I. and Oshimura, M. (2001). Germline transmission of a transferred human chromosome 21 fragment in transchromosomal mice. *J. Hum. Genet.* **46**, 600–603.
104. Raper, S.E., Chirmule, N., Lee, F.S., Wivel, N.A., Bagg, A., Gao, G., Wilson, J.M. and Batshaw, M.L. (2003). Fatal systemic inflammatory response syndrome in a ornithine transcarbamylase deficient patient following adenoviral gene transfer. *Mol. Genet. Metab.* **80**, 148–158.
105. Christ, M., Lusky, M., Stoeckel, F., Dreyer, D., Dieterlé, A., Michou, A.I., Pavirani, A. and Mehtali, M. (1997). Gene therapy with recombinant adenovirus vectors: evaluation of the host immune response. *Immunol. Lett.* **57**, 19–25.
106. Couto, L.B. (2004). Preclinical gene therapy studies for hemophilia using adenoassociated virus (AAV) vectors. *Semin. Thromb. Hemost.* **30**, 161–171.
107. Miller, D.G., Rutledge, E.A. and Russell, D.W. (2002). Chromosomal effects of adeno-associated virus vector integration. *Nat. Genet.* **30**, 147–148.
108. Miller, D.G., Petek, L.M. and Russell, D.W. (2004). Adeno-associated virus vectors integrate at chromosome breakage sites. *Nat. Genet.* **36**, 767–773.
109. Hacein-Bey-Abina, S., von Kalle, C., Schmidt, M., McCormack, M.P., Wulffraat, N., Leboulch, P., Lim, A., Osborne, C.S., Pawliuk, R., Morillon, E., Sorensen, R., Forster, A., Fraser, P., Cohen, J.I., de Saint, B.G., Alexander, I., Wintergerst, U., Frebourg, T., Aurias, A., Stoppa-Lyonnet, D., Romana, S., Radford-Weiss, I., Gross, F., Valensi, F., Delabesse, E., Macintyre, E., Sigaux, F., Soulier, J., Leiva, L.E., Wissler, M., Prinz, C., Rabbitts, T.H., Le Deist, F., Fischer, A. and Cavazzana-Calvo, M. (2003). LMO2-associated clonal T cell proliferation in two patients after gene therapy for SCIDX1. *Science* **302**, 415–419.
110. Cavazzana-Calvo, M., Fischer, A., Hacein-Bey-Abina, S., von Kalle, C., Schmidt, M., Le Deist, F., Wulffraat, N., McIntyre, E., Radford, I., Villeval, J. and Fraser, C.C. (2003). A serious adverse event after successful gene therapy for X-linked severe combined immunodeficiency. *N. Engl. J. Med.* **348**, 255–256.
111. Guhr, A., Kurtz, A., Friedgen, K. and La ser, P. (2006). Current state of human embryonic stem cell research: an overview of cell lines and their use in experimental work. *Stem Cells* **24**, 2187–2191.
112. Guan, K., Nayernia, K., Maier, L.S., Wagner, S., Dressel, R., Lee, J.H., Nolte, J., Wolf, F., Li, M., Engel, W. and Hasenfuss, G. (2006). Pluripotency of spermatogonial stem cells from adult mouse testis. *Nature* **440**, 1199–1203.
113. Shamblott, M.J., Axelman, J., Wang, S., Bugg, E.M., Littlefield, J.W., Donovan, P.J., Blumenthal, P.D., Huggins, G.R. and Gearhart, J.D. (1998). Derivation of pluripotent stem cells from cultured human primordial germ cells. *Proc. Natl. Acad. Sci. U.S.A.* **95**, 13726–13731.
114. Donovan, P.J. and de Miguel, M.P. (2003). Turning germ cells into stem cells. *Curr. Opin. Genet. Dev.* **13**, 463–471.
115. Klimanskaya, I., Chung, Y., Becker, S., Lu, S.J. and Lanza, R. (2006). Human embryonic stem cell lines derived from single blastomeres. *Nature* **444**, 481–485.
116. Hochedlinger, K. and Jaenisch, R. (2006). Nuclear reprogramming and pluripotency. *Nature* **441**, 1061–1067.
117. Takahashi, K. and Yamanaka, S. (2006). Induction of pluripotent stem cells from mouse embryonic and adult fibroblast cultures by defined factors. *Cell* **126**, 663–676.
118. Evans, M.J. and Kaufman, M.H. (1981). Establishment in culture of pluripotential cells from mouse embryos. *Nature* **292**, 154–156.
119. Thomson, J.A., Itskovitz-Eldor, J., Shapiro, S.S., Waknitz, M.A., Swiergiel, J.J., Marshall, V.S. and Jones, J.M. (1998). Embryonic stem cell lines derived from human blastocysts. *Science* **282**, 1145–1147.
120. Orimo, A., Gupta, P.B., Sgroi, D.C., Arenzana-Seisdedos, F., Delaunay, T., Naeem, R., Carey, V.J., Richardson, A.L. and Weinberg, R.A. (2005). Stromal fibroblasts present in invasive human breast carcinomas promote tumor growth and angiogenesis through elevated SDF-1/CXCL12 secretion. *Cell* **121**, 335–348.
121. Hung, S.C., Deng, W.P., Yang, W.K., Liu, R.S., Lee, C.C., Su, T.C., Lin, R.J., Yang, D.M., Chang, C.W., Chen, W.H., Wei, H.J. and Gelovani, J.G. (2005). Mesenchymal

stem cell targeting of microscopic tumors and tumor stroma development monitored by noninvasive in vivo positron emission tomography imaging. *Clin. Cancer Res.* **11**, 7749–7756.

122. Danks, M.K., Yoon, K.J., Bush, R.A., Remack, J.S., Wierdl, M., Tsurkan, L., Kim, S.U., Garcia, E., Metz, M.Z., Najbauer, J., Potter, P.M. and Aboody, K.S. (2007). Tumortargeted enzyme/prodrug therapy mediates long-term disease-free survival of mice bearing disseminated neuroblastoma. *Cancer Res.* **67**, 22–25.

123. Aboody, K.S., Najbauer, J., Schmidt, N.O., Yang, W., Wu, J.K., Zhuge, Y., Przylecki, W., Carroll, R., Black, P.M. and Perides, G. (2006). Targeting of melanoma brain metastases using engineered neural stem/progenitor cells. *Neuro-oncology* **8**, 119–126.

124. Aboody, K.S., Bush, R.A., Garcia, E., Metz, M.Z., Najbauer, J., Justus, K.A., Phelps, D.A., Remack, J.S., Yoon, K.J., Gillespie, S., Kim, S.U., Glackin, C.A., Potter, P.M. and Danks, M.K. (2006). Development of a tumor-selective approach to treat metastatic cancer. *PLoS ONE* **1**, e23–e23.

125. Studeny, M., Marini, F.C., Dembinski, J.L., Zompetta, C., Cabreira-Hansen, M., Bekele, B.N., Champlin, R.E. and Andreeff, M. (2004). Mesenchymal stem cells: potential precursors for tumor stroma and targeted-delivery vehicles for anticancer agents. *J. Natl. Cancer Inst.* **96**, 1593–1603.

126. Zhu, W., Xu, W., Jiang, R., Qian, H., Chen, M., Hu, J., Cao, W., Han, C. and Chen, Y. (2006). Mesenchymal stem cells derived from bone marrow favor tumor cell growth in vivo. *Exp. Mol. Pathol.* **80**, 267–274.

127. Nakayama, N., Lee, J. and Chiu, L. (2000). Vascular endothelial growth factor synergistically enhances bone morphogenetic protein-4-dependent lymphohematopoietic cell generation from embryonic stem cells in vitro. *Blood* **95**, 2275–2283.

128. Kyba, M., Perlingeiro, R.C.R. and Daley, G.Q. (2002). HoxB4 confers definitive lymphoid-myeloid engraftment potential on embryonic stem cell and yolk sac hematopoietic progenitors. *Cell* **109**, 29–37.

129. Kondo, M., Wagers, A.J., Manz, M.G., Prohaska, S.S., Scherer, D.C., Beilhack, G.F., Shizuru, J.A. and Weissman, I.L. (2003). Biology of hematopoietic stem cells and progenitors: implications for clinical application. *Annu. Rev. Immunol.* **21**, 759–806.

130. Burt, R.K., Verda, L., Kim, D.A., Oyama, Y., Luo, K. and Link, C. (2004). Embryonic stem cells as an alternate marrow donor source: engraftment without graft-versus-host disease. *J. Exp. Med.* **199**, 895–904.

131. Park, C., Afrikanova, I., Chung, Y.S., Zhang, W.J., Arentson, E., Fong Gh, G.H., Rosendahl, A. and Choi, K. (2004). A hierarchical order of factors in the generation of FLK1- and SCL-expressing hematopoietic and endothelial progenitors from embryonic stem cells. *Development* **131**, 2749–2762.

132. Doetschman, T.C., Eistetter, H., Katz, M., Schmidt, W. and Kemler, R. (1985). The in vitro development of blastocyst-derived embryonic stem cell lines: formation of visceral yolk sac, blood islands and myocardium. *J Embryol Exp Morphol* **87**, 27–45.

133. Bautch, V.L., Stanford, W.L., Rapoport, R., Russell, S., Byrum, R.S. and Futch, T.A. (1996). Blood island formation in attached cultures of murine embryonic stem cells. *Dev. Dyn.* **205**, 1–12.

134. Yamashita, J., Itoh, H., Hirashima, M., Ogawa, M., Nishikawa, S., Yurugi, T., Naito, M., Nakao, K. and Nishikawa, S. (2000). Flk1-positive cells derived from embryonic stem cells serve as vascular progenitors. *Nature* **408**, 92–96.

135. Kolossov, E., Fleischmann, B.K., Liu, Q., Bloch, W., Viatchenko Karpinski, S., Manzke, O., Ji, G.J., Bohlen, H., Addicks, K. and Hescheler, J. (1998). Functional characteristics of ES cell-derived cardiac precursor cells identified by tissue-specific expression of the green fluorescent protein. *J. Cell Biol.* **143**, 2045–2056.

136. Yang, Y., Min, J.Y., Rana, J.S., Ke, Q., Cai, J., Chen, Y., Morgan, J.P. and Xiao, Y.F. (2002). VEGF enhances functional improvement of postinfarcted hearts by transplantation of ESC-differentiated cells. *J. Appl. Physiol.* **93**, 1140–1151.

137. Min, J.Y., Yang, Y., Converso, K.L., Liu, L., Huang, Q., Morgan, J.P. and Xiao, Y.F. (2002). Transplantation of embryonic stem cells improves cardiac function in postinfarcted rats. *J. Appl. Physiol.* **92**, 288–296.

138. Müller, M., Fleischmann, B.K., Selbert, S., Ji, G.J., Endl, E., Middeler, G., Müller, O.J., Schlenke, P., Frese, S., Wobus, A.M., Hescheler, J., Katus, H.A. and Franz, W.M. (2000). Selection of ventricular-like cardiomyocytes from ES cells in vitro. *FASEB J.* **14**, 2540–2548.

139. Zandstra, P.W., Bauwens, C., Yin, T., Liu, Q., Schiller, H., Zweigerdt, R., Pasumarthi, K.B.S. and Field, L.J. (2003). Scalable production of embryonic stem cellderived cardiomyocytes. *Tissue Eng.* **9**, 767–778.

140. Hidaka, K., Lee, J.K., Kim, H.S., Ihm, C.H., Iio, A., Ogawa, M., Nishikawa, S.I., Kodama, I. and Morisaki, T. (2003). Chamber-specific differentiation of Nkx2.5-positive cardiac precursor cells from murine embryonic stem cells. *FASEB J.* **17**, 740–742.

141. Min, J.Y., Yang, Y., Sullivan, M.F., Ke, Q., Converso, K.L., Chen, Y., Morgan, J.P. and Xiao, Y.F. (2003). Long-term improvement of cardiac function in rats after infarction by transplantation of embryonic stem cells. *J. Thorac. Cardiovasc. Surg.* **125**, 361–369.

142. Pinto, D.O., Kolterud, A. and Carlsson, L. (1998). Expression of the LIMhomeobox gene LH2 generates immortalized steel factor-dependent multipotent hematopoietic precursors. *EMBO J.* **17**, 5744–5756.

143. Lieber, J.G., Webb, S., Suratt, B.T., Young, S.K., Johnson, G.L., Keller, G.M. and Worthen, G.S. (2004). The in vitro production and characterization of neutrophils from embryonic stem cells. *Blood* **103**, 852–859.

144. Eto, K., Murphy, R., Kerrigan, S.W., Bertoni, A., Stuhlmann, H., Nakano, T., Leavitt, A.D. and Shattil, S.J. (2002). Megakaryocytes derived from embryonic stem cells implicate CalDAG-GEFI in integrin signaling. *Proc. Natl. Acad. Sci. U.S.A.* **99**, 12819–12824.

145. Tsai, M., Wedemeyer, J., Ganiatsas, S., Tam, S.Y., Zon, L.I. and Galli, S.J. (2000). In vivo immunological function of mast cells derived from embryonic stem cells: an approach for the rapid analysis of even embryonic lethal mutations in adult mice in vivo. *Proc. Natl. Acad. Sci. U.S.A.* **97**, 9186–9190.

146. Hamaguchi-Tsuru, E., Nobumoto, A., Hirose, N., Kataoka, S., Fujikawa-Adachi, K., Furuya, M. and Tominaga, A. (2004). Development and functional analysis of eosinophils from murine embryonic stem cells. *Br. J. Haematol.* **124**, 819–827.

147. Fairchild, P.J., Nolan, K.F., Cartland, S., GraA a, L. and Waldmann, H. (2003). Stable lines of genetically modified dendritic cells from mouse embryonic stem cells. *Transplantation* **76**, 606–608.

148. Carotta, S., Pilat, S., Mairhofer, A., Schmidt, U., Dolznig, H., Steinlein, P. and Beug, H. (2004). Directed differentiation and mass cultivation of pure erythroid progenitors from mouse embryonic stem cells. *Blood* **104**, 1873–1880.

149. Chang, A.H., Stephan, M.T. and Sadelain, M. (2006). Stem cell-derived erythroid cells mediate long-term systemic protein delivery. *Nat. Biotechnol.* **24**, 1017–1021.

150. Colman, A. (2004). Making new beta cells from stem cells. *Semin. Cell Dev. Biol.* **15**, 337–345.

151. Stoffel, M., Vallier, L. and Pedersen, R.A. (2004). Navigating the pathway from embryonic stem cells to beta cells. *Semin. Cell Dev. Biol.* **15**, 327–336.

152. D'Amour, K.A., Agulnick, A.D., Eliazer, S., Kelly, O.G., Kroon, E. and Baetge, E.E. (2005). Efficient differentiation of human embryonic stem cells to definitive endoderm. *Nat. Biotechnol.* **23**, 1534–1541.

153. D'Amour, K.A., Bang, A.G., Eliazer, S., Kelly, O.G., Agulnick, A.D., Smart, N.G., Moorman, M.A., Kroon, E., Carpenter, M.K. and Baetge, E.E. (2006). Production of pancreatic hormone-expressing endocrine cells from human embryonic stem cells. *Nat. Biotechnol.* **24**, 1392–1401.

154. Hamazaki, T., Iiboshi, Y., Oka, M., Papst, P.J., Meacham, A.M., Zon, L.I. and Terada, N. (2001). Hepatic maturation in differentiating embryonic stem cells in vitro. *FEBS Lett.* **497**, 15–19.

155. Jones, E.A., Tosh, D., Wilson, D.I., Lindsay, S. and Forrester, L.M. (2002). Hepatic differentiation of murine embryonic stem cells. *Exp. Cell Res.* **272**, 15–22.

156. Yamada, T., Yoshikawa, M., Kanda, S., Kato, Y., Nakajima, Y., Ishizaka, S. and Tsunoda, Y. (2002). In vitro differentiation of embryonic stem cells into hepatocyte-like cells identified by cellular uptake of indocyanine green. *Stem Cells* **20**, 146–154.

157. Lin, R.Y., Kubo, A., Keller, G.M. and Davies, T.F. (2003). Committing embryonic stem cells to differentiate into thyrocyte-like cells in vitro. *Endocrinology* **144**, 2644–2649.

158. Ali, N.N., Edgar, A.J., Samadikuchaksaraei, A., Timson, C.M., Romanska, H.M., Polak, J.M. and Bishop, A.E. (2002). Derivation of type II alveolar epithelial cells from murine embryonic stem cells. *Tissue Eng.* **8**, 541–550.

159. Yamada, T., Yoshikawa, M., Takaki, M., Torihashi, S., Kato, Y., Nakajima, Y., Ishizaka, S. and Tsunoda, Y. (2002). In vitro functional gut-like organ formation from mouse embryonic stem cells. *Stem Cells* **20**, 41–49.

160. Brustle, O., Spiro, A.C., Karram, K., Choudhary, K., Okabe, S. and McKay, R.D. (1997). In vitro-generated neural precursors participate in mammalian brain development. *Proc. Natl. Acad. Sci. U.S.A* **94**, 14809–14814.

161. Brustle, O., Jones, K.N., Learish, R.D., Karram, K., Choudhary, K., Wiestler, O.D.,

Duncan, I.D. and McKay, R.D. (1999). Embryonic stem cell-derived glial precursors: a source of myelinating transplants. *Science* **285**, 754–756.

162. Kim, J.H., Auerbach, J.M., Rodríguez-Gómez, J.A., Velasco, I., Gavin, D., Lumelsky, N., Lee, S.H., Nguyen, J., Sánchez-Pernaute, R., Bankiewicz, K. and McKay, R. (2002). Dopamine neurons derived from embryonic stem cells function in an animal model of Parkinson's disease. *Nature* **418**, 50–56.
163. Barberi, T., Klivenyi, P., Calingasan, N.Y., Lee, H., Kawamata, H., Loonam, K., Perrier, A.L., Bruses, J., Rubio, M.E., Töpf, N., Tabar, V., Harrison, N.L., Beal, M.F., Moore, M.A.S. and Studer, L. (2003). Neural subtype specification of fertilization and nuclear transfer embryonic stem cells and application in parkinsonian mice. *Nat. Biotechnol.* **21**, 1200–1207.
164. Tropepe, V., Hitoshi, S., Sirard, C., Mak, T.W., Rossant, J. and van der Kooy, D. (2001). Direct neural fate specification from embryonic stem cells: a primitive mammalian neural stem cell stage acquired through a default mechanism. *Neuron* **30**, 65–78.
165. Bagutti, C., Wobus, A.M., Fa ssler, R. and Watt, F.M. (1996). Differentiation of embryonal stem cells into keratinocytes: comparison of wild-type and beta 1 integrindeficient cells. *Dev. Biol.* **179**, 184–196.
166. Coraux, C., Hilmi, C., Rouleau, M., Spadafora, A., Hinnrasky, J., Ortonne, J.P., Dani, C. and Aberdam, D. (2003). Reconstituted skin from murine embryonic stem cells. *Curr. Biol.* **13**, 849–853.
167. Young, H.E., Duplaa, C., Katz, R., Thompson, T., Hawkins, K.C., Boev, A.N., Henson, N.L., Heaton, M., Sood, R., Ashley, D., Stout, C., Morgan, J.H., Uchakin, P.N., Rimando, M., Long, G.F., Thomas, C., Yoon, J.I., Park, J.E., Hunt, D.J., Walsh, N.M., Davis, J.C., Lightner, J.E., Hutchings, A.M., Murphy, M.L., Boswell, E., McAbee, J.A., Gray, B.M., Piskurich, J., Blake, L., Collins, J.A., Moreau, C., Hixson, D., Bowyer, F.P. and Black, A.C.J. (2005). Adult-derived stem cells and their potential for use in tissue repair and molecular medicine. *J. Cell. Mol. Med.* **9**, 753–769.
168. Serafini, M., Dylla, S.J., Oki, M., Heremans, Y., Tolar, J., Jiang, Y., Buckley, S.M., Pelacho, B., Burns, T.C., Frommer, S., Rossi, D.J., Bryder, D., Panoskaltsis-Mortari, A., O'Shaughnessy, M.J., Nelson-Holte, M., Fine, G.C., Weissman, I.L., Blazar, B.R. and Verfaillie, C.M. (2007). Hematopoietic reconstitution by multipotent adult progenitor cells: precursors to long-term hematopoietic stem cells. *J. Exp. Med.* **204**, 129–139.
169. Petrovic, V. and Stefanovic, V. (2009). Dental tissue–new source for stem cells. *ScientificWorldJournal* **9**, 1167–1177.
170. Fraser, J.K., Wulur, I., Alfonso, Z. and Hedrick, M.H. (2006). Fat tissue: an underappreciated source of stem cells for biotechnology. *Trends Biotechnol.* **24**, 150–154.
171. De Coppi, P., Bartsch, G., Siddiqui, M.M., Xu, T., Santos, C.C., Perin, L., Mostoslavsky, G., Serre, A.C., Snyder, E.Y., Yoo, J.J., Furth, M.E., Soker, S. and Atala, A. (2007). Isolation of amniotic stem cell lines with potential for therapy. *Nat. Biotechnol.* **25**, 100–106.
172. Auriche, C., Carpani, D., Conese, M., Caci, E., Zegarra-Moran, O., Donini, P. and Ascenzioni, F. (2002). Functional human CFTR produced by a stable minichromosome. *EMBO Rep.* **3**, 862–868.
173. Bunnell, B.A., Izadpanah, R., Ledebur Jr, H.C. and Perez, C.F. (2005). Development of mammalian artificial chromosomes for the treatment of genetic diseases: Sandhoff and Krabbe diseases. *Expert Opin Biol Ther* **5**, 195–206.
174. Ren, X., Katoh, M., Hoshiya, H., Kurimasa, A., Inoue, T., Ayabe, F., Shibata, K., Toguchida, J. and Oshimura, M. (2005). A novel human artificial chromosome vector provides effective cell lineage-specific transgene expression in human mesenchymal stem cells. *Stem Cells* **23**, 1608–1616.
175. Goldman, S. (2005). Stem and progenitor cell-based therapy of the human central nervous system. *Nat. Biotechnol.* **23**, 862–871.
176. Klug, M.G., Soonpaa, M.H., Koh, G.Y. and Field, L.J. (1996). Genetically selected cardiomyocytes from differentiating embronic stem cells form stable intracardiac grafts. *J. Clin. Invest.* **98**, 216–224.
177. Li, M., Pevny, L., Lovell-Badge, R. and Smith, A. (1998). Generation of purified neural precursors from embryonic stem cells by lineage selection. *Curr. Biol.* **8**, 971–974.
178. Billon, N., Jolicoeur, C., Ying, Q.L., Smith, A. and Raff, M. (2002). Normal timing of oligodendrocyte development from genetically engineered, lineage-selectable mouse ES cells. *J. Cell. Sci.* **115**, 3657–3665.
179. Gilad, A.A., McMahon, M.T., Walczak, P., Winnard, P.T.J., Raman, V., van Laarhoven, H.W.M., Skoglund, C.M., Bulte, J.W.M. and van Zijl, P.C.M. (2007). Artificial reporter gene providing MRI contrast based on proton exchange. *Nat. Biotechnol.* **25**, 217–219.

180. Gyöngyösi, M., Blanco, J., Marian, T., Trón, L., Petneházy, O., Petrasi, Z., Hemetsberger, R., Rodriguez, J., Font, G., Pavo, I.J., Kertész, I., Balkay, L., Pavo, N., Posa, A., Emri, M., Galuska, L., Kraitchman, D.L., Wojta, J., Huber, K. and Glogar, D. (2008). Serial noninvasive in vivo positron emission tomographic tracking of percutaneously intramyocardially injected autologous porcine mesenchymal stem cells modified for transgene reporter gene expression. *Circ Cardiovasc Imaging* **1**, 94–103.

Chapter 15

Engineered Mammalian Chromosomes in Cellular Protein Production: Future Prospects

Malcolm L. Kennard

Abstract

The manufacture of recombinant proteins at industrially relevant levels requires technologies that can engineer stable, high expressing cell lines rapidly, reproducibly, and with relative ease. Commonly used methods incorporate transfection of mammalian cell lines with plasmid DNA containing the gene of interest. Identifying stable high expressing transfectants is normally laborious and time consuming. To improve this process, the use of engineered chromosomes has been considered. To date, the most successful technique has been based on the artificial chromosome expression or ACE System, which consists of the targeted transfection of cells containing mammalian based artificial chromosomes with multiple recombination acceptor sites. This ACE System allows for the specific transfection of single or multiple gene copies and eliminates the need for random integration into native host chromosomes. The utility of using artificial engineered mammalian chromosomes, specifically the ACE System, is illustrated in several case studies covering the generation of CHO cell lines expressing monoclonal antibodies.

Key words: Artificial chromosomes, ACE system, Recombinant protein production

1. Introduction

There are many methods available to produce recombinant protein expressing cell lines, most of which use plasmid transfection or viral transduction procedures to incorporate DNA sequences containing the gene of interest into cell lines. These processes are limited with regard to the amount of foreign DNA sequence that can be delivered and often result in transfectants with highly variable protein expression due to random integration of the DNA into the host genome. For applications where cloning and delivery of large genomic loci are desired, such as fragments containing long-range genetic elements required for appropriate regulation

Gyula Hadlaczky (ed.), *Mammalian Chromosome Engineering: Methods and Protocols*, Methods in Molecular Biology, vol. 738, DOI 10.1007/978-1-61779-099-7_15, © Springer Science+Business Media, LLC 2011

of gene expression, developmentally regulated multigene loci, or multiple copies of two or more genes in fixed stoichiometry, plasmid or viral vectors may be inadequate. In order to address these limitations bacterial (1, 2) and yeast (3, 4) chromosomal-based cloning vectors have been developed, with carrying capacities for yeast artificial chromosomes in excess of 1 Mb (5). However, similar to conventional plasmid vectors, these vectors still require integration into host mammalian chromosomes for stable maintenance. Such integrations are most often random and the sites of integrations may have profound and unpredictable effects on the expression of recombinant genes (6). In addition, such random integrations may result in inactivation or change in regulation of host genes (7). Furthermore, these methods may necessitate time-consuming amplification events or reinfection to boost the cell's productivity. As a result, the process of generating and selecting a high expressing stable clonal cell line suitable for the clinical and commercial manufacture of biopharmaceuticals can be labor intensive and extremely time consuming.

To increase the speed and efficiency of generating high expressing stable cell lines for the manufacture of recombinant proteins, alternative technologies based on artificial mammalian chromosomes are being considered where genes of interest can be inserted into the chromosome and transfected into industrially relevant cell lines. Compared to traditional techniques, these artificial mammalian chromosomes offer significant advantages for cellular protein production on account of their high carrying capacity and ability to self-replicate without relying on the integration into the host genome. Initially, artificial chromosomes were generated to provide a more stable, safer gene delivery vehicle for gene-based cell therapy (8–19). Several approaches for generating mammalian artificial chromosomes have been reported. For example, in "centromere seeding", artificial chromosomes are assembled de novo in cells from cotransfected DNAs that encode putative centromeres, telomeres, and bacterial drug-resistant marker genes (20–26). In another approach, "minichromosomes" were generated by either fragmenting natural chromosomes via telomere-directed breakage or by identifying naturally occurring fragmented human chromosomes (27, 28). Both approaches generate artificial chromosomes with functional centromeres and telomeres that are stably maintained alongside the host cell's chromosomes. However, although these artificial chromosomes appear attractive as vectors for recombinant protein production these approaches possess practical and technical limitations. Most notably is the inability to isolate and purify the centromere-seeded and fragmented-based artificial chromosomes for transfection into the target cell line. One transfection option that has been considered is to use microcell mediated transfer. Although technically feasible, this technique is very tedious, producing a heterogenous population of microcells containing the artificial chromosomes and resulting in very low transfer efficiencies. Another technically feasible technique has been to generate the artificial chromosomes de novo in the desired target cell line for both centromere seeding and chromosome fragmentation. However, this is an extremely inefficient, labor intensive process resulting in no predictable relationship between input DNA and the de novo chromosomes' composition upon generation, making downstream characterization and quality control difficult and rendering the technique unsuitable for commercial protein production. With respect to centromere seeding, there is also the possibility that rearrangements and gene integrations can occur when the transfected DNA fails to form new chromosomes. Therefore, for practical purposes, a more reliable, reproducible, and quicker method based on using artificial chromosomes had to be developed. To this end, a platform system termed "The ACE System" was developed, which was based on pre-engineered artificial chromosomes called ACEs and allowed for isolation, purification, and delivery of artificial chromosomes along with a constant copy number and recombinant gene expression.

2. ACEs and the ACE System

2.1. Generation of ACEs

Hadlaczky and colleagues discovered that when heterologous DNA is introduced in vitro into the short arms of mouse or human acrocentric chromosomes, which are primarily composed of tandemly repeated ribosomal genes (rDNA) and pericentromeric heterochromatin (commonly referred to as "satellite" or repetitive DNA sequences), large-scale amplification of these sequences were induced (29, 30). Frequently, whole de novo centromeres were subsequently generated, producing dicentric chromosomes. Ensuing breakage of these dicentric chromosomes during mitosis generated new "artificial" chromosomes with functioning centromeres and telomeres, ranging in size from 10 to 360 million base pairs (31–36). These satellite-DNA based mammalian artificial chromosomes are referred to as Artificial Chromosome Expression systems or ACEs (formerly referred to as Satellite DNA-based Artificial Chromosomes or SATACs) (32, 36). These mammalian artificial chromosomes or ACEs contain fully functional centromeres and telomeres and as a result are as mitotically stable as the host chromosomes. Based on their unique nucleotide composition (a predominance of AT base pairs to GC base pairs compared with endogenous host cell chromosomes), the ACEs with or without transgenes may be purified to near homogeneity by high speed flow cell sorting (37) and then transferred to a variety of different cell types (e.g., cell lines, primary cells and adult stem cells etc.) using standard transfection technologies and commercially available transfection agents (33, 38–41). This feature enables the auditioning of alternative lines for improved product quality or quantity, thereby providing an option not typically found in conventional mammalian cell line engineering technologies. In addition, ACEs have been introduced into embryos via a modified microinjection procedure to generate transgenic animals that are able to transmit the ACEs through their germ line for multiple generations without overt changes in the phenotype of the founders or their progeny (42, 43). No chromosomal aberrations were noted in these animals. In addition, there are no known gene-coding sequences on the ACEs that could lead to indirect or direct gene dosage effects. ACEs primarily consist of heterochromatin and pericentric sequences that are thought to be devoid of coding sequences beyond rDNA (29, 30, 32–36). These results gave the first indications that ACEs were stable, non-integrating, and non-deleterious in both in vitro and in vivo.

2.2. The ACE System

Originally ACEs containing target gene(s) or genes of interest were generated de novo. Although feasible, it was a lengthy process to generate a new chromosome for each application. Therefore, to facilitate the rapid and efficient engineering of ACEs with DNA sequences of interest, the ACE System was developed, in which one or more genes could be reliably and reproducibly "loaded" onto an existing ACE and screened for incorporation and expression with relative ease, while still retaining the ability to be purified and transferred to other cell lines. The loading or targeting of the ACE incorporates features of the mechanism used by bacteriophage lambda to integrate itself into the host chromosome of *E. coli* i.e. *att*P donor and *att*B acceptor sites (44).

2.3. Core Components and Targeted Transfection

For mammalian cell line engineering, the ACE System consists of four main components:

1. The Platform ACE: A pre-engineered artificial chromosome containing 50–70 recombinant *att*P acceptor sites, which allows for insertion of multiple copies of DNA sequences (Fig. 1).
2. The Platform ACE Cell Line containing the Platform ACE: A production cell line based for example on a CHO cell line that ideally grows to high cell density and is adapted to suspension serum-free growth conditions.
3. The ACE Targeting Vector (ATV): A plasmid that contains a single *att*B donor site for recombination into the acceptor sites on the Platform ACE, selection marker and the gene(s) of interest along with all genetic elements required for enhanced expression in for example CHO cells.

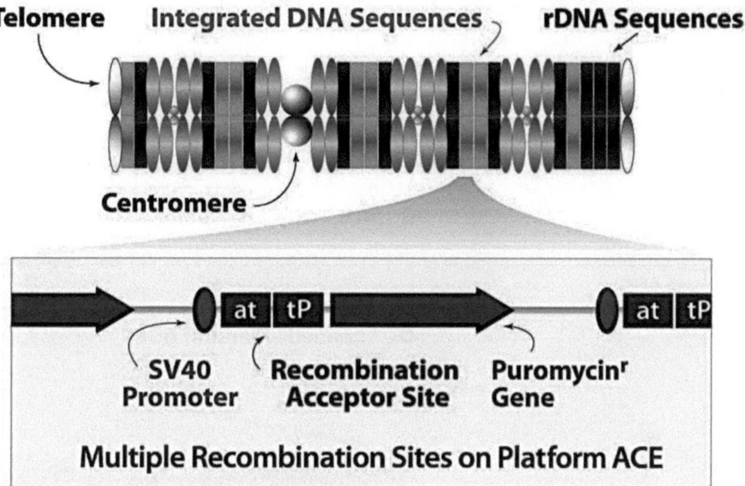

Fig. 1. Schematic representation of Platform ACE. The Platform ACE (depicted as a metaphase chromosome) encodes greater than 50 copies of a recombination acceptor site cassette. Each acceptor site cassette encodes an SV40 promoter, *att*P recombination acceptor site, and the open reading frame of the puromycin resistance gene.

4. The ACE Integrase: A site-specific DNA recombinase that catalyzes the targeting of the ATV onto the Platform ACE.

Targeted transfection or "loading" of the ACE (Fig. 2) is accomplished by cotransfecting the Platform ACE Cell Line with both the ATV containing the gene of interest and ACE Integrase plasmid using standard lipofection methods under adherent conditions. This targeting allows the genes of interest to be localized to a specific genetic environment on the ACE producing an independent expression unit without interference from components of the host chromosome, which is often experienced with the more conventional random integration processes.

Specifically, each recombination acceptor cassette on the Platform ACE consists of a lambda phage *att*P site flanked by a simian virus 40 (SV40) promoter at the 5′-position and an open reading frame sequence encoding the puromycin resistance (puromycin^R) gene at the 3′-position, which confers puromycin resistance to cells carrying the Platform ACE (Fig. 2a). The ATVs (Fig. 2c) encode the bacterial *att*B site upstream of a promoterless secondary drug-selectable marker gene (e.g., zeocin^R, blasticidin^R, neomycin^R or hygromycin^R), which becomes activated by the SV40 promoter when the ATV integrates correctly via recombination between the *att*B site on the ATV and *att*P sites residing on the Platform ACE (Fig. 2d). The ATV also contains the target

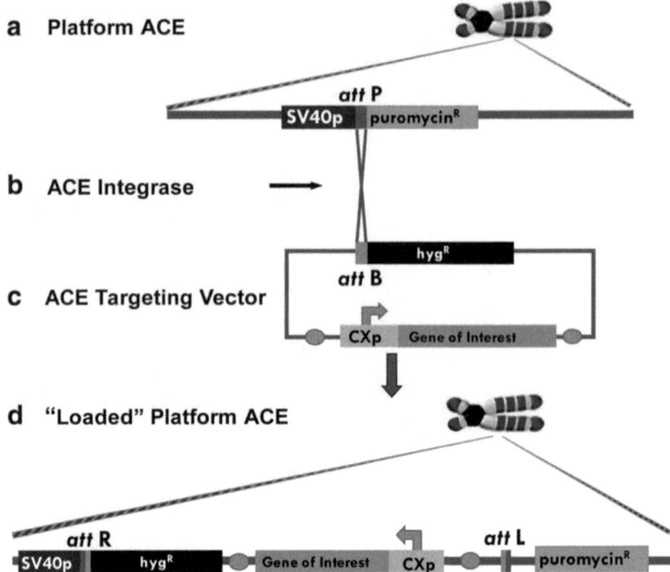

Fig. 2. The ACE System. The Platform ACE (**a**) is a murine artificial chromosome pre-engineered to contain multiple recombination acceptor *att*P sites. The Platform ACE cell line containing the Platform ACE is cotransfected with the ACE Targeting Vector (**c**) or ATV and a DNA plasmid encoding the ACE integrase (**b**). After DNA recombination or ATV loading (**d**) a drug selectable marker (e.g., hygromycin^R) gene is activated and enables efficient identification of cells carrying loaded ACEs. Multiple ATVs may be loaded during one cotransfection. Reprinted with permission from Lindenbaum et al. 2004 (copyright Oxford University Press).

gene cassette, which consists of the gene(s) of interest flanked by insulators and a 5′ upstream CX promoter (chicken β-actin promoter and CMV immediate/early enhancer). Multiple copies of the same gene or more than one gene (e.g., heavy and light chains of an antibody) can be placed into the ATV and loaded on to the Platform ACE. The site-specific recombination is mediated by the transiently expressed ACE Integrase, a mutant version of the lambda phage integrase that has been genetically engineered to function in mammalian cells without bacterial cofactors and catalyze the recombination between donor and acceptor sites (Fig. 2c) resulting in the generation of two new sites *att*R and *att*L. The ACE Integrase reaction is unidirectional and only catalyzes the integration of ATV onto the Platform ACE since it is unable to recognize the new *att*R and *att*L sites and lacks the bacterial cofactors required for excision. The combination of multiple *att*P sites on the Platform ACE and the unidirectional ACE Integrase enables multiple loadings (during a single transfection) or sequential loadings (via multiple transfections) with ATVs. Moreover, the ATV itself has a considerable carrying capacity and has been able to carry payloads exceeding 1.25 Mbp. In addition, the ACE

System targeted integration also increases the efficiency of screening, as only cells in which the ATV has correctly integrated into the Platform ACE are processed. As multiple gene copies are inserted in a single round of loading; very few colonies (100 to 200) have to be screened to identify the high expressing clones.

2.4. Cell Line Generation Using the ACE System

After targeted transfection, identification of integrants is accomplished by growing cells under selective pressure, enabled by the activation of a drug resistance gene encoded on the ATV upon integration into the Platform ACE (Fig. 2). Drug-resistant colonies are then switched to a basic serum-free medium, selected and expanded from 96-well and 24-well cultures to shake flasks and serially screened for growth characteristics and productivity. Since nearly all drug resistant colonies express high levels of recombinant protein, minimal screening is required. The resulting primary transfectants can be used to produce material for research programs (Fig. 3). To stabilize the cell line, selected primary transfectants are single cell subcloned by limiting dilution, expanded to shake flask and subjected to performance testing in terminal shake flask cultures. Candidate clonal cell lines are selected based on growth, yields, and stability of expression and take between 3 and 4 months to generate from ATV transfection. ATV construction can take between 1 and 2 months depending on the target gene DNA source.

To further improve expression, a second transfection or "Double Load" can be carried out on the "Single Load" primary transfectants (Fig. 3). This is possible since not all recombination acceptor sites on the Platform ACE are targeted in a Single Load process. In the Double Load process, Single Load primary transfectants are loaded with a second ATV containing the gene(s) of

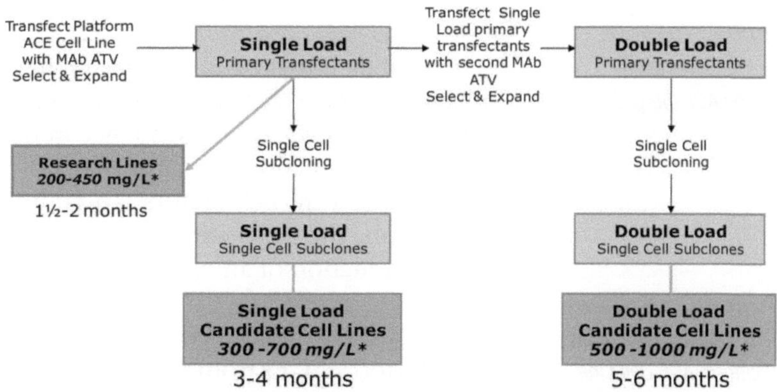

Fig. 3. The ACE System Process Overview. Single-load and double-load maximum batch titers with candidate cell-line generation times from transfection of the Platform ACE cell line.

interest and an alternate drug resistant marker. As with the Single Load, pools of double drug resistant transfectants are generated, screened, and single cell subcloned to identify clonal lines for performance testing in terminal shake flask cultures. Although the Double Load requires an extra one and a half to two months to generate clonal candidate cell lines compared to the Single Load, it has routinely resulted in titers that are 50% higher than those obtained with a Single Load (45). Overall, the ACE System has been used to generate a number of CHO cell lines expressing a variety of different recombinant proteins. For monoclonal antibodies, the Single Load ACE System process routinely produces clonal cell lines with yields of 300–700 mg/L in nonfed batch terminal shake flask cultures, and 500 to >1,000 mg/L for the Double Load ACE System process. Several cell lines have been subject to growth optimization and scale up, in which two to five-fold gain in performance has been noted (45).

2.5. Generation of ACE System Core Components

2.5.1. Generation of the Platform ACE

The original ACE was generated de novo and in vivo in a murine fibroblast cell line (LMTK⁻, American Type Culture Collection, ATTC: CCL-1.3) by the induction of large-scale amplification of "satellite" sequences composing the pericentric heterochromatin. Cells were selected under puromycin conditions and fluorescent in situ hybridization (FISH) was used to determine the presence of the ACE in metaphase spreads. The puromycin resistant cells were subsequently cloned and over 90% of the clones contained stable ACEs. The resulting Platform ACE which contained ~50–70 acceptor *att*P sites were isolated from the LMTK⁻ cells by blocking with colchicine prior to metaphase chromosome harvest. Chromosomes were then stained and the Platform ACEs were sorted by high-speed flow cytometry and could be transferred to other mammalian cell lines using standard lipofection methods (33).

2.5.2. Generation of the ACE Integrase

The lambda integrase gene was amplified by PCR from bacteriophage lambda DNA and a point mutation introduced into the lambda integrase coding sequence in order to allow the resulting integrase to function in mammalian cells without the requirement of bacterial host cell factors. In addition, to increase integrase expression, the ACE Integrase plasmid was further modified by the introduction of an optimized Kozak sequence (33).

2.5.3. Generation of the Platform ACE Cell Line

Using standard lipid-mediated transfection techniques, the purified Platform ACEs from the LMTK⁻ cells were initially transferred to the CHO cell line DG44 (Dr. Lawrence Chasin, Columbia University). Cell lines containing Platform ACEs were selected under puromycin conditions and FISH was used to determine the presence of the ACE in metaphase spreads. The puromycin resistant cells were subsequently cloned and over 95% of the

DG44 clones contained stable ACEs. In a similar fashion the Platform ACEs were then isolated from the Platform ACE containing DG44 cell line and transferred to the CHO cell line CHOK1SV (Lonza Biologics, UK). The resulting puromycin resistant cells went through two rounds of cloning resulting in the generation of the Platform ACE cell line ChK2. The majority of the case studies described below were based on the ChK2 and DG44 Platform ACE cell lines (33, 45–47).

2.5.4. Generation of ATVs Containing the Recombinant Protein Gene

The ACE targeting vectors (ATV) were based on a minimal pUC backbone with the gene of interest contained in an independent transcriptional expression cassette. The gene of interest was under the control of an upstream chimeric promoter/enhancer (CXp: chicken β-actin (CBA) promoter and cytomegalovirus immediate early (CMV-IE) enhancer) and with a downstream simian adenovirus (SV40) polyadenylation sequence. The expression cassette was flanked with six insulator elements (e.g., cHS4, chicken β-globin hypersensitive region 4). In addition, the ATVs contained a resistance gene (hygromycin or zeomycin) downstream of the *att*B integration donor site. In the case of a monoclonal antibody, two expression cassettes were used containing the heavy and light chain genes in either genomic or cDNA form. These two cassettes were then flanked by the insulator elements (Fig. 4) (33, 45).

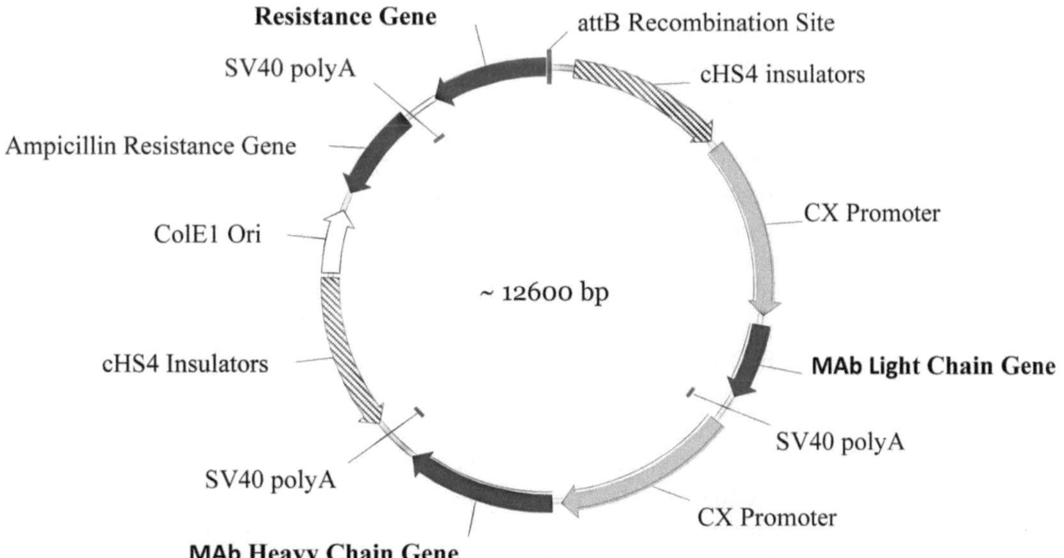

Fig. 4. Schematic of the ACE Targeting vectors (ATV) used to load MAb heavy and light chains onto the Platform ACE. The vector encodes the genes of interest cloned between a CX promoter and the rabbit β-globin polyA signal, with these three elements flanked on both sides by six tandem copies of the chicken β-globin cHS4-core elements. Also encoded in the vector is an *att*B recombination donor site adjacent to a promoterless selectable marker.

3. Protein Production Using Artificial Chromosomes: Case Studies

The following case studies illustrate the usefulness and potential of using engineered mammalian chromosomes in cellular protein production.

3.1. Comparison of the Single and Double Load ACE System Processes

This example focused on comparing the candidate cell lines generated by both the Single and Double Load processes. ChK2 cells were initially targeted with ATVs containing the sequences for both the heavy and light chain of an IgG1 monoclonal antibody (MAb) and the hygromycin resistance gene. Resulting Single Load primary transfectants (SL PT) were expanded from 96-well stage to shake flask stage and expressed the MAb at ~460 mg/L under terminal nonfed shake flask culture in nonoptimized, off-the-shelf, serum-free culture medium (e.g., Invitrogen CD-CHO medium). The top SL PTs were then single cell subcloned by limiting dilution, selected and expanded to shake flask cultures, and subjected to performance testing in terminal nonfed shake flask cultures. The top Single Load single cell subclone (SL ScSc) cell lines reached a viable cell density of over 7.0×10^6 cells/mL with maximum titers of 740 mg/L and specific productivities (Qp) of 14 pg/cell/day. Although single cell subcloning had little effect on the growth profiles of the cells it did result in significantly increasing the MAb expression. This increase in expression was to be expected since the primary transfectants were not completely clonal. In addition to subcloning the SL PTs, they were also subjected to a Double Load with a second ATV containing the heavy and light chain sequences and an alternate secondary drug resistance marker (zeocin). This Double Load process resulted in Double Load primary transfectants (DL PT) with maximum titers reaching 797 mg/L under terminal nonfed shake flask cultures. As with the Single Load process, the top DL PTs were single cell subcloned, resulting in a number of clonal Double Load single cell subclone (DL ScSc) cell lines with MAb maximum titers of up to 1,144 mg/L and Qp of 35 pg/cell/day in terminal nonfed shake flask cultures. Although Double Loading had little overall effect on the growth profiles compared to the parent SL PTs, it did result in almost a doubling of MAb expression. These results demonstrate that it is possible that not all the *att*P acceptor sites were filled during the first transfection (Single Load) and that more copies of the MAb genes can be loaded in the second transfection (Double Load). Table 1 and Fig. 5 compare the range of MAb expression and growth profiles for the top cell lines. It can be seen from Table 1, that the cell lines were generated quite rapidly i.e. 4 months in the case of the SL ScSc and 6 months in the case of the DL ScSc. If only research material is required then the SL PT could be used, which expressed up to

Table 1
Cell line generation using the ACE System. Single Load and Double Load maximum batch titers with candidate cell line generation times from transfection of the Platform ACE Cell Line

	Single load		Double load	
	SL PT	SL ScSc	DL PT	DL ScSc
Range of maximum titer (mg/L)	315–461	456–743	495–797	891–1144
Average maximum titer (mg/L)	415±60	587±101	618±136	976±89
Range of Qp (pg/cell/day)	9–14	10–21	10–35	22–45
Average Qp (pg/cell/day)	11±2	12±4	20±9	27±9
Generation time (months)	2	4	4	6

SL PT→SL ScSc
↓
DL PT→DL ScSc

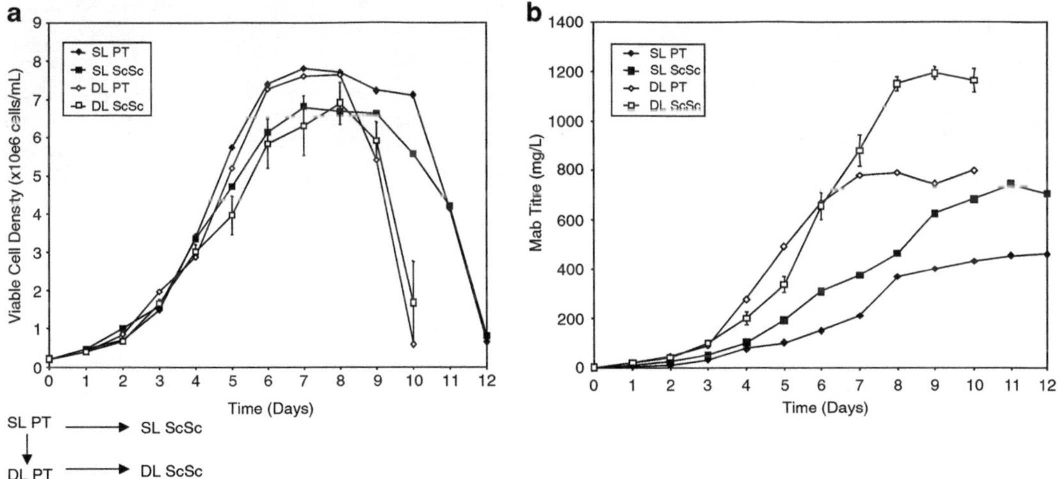

Fig. 5. Examples of the growth profiles for the top MAb candidate cell lines under nonfed, nonoptimized, shake flask conditions. (**a**) Viable cell density verses batch culture time and (**b**) MAb titer versus batch culture time. *Error bars* for DL ScSc are derived from two independent runs.

461 mg/L and were generated in only 2 months. Figure 6 shows the corresponding fluorescent in situ hybridization (FISH) images from the top candidate cell lines. These FISH images show that Single or Double Loading does not appear to affect the integrity of the MAb-ACE with a single intact MAb-ACE containing both heavy and light chain sequences detected in all the metaphase spreads. Moreover, no integration of the MAb sequences onto

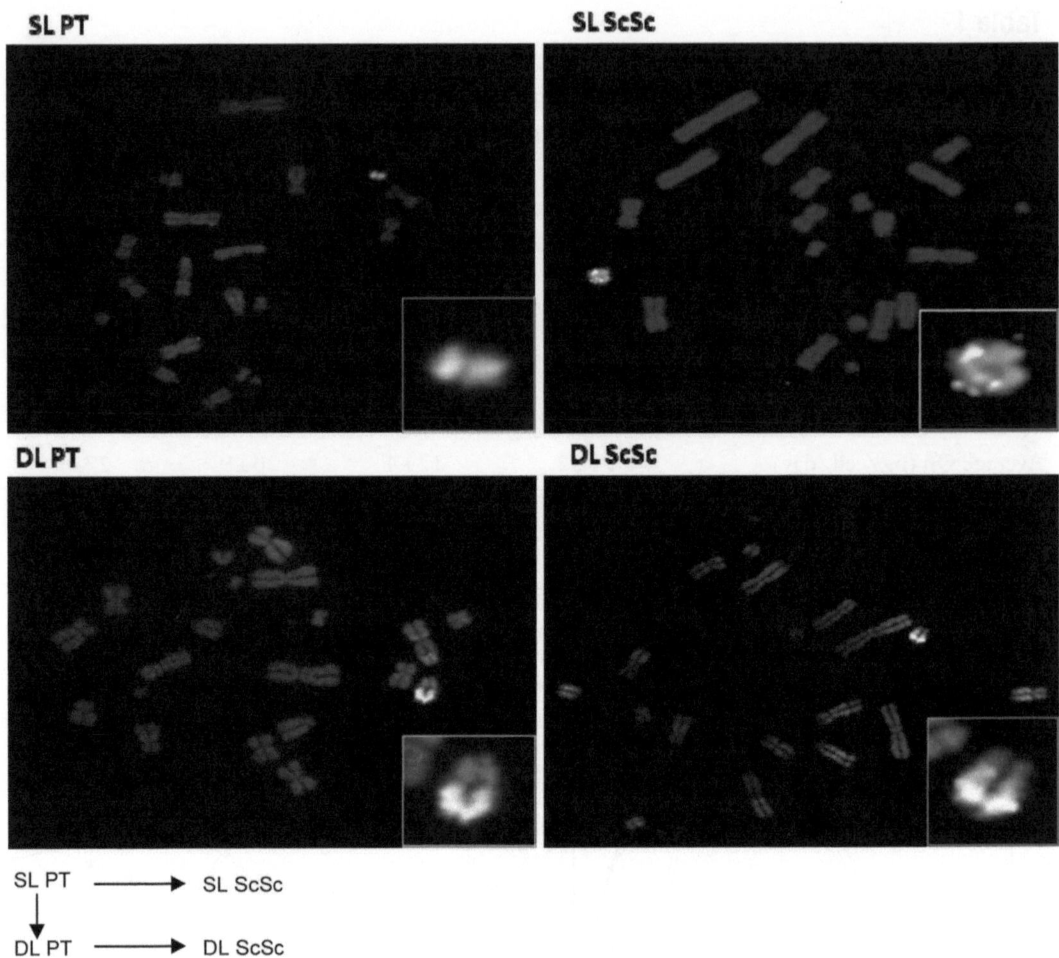

Fig. 6. Examples of FISH analysis for the top MAb candidate cell lines. The FISH image shows a metaphase chromosome spread containing the MAb loaded ACE. Chromosomes were hybridized with a mixture of digoxigenin-labeled mouse major satellite DNA probe to label the ACE (rhodamine-red) and biotin-labeled MAb DNA probe to label the MAb genes (FITC-green). The *inserts* contain enlargements of the loaded ACE.

the host CHO genome was detected over the 6-month period that was required to generate the top DL ScSc.

As part of the ACE System process, the top DL ScSc were subjected to a stability study as part of their performance testing. The final candidate cell lines are selected based on their stability over a minimum number of generations as well as their overall expression. The stability study consisted of maintaining cultures in 125-mL shake flasks and passaging them twice weekly to ~3×10^5 cell/mL under selection. At specific times throughout the study, batch shake flask analysis is carried out to determine growth profiles, maximum titers and Qp as well as FISH analysis. The candidate DL ScSc cell lines were relatively stable under selection with a drop of less than 25% in maximum titer and Qp over 70 generations. FISH analysis showed that the ACE remained intact and stable over the study period. This stability is considered adequate for expansion from a working cell bank to commercial bioreactor scale.

3.2. Fed-Batch Scale-up

The top candidate DL ScSc cell line from Subheading 3.1 was subjected to simple, nonoptimized, fed batch scale-up by supplementing the basal CD-CHO medium with glucose and plant hydrolysates. These cells were evaluated in 500-mL batch shake flasks, 500-mL fed-batch shake flasks, and in 2-L fed-batch bioreactors. The batch shake flasks achieved MAb titers between 0.7 and 1.1 g/L while fed-batch shake flask culture increased the expression to 2.5 and 3.0 g/L, and the fed-batch 2-L bioreactor increased the expression to over 4.0 g/L with an average Qp of 40 pg/cell/day. Figure 7 and Table 2 compare the results of this fed-batch scale up. Stability characterization of the of the top DL ScSc demonstrated that the cell lines were stable up to 96 generations under fed-batch conditions. In addition, there were no product quality issues observed over the 96 generations. This study showed that ACE System derived cell lines perform as well if not better in bioreactors than those cell lines generated by more traditional technologies and exhibit sufficient stability to be used in commercial cellular protein production.

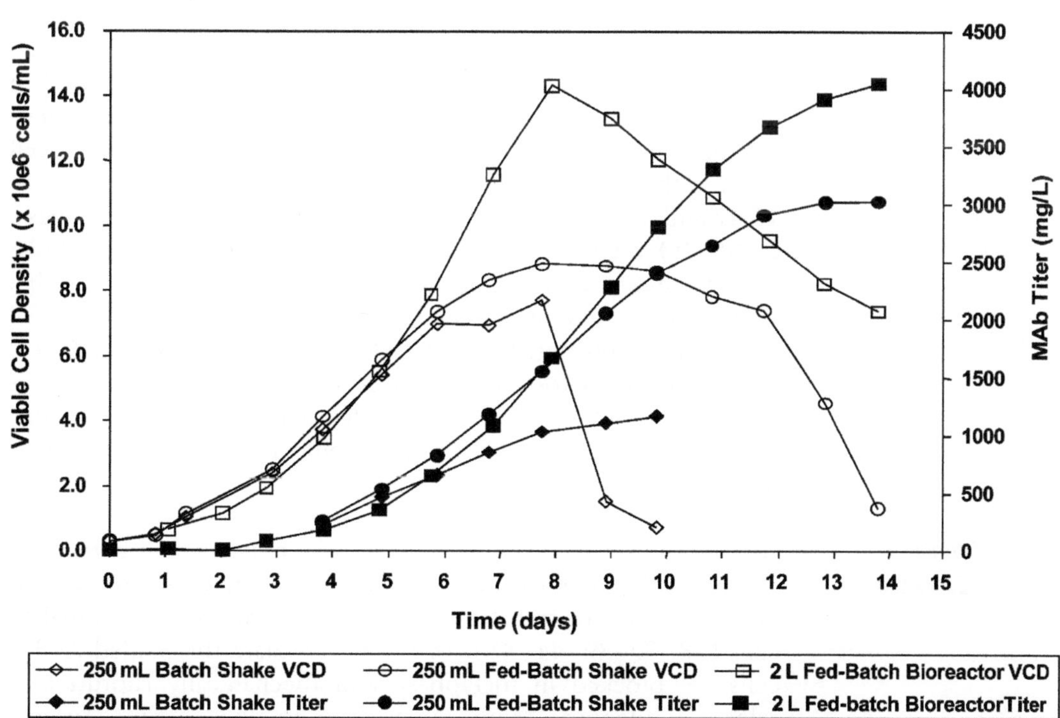

Fig. 7. Scale-up comparison summary for the top DL ScSc. Examples of growth profiles during scale-up from batch shake flask, fed-batch shake flask to 2-L fed-batch bioreactor. Viable cell density (VCD) and MAb titer versus culture time.

Table 2
Summary of fed-batch scale-up for top DL ScSc from Subheading 3.1

Culture	Batch analysis			
	Maximum titer (mg/L)	Viable cell density ($\times 10^6$ cell/mL)	Culture length (Days)	Qp (pg/cell/day)
500 mL Batch shake flask	1,140	7.0	10	45
500 mL Fed-batch shake flask	3,000	9.0	14	41
2.0 L Fed-batch bioreactor	4,060	14.5	14	40

3.3. Rapid Generation of Candidate Cell Lines

In this example, the primary focus was in generating a MAb expressing cell line as quickly as possible to support a toxicity study. It was estimated that ~125 g of material would be required for the study and that a cell line expressing greater than 500 mg/L in a 500-L bioreactor would be sufficient. Therefore, a Single Load process was chosen with minimal screening. Briefly, ChK2 cells were targeted with an ATV containing the sequences for the heavy and light chain of an IgG4 MAb. After transfection, 50 SL PTs were expanded to 96-well plates, 10 were expanded to 24-well plates, and only 2 were selected for shake flask analysis and single cell subcloning. The resulting top candidate SL ScSc had MAb titers of approximately 430 mg/L, was generated in under 3 months from transfection and was stable for over 50 generations under nonselecting medium. This candidate SL ScSc cell line was then subjected to scale-up and simple fed-batch at 1.6-L, 15-L and 500-L bioreactor scale by supplementing the basal CD-CHO medium with glucose and plant hydrolysates. MAb titers were over doubled in the 1.6-L fed-batch bioreactor with a significant increase in Qp and culture time. The MAb titer fell slightly to ~660 mg/L when scaled up to the 15-L and 500-L fed-batch bioreactors. Finally, over 140 g of purified MAb was recovered from the 500-L fed-batch bioreactor. This demonstrated that a commercial cell line could be developed in under 3 months using the ACE System.

3.4. Auditioning of Cells for Expression

Among the key features of the ACE System is the ability to purify ACEs (loaded with the gene(s) of interest) using high-speed flow cytometry and to transfer the ACEs to a variety of mammalian cells (33, 37–39, 41, 48). In order to isolate the loaded ACEs, cells are blocked in metaphase and mechanically ruptured to release condensed chromosomes prior to flow sorting. To sort the chromosomes, they are stained with Hoechst 33258 and chromomycin-A3, which bind preferentially to AT- and GC-base pairs respectively. Platform ACEs are composed of greater than 350,000 copies of the AT-rich 234-bp mouse major satellite sequences, and hence bind more Hoechst 33258 and less chromomycin-A3 than the endogenous host cell chromosomes. The dual-stained loaded ACEs can be readily distinguished and separated from the host chromosomes at sort rates exceeding 1 million ACEs/hour/sorter and at purities exceeding 99% (40). These purified loaded ACEs could then be transferred to other cell lines using standard lipofection techniques (33, 38–41). After ACE transfer, stable production cell lines are generated in a similar manner to the Single Load process previously described. This ability to purify and transfer loaded ACEs, allows host cells to be rapidly auditioned for improved quality of the product (e.g., desired glycosylation pattern) or enhanced quantity (e.g., improved growth or expression). In addition, unlike the more common random integration techniques, by transferring loaded ACEs from one cell line to another ensured that the gene remains in an identical genetic environment from cell line to cell line. In this manner, any differences in expression can be attributed to phenotypic properties of the host cell rather than the transfection technique, gene copy number, or the location of the gene in the host chromosome.

To demonstrate this unique auditioning feature of the ACE System, loaded MAb-ACEs were isolated from the top CHOK1SV based DL ScSc generated in Subheading 3.1 (46) and transferred to three different CHO cell lines namely CHOK1SV, CHO-S, and DG44 (47). This MAb-ACE donor cell line had a maximum MAb titer of 975 mg/L and a Qp of 23 pg/cell/day in terminal, nonfed, nonoptimized shake flask culture. Transfecting the three different CHO cell lines with the same MAb-ACEs ensured that all resulting cell lines would contain the same number of MAb genes in the same genetic environment allowing a direct comparison to be made between the three CHO host cell lines. The resulting ScSc showed large differences in overall MAb expression between the three CHO cell lines with the CHOK1SV based cell lines consistently expressing higher MAb levels (average maximum titer 1,218 mg/L and average Qp 34.5 pg/cell/day) compared to the CHO-S based cell lines (average maximum titer 167 mg/L and average Qp 4.5 pg/cell/day) and the DG44 based cell lines (average maximum titer 452 mg/L and average Qp 7.4 pg/cell/day). The MAb expression of the top CHOK1SV based ScSc remained very similar to the original MAb-ACE donor cell line. Figures 8 and 9 compare the growth profiles and FISH analysis respectively of the top ScSc for each CHO cell line. As can be seen from Fig. 8a, the maximum viable cell densities were similar for all subclones (~6×10^6 cells/mL) although the batch culture time varied slightly between ScSc. Figure 9 shows that a single intact MAb-ACE was detected in all the metaphase spreads

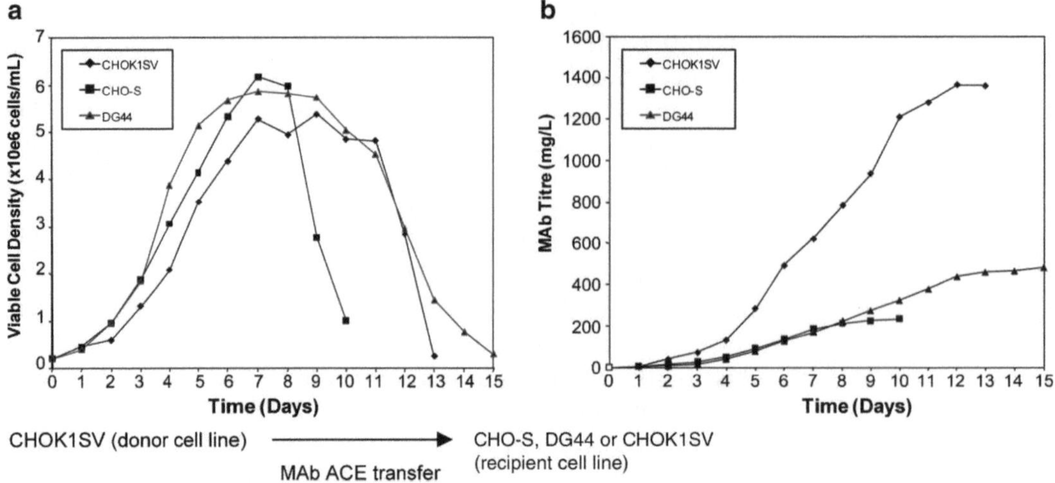

Fig. 8. Examples of the growth profiles for the top MAb expressing ACE transfer ScSc under nonfed, nonoptimized, shake flask conditions. (**a**) Viable cell density verses batch culture time and (**b**) MAb titer versus batch culture time.

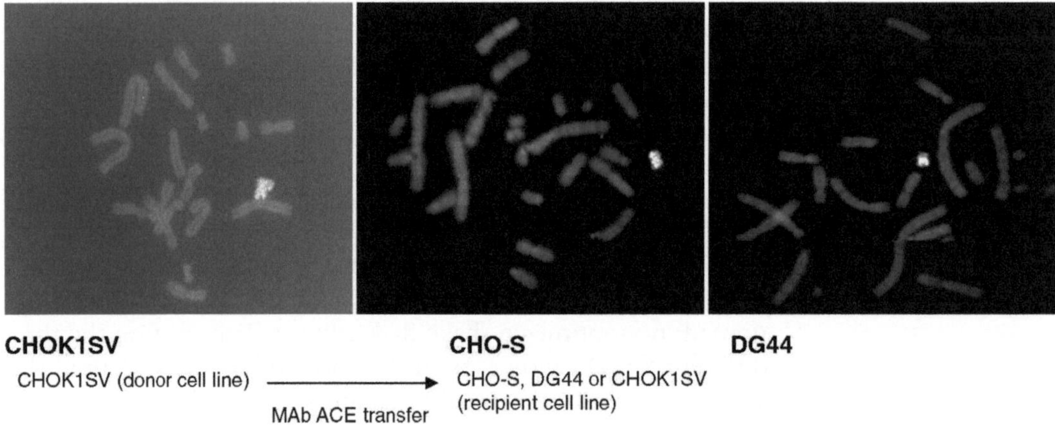

Fig. 9. Examples of FISH analysis for the top MAb expressing ACE transfer ScSc. The FISH images show metaphase chromosome spreads containing the MAb ACE. Chromosomes were hybridized with a mixture of digoxigenin-labeled mouse major satellite DNA probe to label the ACE (rhodamine-red) and biotin-labeled MAb DNA probe to label the MAb genes (FITC-green).

analyzed by FISH for each CHO cell line and that the MAb genes remained localized to the ACE over the 3 month period that was required to generate the top ScSc. The dramatic differences in MAb expression (Fig. 8b) can, most likely, be attributed to the cell line rather than the copy number or location of the MAb genes since the FISH analyses for each CHO based subclone were similar.

In order to investigate further whether differences in MAb expression between the three CHO host cell lines were due to changes in the MAb-ACE (e.g., changes in gene copy number) during the MAb-ACE transfer from the CHOK1SV based donor

Table 3
Batch analysis comparison of top expressing MAb back transfer CHOK1SV based subclone with original MAb ACE donor cell lines

Cell line	Batch analysis Maximum titer (mg/L)	Viable cell density ($\times 10^6$ cell/mL)	Culture length (Days)	Qp (pg/cell/day)
Original CHOK1SV based MAb ACE donor DL ScSc cell line	975	7.8	11	23
CHO-S based MAb-ACE donor subclone cell line	188	11.6	9	2.8
CHOK1SV based subclone	902	8.0	11	21

CHOK1SV (donor cell line) ——MAb ACE transfer——→ CHO-S ——MAb ACE transfer——→ CHOK1SV

cell line to the three recipient CHO host cell lines, the MAb-ACE was purified from a low MAb expressing CHO-S based subclone (maximum MAb titer of 188 mg/L and a Qp of 2.8 pg/cell/day) and transferred back to fresh CHOK1SV cells. The back transfer of the MAb ACE from the CHO-S based subclone to CHOK1SV resulted in generating a CHOK1SV cell line that returned the expression of the MAb to that of the original CHOK1SV based donor cell line and was over five times that of the CHO-S based subclone. Table 3 compares the batch analysis of the top back transfer CHOK1SV subclone with its CHO-S based donor subclone and the original CHOK1SV based donor cell line. This comparison clearly shows that returning the MAb-ACE into a CHOK1SV environment restores the growth profile and MAb expression to that of the original CHOK1SV based donor cell line.

From these results, it can be concluded that the MAb-ACE had remained intact (with unchanged copy number) during the transfer and that the differences in MAb expression between the different CHO cell lines was due to the cell phenotype and not due to gene copy number or location and that there are significant differences in recombinant protein expression between these CHO cell types. Using ACE transfer, therefore, provides a rapid and unique technique to audition cell lines allowing a direct comparison to be made between cell lines based on cell type alone without the confounding effects of recombinant gene copy number and genetic environment.

3.5. Multiple Gene Expression: Potential for Metabolic Engineering

Another feature of the ACE System is the ability to integrate several copies of different genes onto the same ACE either through transfection with an ATV containing more than one gene or through double loading with ATVs containing different genes. To highlight this novel feature the DG44 CHO Platform ACE cell line was loaded with both green fluorescent protein (GFP) and human erythropoietin (EPO) (33). The Platform ACE was initially loaded with an ATV encoding GFP, resulting in a cell line expressing GFP at levels detectable by fluorescence microscopy and flow cytometry. This GFP expressing cell line was then loaded with a second ATV encoding EPO. The resulting cell lines maintained parental GFP expression levels and also expressed EPO at levels greater than 400 IU/10^6 cells/day as measured by ELISA. The capacity of the ACE System to stably express multiple different genes provides the basis for potential metabolic engineering applications. For instance, genes encoding growth factors, anti-apoptotic factors, or factors affecting post-translational modifications or protein secretion could be sequentially loaded onto an ACE expressing a product gene, thereby enhancing the growth characteristics of that cell line for clinical and commercial manufacture in a bioreactor. Alternately, a Platform ACE could be loaded with multiple copies of metabolic factors, sorted and purified, and transferred into a preexisting production cell line.

4. Conclusion

There are several problems in use of the current techniques to generate mammalian cell lines expressing recombinant proteins. These techniques are mainly based on random integration of the gene of interest into the host genome through plasmid transfection or viral transduction, which can result in the generation of transfectants with highly variable recombinant protein expression. Also these techniques have a limited DNA carrying capacity, which prevents the transfer of multiple or large complex genes. To identify the high expressing clones requires extensive screening of large numbers transfectants with no guarantee of identifying a stable cell line. Furthermore, these methods may necessitate time-consuming amplification events or reinfection to boost the cell's productivity. As a result, the process of generating and selecting a high expressing stable clonal cell line suitable for the clinical and commercial manufacture of biopharmaceuticals can be labor intensive and extremely time consuming. Other techniques have been generated to targeting the genes of interest to specific "hot spots" on the host genome in order to reduce gene silencing and the number of copies required for high protein expression. These have met with some success although they still do not deal with the potential interference effects of integrating foreign genes onto the host genome. Technologies that can accommodate large DNA payloads and do not require integration into the host genome for long-term stable maintenance would be advantageous, in terms of both more predictable gene expression and noninterference in host cell functions. In addition, such systems would also be of great utility in gene therapy applications where safety and stability considerations are paramount. To address these issues, artificial mammalian chromosomes have been considered. Compared to traditional methodologies, using artificial mammalian chromosomes offer significant advantages for cellular protein production as well as animal transgenesis, and gene-based cell therapy applications on account of their high carrying capacity and ability to self-replicate without relying on integration into the host genome. Probably the most successful artificial mammalian chromosome technique applied to cellular protein production has been the artificial chromosome expression system or ACE System.

The ACE System, which consists of the Platform ACE, Platform ACE Cell Line, ACE targeting vector (ATV), and ACE Integrase, was developed as a modular platform technique that would enable a wide-ranging variety of genes to be rapidly introduced into mammalian cells. The success of the system is based on the unique characteristics of the Platform ACE or artificial chromosome that allows targeted transfection through the action of the specifically designed ATVs and ACE Integrase and through the use of the lambda bacteriophage *att*P and *att*B acceptor and donor sites. This targeting allows the genes of interest to be localized to a specific genetic environment without interference from components of the host chromosome. The design of the ATVs allows considerable flexibility in the size, number and variety of genes that can be targeted to the Platform ACE. In the case of monoclonal antibodies it is usual to have both heavy and light chains contained in one ATV, although it is quite possible for the ATV to contain more than one copy of each heavy and light chain gene and it appears that there is no limitation to the gene carrying capacity of the ATV. The Platform ACE contains natural centromeres and telomeres that allow them to be stably maintained side by side with the host chromosomes. The Platform ACE large payload capacity and multiple acceptor sites allow the integration of high copy numbers of genes with resulting high gene expression without the need for amplification. In addition, once loaded, the Platform ACEs can be easily isolated to highly purified yields and subsequently transferred into numerous other cell types. This provides a routine technique requiring minimal effort to audition cell lines by ensuring that cells are transfected with the same number of genes located in the same genetic environment. To demonstrate the utility of the ACE system, the ACE System has been used to generate stable clonal cell lines that express mainly MAbs that were generated in 3–6 months with minimal screening. These candidate cell lines had MAb titers in terminal nonfed shake flask cultures of 300–1,000 mg/L with Qp of 20–40 pg/cell/day. These cell lines had a stable gene expression for over 96 generations and were suitable for media and growth optimization and scale up. In addition to the advantages of the ACE System, it has the potential to be developed further to reduce cell line generation times and to increase final titers. For example, the modular nature of the ACE System allows the ACES and ATV to be improved independently. Where, for example, new ACEs can be developed with more acceptor sites and new ATVs can be developed with new promoters and insulators. Platform ACE Cell Lines based on mammalian cells other than CHO cells may be developed (e.g., human cell lines such as HEK293 or PER.C6) to determine whether improvements to cell line expression, stability, or product quality can be made. Cell line generation times can be reduced by optimizing screening, improving transfection, and overlapping process steps. Furthermore, the ability to generate cell lines that express several different genes may allow the ACE System to be used to insert genes that are not normally expressed by the host cell, but may be required to render the recombinant protein active through, for example, post-translational modification. With its advantages of performance, speed and versatility, the ACE System provides an attractive and practical alternative to conventional methods for cell line generation and cellular protein production.

References

1. Sternberg, N. (1990) Bacteriophage P1 cloning system for the isolation, amplification, and recovery of DNA fragments as large as 100 kilobase pairs *Proc. Natl. Acad. Sci. U.S.A.* **87**, 103–107.
2. Ioannou, P. A., Amemiya, C. T., Garnes, J., Kroisel, P. M., Shizuya, H., Chen, C., Batzer, M. A. and de Jong, P. J. (1994) A new bacteriophage P1-derived vector for the propagation of large human DNA fragments. *Nat. Genet.* **6**, 84–89.
3. Murray, A. W. and Szostak, J. W. (1983) Construction of artificial chromosomes in yeast *Nature*, **305**, 189–193.
4. Burke, D. T., Carle, G. F. and Olson, M. V. (1987) Cloning of large segments of exogenous DNA into yeast by means of artificial chromosome vectors. *Science* **236**, 806–812.
5. Chumakov, I. M., Rigault, P., Le, G. I., Bellanne-Chantelot, C., Billault, A., Guillou, S., Soularue, P., Guasconi, G., Poullier, E., Gros, I. *et al.* (1995) A YAC contig map of the human genome *Nature* **377**, 175–297.
6. Al Shawi, R., Kinnaird, J., Burke, J. and Bishop, J. O. (1990) Expression of a foreign gene in a line of transgenic mice is modulated by a chromosomal position effect *Mol. Cell Biol.* **10**, 1192–1198.
7. Costantini, F., Radice, G., Lee, J. L., Chada, K. K., Perry, W. and Son, H. J. (1989) Insertional mutations in transgenic mice *Prog. Nucleic Acid Res. Mol. Biol.* **36**, 159–169.
8. Huxley, C. (1994) Mammalian artificial chromosomes: a new tool for gene therapy *Gene Ther* **1**, 7–12.
9. Ascenzioni, F., Donini, P., and Lipps, H. J. (1997) Mammalian artificial chromosomes-vectors for somatic gene therapy *Cancer Lett* **118**, 135–142.
10. Vos, J. M. (1998) Mammalian artificial chromosomes as tools for gene therapy *Curr Opin Genet Dev* **8**, 351–359.

11. Brown, W. R., Mee, P. J., and Hong, S. M. (2000) Artificial chromosomes: ideal vectors? *Trends Biotechnol* **18**, 218–223.
12. Lipps, H. J., Jenke, A.C., Nehlsen, K. et al. (2003) Chromosome-based vectors for gene therapy *Gene* **304**, 23–33.
13. Cooke, H. (2001) Mammalian artificial chromosomes as vectors: progress and prospects. *Cloning Stem Cells* **3**, 243–249.
14. Lewis, M. (2001) Human artificial chromosomes: emerging from concept to reality in biomedicine *Clin Genet* **59**, 15–16.
15. Lipps, H. J., and Bode, J. (2001) Exploiting chromosomal and viral strategies: the design of safe and efficient non-viral gene transfer systems *Curr Opin Mol Ther* **3**, 133–141.
16. Grimes, B. R., Warburton, P.E., and Farr, C. J. (2002) Chromosome engineering: prospects for gene therapy. *Gene Ther* **9**, 713–718.
17. Grimes, B. R., Rhoades, A. A., and Willard, H. F. (2002) Alpha-satellite DNA and vector composition influence rates of human artificial chromosome formation *Mol Ther* **5**, 798–805.
18. Larin, Z., and Mejia, J. E. (2002) Advances in human artificial chromosome technology. *Trends Genet* **18**, 313–319.
19. Saffery, R., and Choo, K. H. (2002) Strategies for engineering human chromosomes with therapeutic potential *J Gene Med* **4**, 5–13.
20. Harrington, J. J., Van Bokkelen, G., Mays, R. W. et al. (1997) Formation of de novo centromeres and construction of first-generation human artificial microchromosomes. *Nat Genet* **15**, 345–355.
21. Warburton, P. E., and Cooke, H. J. (1997) Hamster chromosomes containing amplified human alpha-satellite DNA show delayed sister chromatid separation in the absence of de novo kinetochore formation *Chromosoma* **106**, 149–159.
22. Ikeno, M., Grimes, B., Okazaki, T. et al. (1998) Construction of YAC-based mammalian artificial chromosomes *Nat Biotechnol* **16,** 431–439.
23. Henning, K. A., Novotny, E. A., Compton, S.T. et al. (1999) Human artificial chromosomes generated by modification of a yeast artificial chromosome containing both human alpha satellite and single-copy DNA sequences *Proc Natl Acad Sci USA* **96**, 592–597.
24. Ebersole, T. A., Ross, A., Clark, E. et al. (2009) Mammalian artificial chromosome formation from circular alphoid input DNA does not require telomere repeats *Hum Mol Genet* **9**, 1623–1631.
25. Mejia, J. E., Willmott, A., Levy, E. et al. (2001) Functional complementation of a genetic deficiency with human artificial chromosomes *Am J Hum Genet* **69**, 315–326.
26. Mejia, J. E., Alazami, A., Wilmott, A. et al. (2002) Efficiency of de novo centromere formation in human artificial chromosomes *Genomics* **79**, 297–304.
27. Carine, K., Solus, J., Waltzer, E. et al. (1986) Chinese hamster cells with a minichromosome containing the centromere region of human chromosome 1 *Somat Cell Mol Genet* **12**, 479–491.
28. Farr, C. J., Stevanovic, M., Thomson, E. J. et al. (1992) Telomere-associated chromosome fragmentation: applications in genome manipulation and analysis *Nat Genet* **2**, 275–282.
29. Hadlaczky, G., Praznovszky, T., Cserpan, I., Kereso, J., Peterfy, M., Kelemen, I., Atalay, E., Szeles, A., Szelei, J., Tubak, V. *et al.* (1991) Centromere formation in mouse cells cotransformed with human DNA and a dominant marker gene *Proc. Natl. Acad. Sci. U.S.A* **88**, 8106–8110.
30. Praznovszky, T., Kereso, J., Tubak, V., Cserpan, I., Fatyol, K. and Hadlaczky, G. (1991) De novo chromosome formation in rodent cells *Proc. Natl. Acad. Sci. U.S.A* **88**, 11042–11046.
31. Christ, M., Lusky, M., Stoeckel, F. et al. (1997) Gene therapy with recombinant adenovirus vectors: evaluation of the host immune response *Immunol Lett* **57**, 19–25.
32. Hadlaczky, G. (2001) Satellite DNA-based artificial chromosomes for use in gene therapy *Curr. Opin. Mol. Ther.* **3**, 125–132.
33. Lindenbaum, M., Perkins, E., Csonka, E., Fleming, E., Garcia, L., Greene, A., Gung, L., Hadlaczky, G., Lee, E., Leung, J., MacDonald, N., Maxwell, A., Mills, K., Monteith, D. Perez, C.F., Shellard, J., Stewart, S., Stodola, T., Vandenborre, D., Vanderbyl, S. and Ledebur, H.C. (2004) A mammalian artificial chromosome engineering system (ACE System) applicable to biopharmaceutical protein production, transgenesis and gene-based cell therapy *Nucleic Acids Research* **32** (21), e172.
34. Csonka, E., Cserpan, I., Fodor, K., Hollo, G., Katona, R., Kereso, J., Praznovszky, T., Szakal, B., Telenius, A., de Jong, G. *et al.* (2000) Novel generation of human satellite DNA-based artificial chromosomes in mammalian cells *J. Cell Sci.* **113**, 3207–3216.
35. Hollo, G., Kereso, J., Praznovszky, T., Cserpan, I., Fodor, K., Katona, R., Csonka, E., Fatyol, K., Szeles, A., Szalay, A. A. *et al.*

(1996) Evidence for a megareplicon covering megabases of centromeric chromosome segments *Chromosome Res.* **4**, 240–247.

36. Kereso, J., Praznovszky, T., Cserpan, I., Fodor, K., Katona, R., Csonka, E., Fatyol, K., Hollo, G., Szeles, A., Ross, A. R. *et al.* (1996) De novo chromosome formations by large-scale amplification of the centromeric region of mouse chromosomes *Chromosome Res.* **4**, 226–239.
37. De Jong, G., Telenius, A. H., Telenius, H., Perez, C., Drayer, J., and Hadlaczky, G. (1999) Mammalian artificial chromosome pilot production facility: large-scale isolation of functional satellite DNA-based artificial chromosomes *Cytometry* **35**, 129–133.
38. Perez, C.F., Vanderbyl, S.L., Mills, K.A., and Ledebur, H.C. (2004) The ACE System: A versatile chromosome engineering technology with applications for gene-based cell therapy. *Bioprocessing Journal* **3**, 61–68.
39. De Jong, G., Telenius, A., Vanderbyl, S., Meitz, A. and Drayer, J. (2001) Efficient in-vitro transfer of a 60-Mb mammalian artificial chromosome into murine and hamster cells using cationic lipids and dendrimers *Chromosome Res.* **9**, 475–485.
40. Vanderbyl, S., MacDonald, N. and de Jong,G. (2001) A flow cytometry technique for measuring chromosome-mediated gene transfer. *Cytometry* **44**, 100–105.
41. Vanderbyl, S., MacDonald, G. N., Sidhu, S., Gung, L., Telenius, A., Perez, C. and Perkins, E. (2004) Transfer and stable transgene expression of a mammalian artificial chromosome into bone marrow-derived human mesenchymal stem cells *Stem Cells* **22**, 324–33.
42. Co, D. O., Borowski, A. H., Leung, J. D., van der Kaa, J., Hengst, S., Platenburg, G. J., Pieper, F. R., Perez, C. F., Jirik, F. R. and Drayer, J. I. (2000) Generation of transgenic mice and germline transmission of a mammalian artificial chromosome introduced into embryos by pronuclear microinjection *Chromosome Res.* **8**, 183–191.
43. Monteith, D. P., Leung, J. D., Borowski, A. H., Co, D. O., Praznovszky, T., Jirik, F. R., Hadlaczky, G. and Perez, C. F. (2004) Pronuclear microinjection of purified artificial chromosomes for generation of transgenic mice: pick-and-inject technique. *Methods Mol. Biol.* **240**, 227–242.
44. Landy, A. (1989) Dynamic, structural, and regulatory aspects of lambda site-specific recombination. *Ann. Rev. Biochem.* **58**, 913–949.
45. Kennard, M. L., Goosney, D. G., and Ledebur, H. C. (2007) Generating stable, high-expressing cell lines for recombinant protein manufacture *BioPharm International* **March**, 52–59.
46. Kennard, M. L., Goosney, D. G., Monteith, D., Zhang, L., Moffat, M., Fischer, D., and Mott, J. (2009) The generation of stable, high MAb expressing CHO cell lines based on the Artificial Chromosome Expression (ACE) technology *Biotech. and Bioeng.* **104(3)**, 540–553.
47. Kennard, M. L., Goosney, D. G., Monteith, D., Roe, S., Fischer, D., and Mott, J. (2009) Auditioning of CHO Host Cell Lines Using the Artificial Chromosome Expression (ACE) Technology *Biotech. and Bioeng.* **104(3)**, 526–539.
48. Vanderbyl, S., Sullenbarger, B., White, N., Perez, C.F., MacDonald, G.N., Stodola, T., Bunnell, B.A., Ledebur Jr., H.C., and Lasky, L.C. (2005) Transgene expression after stable transfer of a mammalian artificial chromosome into human hematopoietic cells *Experimental Hematology* **33**, 1470–1476.

Index

A

Gyula Hadlaczky (ed.), *Mammalian Chromosome Engineering: Methods and Protocols*, Methods in Molecular Biology, vol. 738, DOI 10.1007/978-1-61779-099-7, © Springer Science+Business Media, LLC 2011

D

E

F

I

J

K

L

M

Q

R

S

T

U

V

W

X

Y

Z

MIX
Papier aus verantwortungsvollen Quellen
Paper from responsible sources
FSC® C105338

If you have any concerns about our products,
you can contact us on
ProductSafety@springernature.com

In case Publisher is established outside the EU,
the EU authorized representative is:
Springer Nature Customer Service Center GmbH
Europaplatz 3, 69115 Heidelberg, Germany

Printed by Libri Plureos GmbH
in Hamburg, Germany